Christian Jooss
Self-organization of Matter

Also of interest

Electrons in Solids.
Mesoscopics, Photonics, Quantum Computing, Correlations, Topology
H. Bluhm, T. Brückel, M. Morgenstern, G. Plessen, C. Stampfer, 2019
ISBN 978-3-11-043831-4, e-ISBN 978-3-11-043832-1

Statistical Physics
Michael V. Sadovskii, 2019
ISBN 978-3-11-064510-1, e-ISBN 978-3-11-064848-5

Vom Energieinhalt ruhender Körper.
Ein thermodynamisches Konzept von Materie und Zeit
Grit Kalies, 2019
ISBN 978-3-11-065556-8, e-ISBN 978-3-11-065696-1

Spintronics.
Theory, Modelling, Devices
Tomasz Blachowicz, Andrea Ehrmann, 2019
ISBN 978-3-11-049062-6, e-ISBN 978-3-11-049063-3

Christian Jooss

Self-organization of Matter

A dialectical approach to evolution of matter in the microcosm and macrocosmos

DE GRUYTER

Author
Prof. Dr. Christian Jooss
University of Goettingen
Institute of Materials Physics
Friedrich-Hund-Platz 1
37077 Göttingen
cjooss@gwdg.de

ISBN 978-3-11-064419-7
e-ISBN (PDF) 978-3-11-064420-3
e-ISBN (EPUB) 978-3-11-064431-9

Library of Congress Control Number: 2020931596

Bibliographic information published by the Deutsche Nationalbibliothek
The Deutsche Nationalbibliothek lists this publication in the Deutsche Nationalbibliografie;
detailed bibliographic data are available on the Internet at http://dnb.dnb.de.

© 2020 Walter de Gruyter GmbH, Berlin/Boston
Cover image: Grafissimo / E+ / Getty Images
Typesetting: Integra Software Services Pvt. Ltd.
Printing and Binding: LSC Communications, United States

www.degruyter.com

Foreword to the English edition

"A Crisis at the Edge of Physics" was the headline in the New York Times on June 7th, 2015. "Supersymmetry and the Crisis in Physics "was the lead story in the May 2014 issue of *Scientific American*. The physics crisis has broken out openly and none of the particles predicted by "unified matter theories" for decades have been found experimentally with particle accelerators. The physical worldview prevailing today is directed against a deeper understanding of the structure and development processes of matter. Instead, it seeks the essence of matter in mathematical symmetries of microscopic building blocks of matter that stand above nature. "Lost in mathematics" criticizes the physicist S. Hossenfelder [Hossenfelder 2018]. Instead of experimental observations, particle physics is guided by the "mathematical beauty of equations," Hossenfelder says.

However, the crisis in physics that has arisen not only affects individual disciplines, such as particle physics. In cosmology, the Big Bang theory claims the "creation of matter and energy from nothing in an initial singularity." It is based solely on the interpretation of the redshift of the light of distant celestial bodies as a general "expansion of space" and is detached from the observable development of matter. In the underlying General Theory of Relativity, gravity was geometrized, declared a property of a curved "empty" space, a "nothing." The "origin of matter from nothing" in the Big Bang, the "substitution of matter by the geometry of empty space" in gravity, and the interpretation of subatomic particles and quantum fields in particle physics as "excitations of nothing." The open crisis of different areas of physics has a common cause in epistemology.

This epistemological crisis of physics arose despite tremendous progress in individual topics. It began already at the turn of the 19th to the 20th century, when under the influence of idealistic philosophies, especially empirical criticism, pragmatism and neopositivism, the claim of physics, to recognise and understand matter as a reality that exists objectively and independently of human consciousness, was abandoned. Max Planck stated already in the early 1930s in his lecture: "Positivism and the Real External World" with concern: *"Even this (physics) has not been spared from the general crisis. A certain uncertainty has arisen in its field, and opinions on epistemological questions sometimes diverge considerably. Its previously generally accepted principles, even causality itself, are sometimes thrown overboard."* [Planck 1949 p. 228]

Every scientist works on the foundation of a worldview which affects the selection and methods of their experiments, their cognitive process and their conclusions. Worldviews are a system of theoretical views and judgments about nature and society and interact in many ways with the methodology of scientific work. Despite their diversity, however, they must all be differentiated according to how they answer the fundamental question of the relationship between being and consciousness: whether they belong to the materialistic direction according to which being, objective reality, is primary, can be reflected by human consciousness and exists independently of it;

https://doi.org/10.1515/9783110644203-202

or whether they belong to the idealistic direction, whereby complexes of sensation, ideas or principles are primary, standing above reality, which is thus secondary.

The emergence of opposing worldviews and their effects in the natural sciences have their origins in the division of society into classes. Every ideology bears the mark of a class. Thus, with its emergence at the beginning of modern times in the ideological struggle of the bourgeoisie against idealistic feudal scholasticism, modern natural science first produced a native materialism based on the unity of theory and practice.

As revolutionary as this birth phase of modern natural science was, its further development in the turn from the 19th to the 20th century was not able to correctly interpret the new findings of the time, as a result of its metaphysical mechanical concept. Dialectics as a theory and research method of development processes had to, and did, find its way into natural science: Kant predicted brilliantly the idea of the birth and death of solar systems in space. Darwin developed the theory of evolution, the theory of the development of living beings from lower to higher levels. The materialistic inversion of the systematic exposition of dialectics by Hegel and the observation of class struggles in the 19th century formed a basis for Karl Marx's and Friedrich Engels' dialectical-materialistic view of development in nature and society. As a scientific worldview of the working class, it not only enriched humanity's efforts to liberate itself from capitalist exploitation and oppression, but also the thinking and research of progressive scientists worldwide.

At the turn of the 20th century, new discoveries in the natural sciences (electricity, radioactivity, effect of gravitation on light, etc.) called for these to be interpreted in a comprehensive, dialectical and materialist approach. Although even today many natural scientists spontaneously lean toward materialism, the crisis of bourgeois idealist ideology penetrated into the natural sciences: neopositivism and pragmatism became the dominant variant of idealistic epistemology in science. Although it enables individual insights, it fragments the necessary knowledge of the overall context of nature and separates mathematical theory from reality in theoretical interpretation. Thus, the concept of mathematical models becomes primary vis-à-vis matter. However, modern mathematics with its tremendous effectiveness can only support the theoretical understanding of matter and achieve quantitative predictions if it is understood as an approximate reflection of objective reality and if it is consciously applied in this sense, i.e., with the awareness of its limits.

This book *Self-Organisation of Matter* critically examines the effect of physical idealism in the theoretical understanding of structure and evolution of matter, from quantum fields and development of subatomic building blocks to cosmology, the evolution of galaxies, and the gigantic structures of galactic superclusters. In particular, the properties and development processes of the structural planes of matter below the atomic building blocks and their relationship to the macrocosm are investigated.

In contrast to physical idealism, which replaces microscopic systems of matter by a "vacuum," a "void" that form multi-dimensional mathematical spaces, in which supposedly "superstrings" or "membrane worlds" are formed, the book consistently takes a materialistic point of view: the various quantum fields and their excitations in the form of subatomic particles are combined with the theory of the superfluid quantum æther. To this end, it builds on the progress of knowledge in the understanding of systems of condensed matter and their possibility of different quantum-fluid states, for example, superfluid states. In fact, the Higgs boson, the only new particle that was probably found at the CERN particle accelerator in 2012, was predicted back in 1964 from the theory of superfluidity [Higgs 1964]. However, this is swept under the carpet in the idealistic interpretation as the "God particle."

The most advanced epistemological theory and method, which is best suited to the complexity of scientific questions, is materialistic dialectics. It is by no means complete once and for all, but must constantly absorb new insights wrested from nature and society, and in so doing, raise itself further and further to the level of the all-embracing and systemic consideration of the overall context of development.

This requires:

- The critical and materialistic analysis and synthesis of a plethora of insights from modern science, combined with the critique and dialectical negation of idealistic interpretations, especially in the context of neopositivism.
- A qualitative comparative method of investigating the development processes of various forms of matter as a materialistic basis for an approximate quantitative description using mathematical-physical theories.
- The determination of the inner driving forces, opposing forces and contradictions that are decisive for development processes and their unfolding, depending on external conditions and influences.
- The generalization of new dialectical-materialistic concepts of the development of matter systems, especially their development through self-organization and the role of the interaction of the development of different systems in the development of the cosmos.
- A materialistic understanding of the nature of natural laws as emergent, as evolving and changing with the structure of matter.

This book is intended to be a contribution to the discussion, a polemic, and a suggestion for a work program for research, and by no means a complete theory and method. It is based on studies and lectures over a long period of time in constant critical discussion with numerous colleagues. An important starting point was the work of W. Dickhut on materialistic dialectics in natural science [Dickhut 1987]. Further important suggestions came from G. E. Volovik on the critical investigation of models of particle physics in the light of quantum fluids [Volovik 2003], from F. Selleri, H. Preston, and F. Potter on the critical analysis of relativity theory [Selleri 1998, 2004; Preston & Potter 2006], from J. Lutz on development processes of the galaxies [Lutz 1991], as well as from L. Landau's dialectical approach to physics in the socialist period of the former Soviet Union. I would like to thank K. Arnecke, F. Hessmann, H.-U. Jüttner, J. Lutz, W.-D. Rochlitz, C. Volkert, and R. Wolk for their critical remarks during the final editing of the book and others for valuable critiques and corrections, which were included in the english edition. Their mention does not mean in any way that they agree with the entire contents of the book.

Göttingen, December 2019
Christian Jooss

Contents

1 Two conflicting directions in modern natural science

Research into the structure and motion of matter in the microcosm and macrocosm has made enormous progress in the 20th and the early 21st centuries.

The world around us produces a vastly complex variety of physical systems. In the microcosm, matter consists of molecules with their atoms, which in turn consist of different subatomic building blocks, such as atomic nuclei, electrons, and photons. It has now been proven that even the protons and neutrons as building blocks of atomic nuclei consist of quarks and gluons, which move like a "bubble" in a surrounding gluon fluid. In the macrocosm, our Earth is part of the solar system, which together with billions of other stars forms our galaxy, the Milky Way. This, in turn, is part of galaxy clusters, galaxy superclusters, and vast cosmic structures of matter.

A huge amount of experimental material has collected showing that the different forms of matter form a system of structural levels that build on each other and interact with one another in their development, with length scales ranging from less than femtometres (10^{-15} m as the classical radius of the electron) to larger than billions of light years (10^{25} m as the size of the Great Wall of galaxy superclusters); see Figure 1.

Their motions, transformations, and developments run on time scales from less than one femtosecond (10^{-15} s for electronic transitions) to more than tens of billions of years (over 10^{18} for evolutionary galaxy processes). The question of the theoretical interpretation, the nature, the origin, and the laws of development of these material structures is the subject of a lively and controversial scientific and social debate. In the multitude of theories and views, there are two opposing views: the dialectical-materialistic view, according to which matter exists objectively and independently of our consciousness and is itself subject to an infinite development that encompasses all structural levels of matter. At every structural level, it produces qualitatively new characteristics and types of motion that can be researched through observation and experimentation and theoretically understood through scientific work.

In contrast, the idealistic theory that is socially dominant today means a historical regression, according to which matter can ultimately be traced back to geometry, principles, ideas, and a "world formula" that stands above matter. According to it, matter originated from nothing in a "big bang" and in a linear chain of development, will finally end gradually in a heat death, unless a new big bang continues the development. Such hypotheses are detached from any practical experience gained through scientific experiments; they are based on the separation of theory and experiment.

https://doi.org/10.1515/9783110644203-001

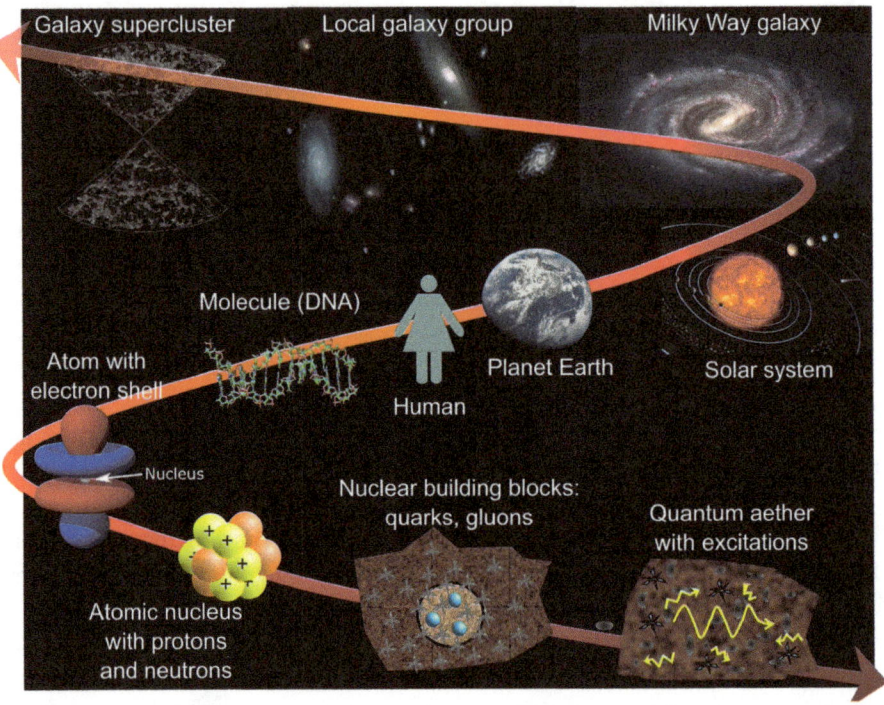

Figure 1: Structural levels of matter in the microcosm and macrocosm (not to scale).

1.1 The replacement of matter by geometry and world formulas

This predominant world view wants to trace the structure and development of all matter in the cosmos back to the final elementary building blocks, which are located in an "empty space," the "vacuum." The unity of matter is sought in geometric structures, the "strings" as vibrating energy filaments in multi-dimensional mathematical spaces.

The geometrization of properties of matter by curved spaces is also a component of corresponding interpretations of Einstein's relativity theories of gravity and rapid motion. The string theorist Brian Greene describes in his book *The Fabric of the Cosmos* the idealistic pole of theories of the structure of matter which is widespread today:

> According to superstring theory, each particle consists of a tiny energy string, about 100 billion times smaller than a single atomic nucleus (far too small for our experimental techniques today) and shaped like a tiny string. Just as a violin string can have different vibrational patterns, each corresponding to a different note, so the filaments of superstring theory have different vibrational patterns. These vibrations, however, do not correspond to different notes, but rather, as the theory interestingly claims, to different particle properties. [Greene 2004, p. 33]

The theory of the origin of matter from immaterial "energy strings" is not the result of experimental observations and their generalization by induction. The superstring theory arises instead deductively from a mathematical construct combining two theories in the search for a "world formula":

> As we will see, the fusion of general relativity and quantum mechanics proposed in superstring theory makes mathematical sense only if we impose another revolution on our idea of space-time. Instead of the three spatial dimensions and the one temporal dimension of our everyday experience, superstring theory demands nine spatial dimensions and one temporal dimension. In an even more viable version of superstring theory, known as "M theory," there are even ten spatial dimensions and one temporal dimension – a cosmic substrate with a total of eleven space-time dimensions. (...) And the space offered by the large additional spatial dimensions could open up even more remarkable possibilities: other, nearby worlds, not nearby in ordinary space, but nearby in the extra-dimensional worlds of which we have not yet noticed anything.
>
> [Greene 2004, pp. 34–35]

The replacement of matter by the geometry of curved empty spaces is methodically and ideologically related to the views of the ancient idealist Plato, who rejected the materialistic atomic hypothesis and replaced matter with idealized geometric bodies. Today, the geometrization of matter in the microcosm is an essential theoretical basis of big-bang cosmology. A big bang can only be calculated by the geometrization of gravity.

According to the Big Bang theory, although before the Big Bang there was no matter, space or time, there were already "laws of nature" in the form of a world formula and "initial conditions," from which the real manifold world with all its different forms of matter is supposed to have emerged in the last 14 billion years. The origin of all matter is traced back to an absolute idea standing above matter. The idealistic processing of the discoveries has led to a deep crisis of the physical view of the world as a whole, despite all the advances in knowledge in detail.

1.2 The discovery of a new structural level of matter at the beginning of the 20th century

An essential starting point of the crisis of modern physics is the absolute rejection of the concept of æther, not only its unsustainable mechanical aspects, at the beginning of the 20th century. Exactly at the time when, with the discovery of quantum effects and the increase in the mass of the electron at high speeds, the first effects of deeper structural levels of matter became experimentally visible, the material causes of these effects were negated under the influence of the spreading positivism and idealism. From the justified criticism of too simple and naïve concepts of the æther developed the special and general theory of relativity as well as Copenhagen interpretation of quantum theory.

The mechanical concept of æther as fluid, which "surrounds" corpuscular matter and was supposed make itself noticed by mechanical forms of motion like æther winds and elastic æther vibrations (light), had become untenable because it led to blatant contradictions to reality. Instead of finding a dialectic concept of æther, however, the baby was thrown out with the bathwater. The replacement of matter by the assumption of an "empty space," a "vacuum," a "nothing," provided the fundamental basis for the geometrization of matter.

Despite the abolition of the æther from the worldview of physics, at the same time, comprehensive material has accumulated from particle, quantum, and relativistic physics that there is no "empty space," no "vacuum." It is teeming with unstable particles, the zero-point fields. The "vacuum" can be polarized electrically and magnetically – yes, it is even possible to melt the "vacuum" (according to the prescribed language shaped by the positivist world view) and to transform it into another state of matter by means of phase transformations. The "zoo of elementary particles" known today shows that these particles are not "mass points in empty space," but excitation states of a deeper structural level of matter. Even though these facts are known to many scientists, these contradictions are not openly discussed.

Thus the prevailing doctrine includes the constantly regurgitated statement "Einstein's theory of relativity proves that there is no æther." Albert Einstein, however, formulated in a letter to H. A. Lorentz in 1919 that he would have abolished the idea of the velocity of æther:

> "It would have been more correct, if in my earlier publications I would have limited myself to emphasizing the nonreality of æther velocity, instead of asserting the nonexistence of æther at all. For I realize that the word "æther" means nothing more than that space must be understood as a medium of physical qualities.   [Kostro 2000, p. 189]

The idea of æther velocity was based on the mechanical worldview, which contrasted particles and æther. Einstein at times saw his theories as an expression of a dynamic æther, whose motions are determined by spatial distribution and motion of masses and which affects the internal atomic motions (slowing of clocks, the redshift of light, length-contraction) in various ways [ibid].

A decisive contribution to the consolidation of physical idealism has been made by the fact that the deterministic laws of mechanics applicable to the local motions of macroscopic bodies have given way to statistical laws of motion in the microscopic world.

A number of physicists, including Albert Einstein, denied the objective reality of the novel laws of motion of particles in fluctuating quantum fields in their justified criticism of the widespread idealistic Copenhagen interpretation of quantum physics. The quantum physicist Anton Zeilinger writes:

> "Albert Einstein was apparently unhappy throughout his life about the new role of randomness in quantum physics. He expressed this by stressing 'God does not play dice!' Niels Bohr then answered that he should stop telling the Lord God what to do. From our new view of quantum physics as a science of information – as a science of what can be known in

principle – a very natural explanation of randomness follows. And it further follows that this randomness is necessary and unavoidable and cannot be avoided, as Einstein hoped."

[Zeilinger 2003, p. 46]

In fact, quantum theory, with its statistical laws, is the expression of objective and real random fluctuations in the location and velocity of microscopic particles. They have their origin in the excitations of the zero-point field. The random motion of particles in the statistically fluctuating zero-point field leads to qualitatively new laws of motion, which are, among other things, connected with the development of matter-wave fields. Randomness and regularity are inextricably interdependent.

However, Zeilinger does not regard the laws of quantum theory as independent of the researching consciousness, but as the result of what an observer who possesses only "classical measuring instruments" can "know" about the microscopic world. This subjectivist view does not interpret matter waves as real waves, but as "tools for our thinking" [Zeilinger 2003, p. 191] and has today become a decisive obstacle to the theoretical understanding of microscopic processes. Zeilinger's critique of Einstein's rejection of the objectively existing randomness is therefore correct, but not the subjectivist interpretation as something that arises from limited knowledge. The objectively existing dialectic of necessity (regularity) and randomness is of great importance for the understanding of any complex development. In quantum physics, statistical laws are precisely the expression of an objectively existing self-organization process through the mutual interaction of particle motion and randomly fluctuating motions of the zero-point field. This self-organization process leads to the formation of matter waves that produce stable motion patterns while maintaining the energy and momentum of the particle.

1.3 Paradoxical contradictions between microphysics and big-bang cosmology

The replacement of matter with space-time geometry leads to paradoxical contradictions. Although, today, zero-point fields which penetrate the entire "empty space" are experimentally verifiable, they must be declared "virtual matter" under the effect of positivist ideology. According to this, it is present only virtually, but not real. Otherwise, there would be a blatant contradiction in the energy density of the zero-point field. Its energy density, determined from experiments and models of quantum physics, is 120 orders of magnitude greater than is compatible with a big-bang model of an "expanding empty space." In the current Big Bang cosmology, instead, the "empty space" between visible forms of matter is currently arbitrarily filled with dark matter and dark energy in order to save the geometric theory of gravity on length scales of galaxies and superclusters and bring it into line with the observation of a flat space in the cosmos. The separation of theory from experimental observation goes so far that speculative hypotheses are advanced that only 1% of the total matter in the cosmos consists of the building blocks known to us, the atoms with their sub-atomic building blocks. Not that one could rightly assume that there are still unlimited new forms of matter that have not yet been recognized! No, 99% of the cosmos is claimed to be dark matter and energy, introduced to save the Big Bang model.

But in theoretical quantum physics there are also great theoretical difficulties, which consist of the fact that the energy density of the zero-point field ("the vacuum energy") rises rapidly at smaller length scales and leads to infinite values, which are treated with complicated mathematical techniques (what is called "renormalization") in order to be able to calculate correct values of such simple quantities as the electrical charge, or the mass of the electron, at all. Interpreted materialistically, the complex mathematics of quantum field theory is a direct reflection of the fact that even "elementary particles," like the electron, cannot be understood as "punctiform structure-less objects," but represent complex matter structures.

1.4 The anthropic principle: from superstrings to creationism

Instead of systematically investigating the structure and laws of motion of the deeper structural levels of matter more and more deeply, through the interaction of experiment and theory, the arbitrary replacement of particles by one-dimensional energy strings reinforces the paradoxical contradictions. Superstring theory does not even explain the masses or intrinsic angular momenta of the simplest subatomic building blocks. This is not about details: the idea that the different masses of the electron and the proton can be explained with different vibrational states of the string leads to the calculation of masses that are 10^{19} times too large.

The former director of the particle accelerator in Stanford, Burton Richter, has the feeling that superstring theory is a dead end. In a commentary for *Physics Today,* "Theory in Particle Physics: Theological Speculation versus Practical Knowledge" he discusses the latest versions of string theory, from which countless different universes with the most varied properties of the "elementary particles" can be calculated. However,

> "no solution that looks like our universe has been found."   [Richter 2006, S. 8]

As a way out, the string theorists refine their constructions and flee into a fine tuning of "fundamental natural constants." These include quantities such as the mass and electrical charge of the electron or proton as well as the strength of certain natural forces. They largely determine the structure of atoms and the structure of macroscopic forms of matter. Why are these quantities just as they are observed in experiments? Instead of understanding them materialistically as an expression of the self-organization of matter, a "fine tuning" is claimed. And who is supposed to have fine-tuned it? What is called the "anthropic principle" claims that "fine tuning" is dictated by the condition that organic matter, life, and ultimately human consciousness could form. Burton Richter writes quite aptly about this:

> The anthropic principle is an observation and not an explanation. To believe otherwise would mean that the development of mankind in a late stage of the universe is the cause of the natural constants being set to the correct values at the beginning. If you believe that, you are a creationist.   [ibid]

The combination of the world formula and big-bang models is an expression of the replacement of science by religion. George Smoot, who received the Nobel Prize for Physics in 2006 for his study of cosmic background radiation as "confirmation of the theory of the Big Bang," answers the question of what triggered the Big Bang:

> (...) Stephen Hawking's universe, for example, needs no cause at all, it emerges from nothing and draws its power from itself. Somehow a situation was necessary where basic energy and matter are present. If the correct physical laws were added, this would automatically result in a Big Bang and a phase of cosmic inflation. The question would only be, is a God responsible for it? The answer is, of course, uncertain, as it always is when one asks how everything began and why and what caused it. It all boils down to this: someone or something must have made the initial conditions possible and set the experiment in motion, right? So, the question is: is this universe just a calculation made by someone? (...) I suspect that it will be difficult to rule out or prove the existence of a God.   [Smoot 2006, p. 41]

All that remained to be clarified was whether the "fine-tuning of natural constants" was carried out by a creator with or without a beard. The metaphysical principle, which introduces an external impulse for every development, forms the opposite pole to the scientific development theory of matter and ideologically to dialectical materialism.

1.5 The penetration of the positivist world view into physics

To understand the effect and dominance of physical idealism in modern science and research today, we must return to its origins. Positivism was able to spread as its main form in natural science and to dominate it largely ideologically because it combines materialistic knowledge in individual questions with a subtle attack on the ability of natural science to understand matter ever more deeply and generally.

Positivism originally goes back to the Frenchman August Comte, who in 1830 declared in his book *Cours de Philosophie positive* as a reaction to the materialism of the French Revolution:

> *"The positive explanations do not offer causes which produce the phenomena one only examines the circumstances under which they arose and connects them through the relationship ... among themselves."*   [Comte 1830 p. 8]

One of the most influential representatives of positivism at the end of the 19th century was the physicist Ernst Mach, who contributed to bringing this philosophy into physics. In 1883, he wrote in his main work *The Science of Mechanics: A Critical and Historical Account if its Development*, about the source of knowledge in physics:

> "Nature is composed of the elements given by the senses. The natural man, however, first identifies certain complexes of these elements which occur with relative stability and which are more important to him. The first and oldest words are names for "things." (…) The sensations are also not "symbols of things." Rather, the "thing" is a thought symbol for a complex of sensations of relative stability. Not the things (bodies), but colors, sounds, pressures, spaces, times (what we usually call sensations) are real elements of the world."   [Mach 1883, p. 457]

Thus, the world exists only because human beings distinguish between different sensory complexes through thought symbols. According to Mach, one's sensations and thoughts are therefore not a reflection, an illustration of objective reality, but the other way round: reality is a product of human consciousness.

The essence of positivism's attack on materialism is the assertion that matter is something indefinite, something abstract, something that cannot be verified by any experience and practice, and thus something "metaphysical," that is, something outside experience. The replacement of the concept of "matter" by that of "experience" is, therefore, the only scientifically permissible one. Even if today no physicist would seriously doubt the objective and real existence of light as electromagnetic radiation with different wavelengths, which are reflected in the retina and in the brain of man in different colors, Mach's epistemology is the ideological starting point of a whole series of idealistic dogmas in modern physics. This ranges from the idealistic Copenhagen interpretation of quantum mechanics to the rejection of a material carrier of electromagnetic fields to the questioning of the objectivity of space and time.

The excellent book *Materialism and Empirio-Criticism* by V. I. Lenin of 1908 was the first to thoroughly examine the penetration of positivism into modern physics and to further develop materialist epistemology in the critique of "Machism." The reflection theory of dialectical materialism states: *"For the materialist, on the contrary, the world is richer, more alive, more diverse than it seems, because every step of scientific development discovers new sides in it. For the materialist, our sensations are images of the only and last objective reality – the last one not in the sense that it is already completely recognized, but in the sense that there is no other beside it exists."* [Lenin 1908, p. 123]

Lenin also examines the crisis of modern physics in its early days. *"The essence of the crisis of modern physics consists in the destruction of the old laws and basic principles, in the abandonment of the objective reality existing outside consciousness, i.e. in the replacement of materialism by idealism and agnosticism. 'Matter has disappeared' – this is how one can express the fundamental difficulty created by this crisis, which is typical of many individual questions."* [Ibid, p. 257]

The present crisis of physics did not arise from the fact that a certain historical stage of development of knowledge was in a crisis. Thus, at the end of the 19th century, the laws of mechanics had to be extended and enriched by new laws of electrodynamics, relativity, quantum physics and so on. Thomas Kuhn therefore, describes in his book *The Structure of Scientific Revolutions* [Kuhn 1962] the emergence of crises in the natural sciences as the result of new discoveries that contradict given paradigms and are discussed as "anomalies."

However, the appearance of such contradictions and their resolution through theoretical extensions or, where necessary, corrections that can also assume the dimension of a scientific revolution, is a normal driving force of natural science. In contrast, today's crisis, was caused by turning away from materialistic epistemology and worldview. It went hand in hand with the scientific revolution of physics in the 20th century, but has epistemological causes that ultimately stem from the development of crises of capitalist society.

1.6 Popper's attack on the inductive method

The influence of positivist ideology in science has continued to be refined since the early 20th century. In the 1930s to 1950s, positivism reoriented itself, especially against the objective reality of the laws of development of nature, society, and consciousness. One of its main representatives, Karl R. Popper, tried to build a dam against the materialistic knowledge of nature. In his influential book *The Logic of Scientific Discovers* he denied the possibility of deriving laws of nature from the observation of reality by means of the inductive method:

> In our opinion, however, there is no induction. The conclusion of the special statements on the theory verified by "experience" is logically inadmissible, theories are thus never empirically verifiable. If we want to avoid the positivistic error of excluding the scientific-theoretical systems by the demarcation criterion, we must choose this in such a way that even sentences that are not verifiable can be recognized as empirical.
>
> Now, however, we only want to recognize as empirical a system that is capable of verification through "experience." This consideration suggests the idea of proposing as a differentiation criterion, not the verifiability, but the falsifiability of the system; in other words: although we do not demand that the system can finally be positively marked by empirical-methodological means, we demand that the logical form of the system makes it possible to mark it negatively by means of methodological verification: an empirical-scientific system must be able to fail because of experience. [Popper 1935, p. 13]

The existence of theories, concepts and conceptions should be accepted without questioning the method of their realization:

> "We have characterized the activity of the scientific researcher initially as setting up and examining theories." [ibid, p. 11]

The complete rejection of the inductive method, which is a component of centuries of successful natural science, establishes the epistemological freedom to "invent" all possible systems and laws by deduction alone, which are all equal and true until they are falsified. Superstrings, rolled up multi-dimensional spaces, Big Bang – everything is scientifically allowed as long as the theory delivers individual statements that agree with individual measurement data. But in fact, materialistic insights can only develop in the fundamental unity of theory and practice as a growing infinite process of cognition. The dialectic of inductive and deductive methods is fundamental for the examination by the natural sciences of ever new stages and types of motion of matter, and its systemic overall context. The confirmation of a theory by its application in practice is the criterion of truth. A truth that will never be complete because there is an infinite dialectical process of bringing relative human knowledge closer to the absolute truth, i.e., the objective reality.

1.7 Idealistic rejection of laws of development

Popper's main attack in his philosophical pamphlet *The Poverty of Historicism* is directed at the possibility of a scientific theory of the development of matter, of life, or even of a society:

> But can there be a law of evolution? (. . .) I am of the opinion that this question must be answered with "no" and that the search for the law of the "irrevocable order" of development can by no means fall within the scope of the scientific method, regardless of whether it is biology or sociology. My reasons for this are very simple. The development of life on earth and human society is a unique historical process. We may assume that such a process takes place according to a range of different causal laws, such as the laws of mechanics, chemistry, heredity and segregation, natural selection, and so on. Its description, however, is not a law, but only a singular historical statement. [Popper 1965, p. 85 ff]

A complete regression over a hundred years after Darwin's ground-breaking theory of evolution! As if the interaction of various lawful processes and coincidences is not always the basic condition for the development of complex forms of matter such as life, which occurs lawfully under appropriate conditions and is not a singular historical happenstance, even if the concrete course is subject to many chance conditions. Dialectical materialism also rejects the existence of a "law of the 'irrevocable order' of development," for such a law would be metaphysical in content.

In his polemical book, Popper denies even the existence of laws of development of solar systems from dust clouds that were already generally accepted at that time. He recognizes only the lawful character of eternally "repeating celestial mechanics" [ibid, p. 89] – so much for his corruption of "development" as "irrevocable order." Popper thus gives the epistemological justification for the disintegration of natural science into special fields and the detachment of theory from reality. The rejection of the existence of laws of development is at the core of the ideological superstructure of scientific theory in a capitalist society shaken by crisis developments since the beginning of the 20th century. Although many natural scientists possess a natural relationship to the materialistic world view (attacked by positivism as "naive realism"), natural scientists cannot free themselves from the ideological effects of crises without becoming aware of these questions.

1.8 About the character of natural laws

In connection with the natural laws, Hegel writes in the *Science of Logic*: *"The law is the reflection of appearance into self-identity; (...) This identity, the substrate of appearance, which constitutes law, is appearances's own moment; (...) The law, therefore, is not beyond appearance but is immediately present in it; the kingdom of laws is the restful [emphasized by Hegel] copy of the concretely existing or appearing world."* [Hegel 1813, p. 4441]

And Lenin comments: *"This is a remarkably materialistic and remarkably appropriate (with the word "quiescent") determination. Law takes the quiescent – and therefore law, every law, is narrow, incomplete, approximate. (...) Law is the reflection of the essential in the movement of the universe. The appearance is richer than law."* [Lenin 1914, p. 161]

All laws of nature, as good as they have proven themselves in explaining and solving one or the other problem, always carry the moment of simplification within them. The rejection of a scientific theory of the development of matter by positivism is absurdly accompanied by the elevation of individual approximation models to dogmas, which could be derived from symmetries, mathematical superstructures, or even a "world formula." Just the search for such a world formula is nothing but the search for a final truth, unchangeable for all eternity, which is diametrically opposed to the rich knowledge of physics about the mutual conditionality of the development of matter systems and their dialectical laws of motion.

A striking criticism of the reductionist explanation of natural laws via a "world formula" by means of properties of elementary "basic building blocks" is made by Physics Nobel Laureate Bob Laughlin:

> The natural world is governed both by essential building blocks/elements and by powerful principles of organization that emerge from them. These principles are transcendent in the sense that their validity remains even if the essential elements change somewhat. (...) In other words, the laws of nature that we perceive develop through collective self-organization and really do not need knowledge of their components to be understood and applied.
>
> [Laughlin 2005, p. IX]

In fact, qualitatively new properties emerge in the collective interaction of their components. Structure and motion of the individual components become part of the collective system behavior. Even a small increase in temperature can transform a solid with shear elasticity into a liquid state, for example, ice into liquid water. The shear stiffness suddenly collapses at the transition temperature and new hydrodynamic laws of liquids replace the elastic laws of the solid state. Other forms of matter such as electrons, neutrons, atomic nuclei, the magnetic flux or even a collective of soap bubbles can also assume gaseous, liquid, or solid states of aggregation, with corresponding mechanical or hydrodynamic laws of motion.

These are collective orders, or motions of building blocks of matter that produce such macroscopic properties with specific laws.

The manifold structures of matter cannot be reduced to a "primordial matter particle," nor can the diversity of natural laws be reduced to a "primordial law." Just as in biological evolution, neither the chicken nor the egg were present first, but both arose in the process of species development, so forms of matter develop in the microcosm and macrocosm with their laws of motion in a reciprocal, dialectical process. On different structural levels of matter, new qualitative laws emerge. Also, the dialectic of nature is not a "primordial law," but must be found in concrete reality and develops new aspects with each stage of development of matter.

1.9 Development from within through self-organization

In the 1930s, the Soviet physicist Lev Landau, among others, provided impulses for the study of the laws of self-organization of complex systems. In his theory of phase transitions, he generalized insights from various interacting many-particle systems (atoms, magnetic moments, electrons, etc.).

He explained how, despite chaotic individual motions of the individual particles, orders spontaneously occur in the form of phases determined by the appearance of new collective quantities (elasticity, ferromagnetism, superconductivity, etc.). In the 1950s, impulses came from the development of cybernetics as the science of self-control of systems by feedback. Even seemingly chaotic systems, such as molecules with a thermal random motion in gases and liquids, develop orders in the driven state far away from equilibrium. Examples are the transition from diffusion to laminar or turbulent flow or the transition from heat transport by diffusion to convection movement in ordered cells.

In the 1970s, Nobel Laureate Ilyia Prigogine developed the theory that the minimization of energy dissipation has a regulating effect in the development of complex structures far from equilibrium, based on experimental investigations of biochemical processes. He pointed out that, depending on the degree of nonequilibrium, a number of patterns occur in chemical reactions which are sharply separated by instabilities [Prigogine 1977].

He concludes for the development of the cosmos:

"The universe – just like the origin of life – can only be the result of a succession of instabilities."   [Prigogine & Stengers 1993, p. 296]

Consistently thought through to the end, this would be a fundamental critique of the Big Bang theory. Although he could not completely detach himself from it, he developed the consideration that the "quantum vacuum" in a phase transition transforms into developed forms of matter and gravity. [ibidem, p. 300]

Important contributions to the generalization of dialectical laws of self-organization were made in the 1980s by physicist Herrmann Haken. On the basis of the formation of ordered structures of matter, for example, in the generation of laser light, or in the stable flow cells, he generalized that the formation of opposing collective types of motion, their competition, and their interaction, is a basic law of development through self-organization [Haken 1981, 2004]. The dialectic of randomness and necessity is essentially expressed in self-organization. For example, in stable dynamic states of the convection cell, the collective ordered movement determines the individual movement, "the order enslaves the individual parts."

In contrast, for the erratic transition between different patterns of convection cells, random fluctuations, and their interaction play a decisive role in destabilizing one order and stabilizing another.

A Nobel Laureate in Physics, Hannes Alfvén, developed a theory of the electrodynamic development of stars and galaxies out of themselves [Alfvén 1984] based on the investigation of the laws of self-organization in plasmas. This theory was taken up by the US physicist Eric Lerner in his critical book *The Big Bang Never Happened* [Lerner 1992]. However, they only consider the electromagnetic structure formation of matter and neglect the existence of other structural levels of matter. The physicist Josef Lutz developed a fundamental critique of the Big Bang theory [Lutz 1991] based on a summary of the experimental observation of structure and development of different matter systems in the macrocosm. Building on the work of V. A. Ambarzumjan [Ambarzumjan 1976], he showed that, similarly to the evolutionary process of stars, galaxy systems are subject to an evolutionary process in which the galaxies go through different phases of their evolution.

What is still pending today, however, is the investigation of the overall connection and the interdependence of these development processes in the macrocosm with those in the microcosm.

1.10 Why a new dialectic concept of the æther is necessary

A dialectical-materialistic theory of the development of matter in the cosmos today therefore requires a systemic view that includes all structural levels of matter and their interactions, and at the same time, the coining of new appropriate dialectical terms. This affects, in particular, the "æther". Neither the mechanical concept of æther, nor that of the "vacuum state," nor the concept of the field, are suitable for the designation of this new structural level of matter.

Fields as expressions of excitations, tensions or orders are always only expressions of material processes. A new concept of æther is necessary precisely to find the unity of the different fields (electromagnetic fields, fields of gravity, strong and weak nuclear force, matter waves etc.) within a new structural level of matter. Even though the concept of a "vacuum" has undergone a subtle change of meaning in recent decades toward the "ground state of a many-particle system," "vacuum" is ultimately linked ideologically with the idea of an "emptiness," and its motions as "excitations of the void." Therefore, the new term "quantum æther" is used in this book.

This term ties in with the progressive role of the concept of æther in the history of science [Steimle 1998]. In contrast to the mechanical concepts of the 19th century, however, the revolutionary findings from the research into quantum fluids are taken into account by the addition of "quantum."

Other terms are sometimes used for this purpose, such as quantum vacuum, Planck medium [Volovik 2003], Dirac sea [Dirac 1930], or continuous matter [Dickhut 1987]. The term "Dirac Sea " is historically a first approximation from relativistic quantum theory to this new stage of matter. However, it is too closely linked to the special model of Dirac. The term "continuous matter," first proposed by W. Dickhut in the 1940s [Dickhut 1987], is a general antithesis to the term "particle." In his remarkable study *Materialistic Dialectics and Bourgeois Natural Science*, he processed the findings of quantum physics into a dialectical theory of the development of particles from the continuum, criticizing both idealistic interpretations and mechanical ideas associated with the concept of æther. Continuous matter is not a physical expression for a specific substance but a fundamental epistemological concept in the dialectic of the reciprocal change of continuous and discrete forms of matter. It means that particles are concentration points, or development products, of matter and that there is no "empty space without matter" between the particles.

With the study of quantum fluids, an understanding has developed that a collective of discrete particles can form a continuum in which their individual motions cancel each other out completely in the collective. And vice versa, that quantized particles result from certain motions of continua. On the basis of these findings, it can be assumed that the quantum æther also has discrete and continuous aspects, and is only a further stage of matter as it progresses to ever smaller systems in the microcosm. Thus it not only produces different quantum fields and particles as development products, but also consists of different condensates with their forms of movement. This is proven, for example, by the occurrence of different states of aggregation of the quantum æther (Dirac phase, electroweak phase and quark-gluon plasma).

1.11 Dialectics of self-organization of matter

The struggle between the two directions in modern science is reaching a decisive phase. With the development of ever new observation instruments and methods, natural science is penetrating ever deeper into the microcosm and the macrocosm. Every structural level of matter, once regarded as "elementary," turned out to be composed of new, deeper layers of matter. Every structure declared the largest in the macrocosm has been replaced by the discovery of even larger structures. Without a conscious processing of the individual findings guided by a progressive worldview, however, the crisis of physics cannot be overcome.

The knowledge of the unlimited forms of matter and its motions, transformation, and development from within itself brilliantly confirms and enriches the dialectical-materialistic basic conception of Marx and Engels that the development of matter is driven by the struggle and unity of opposites and their interpenetration and is unending. Engels explains in *Dialectic of Nature*:

> "The whole of nature accessible to us forms a system, an overall context of bodies, and by bodies, we mean all material existences from the star to the atom, even to the æther particle, as far as its existence is admitted. The fact that these bodies are connected already implies that they interact with each other, and their mutual influence is motion. It is already evident here that matter is unthinkable without motion. And if matter continues to confront us as something given, both uncreatable and indestructible, then it follows that motion is also as uncreatable as indestructible. This conclusion became irrefutable as soon as the universe was recognized as a system, as a connection of bodies." [Engels 1885, p. 355]

The dialectical method also continues to develop with new revolutionary upheavals in the progress of human knowledge. New dialectical laws of motion must consciously be generalized for a theory of the development of matter in the cosmos. Such a higher level of scientific-dialectical thinking related to the overall context requires social conditions in which the upcoming tasks of mankind are solved in the interest of the common good by a collective combination of theory and practice. In other words, a socialist society, in which production, way of life and natural science can develop according to plan, in the unity of man and nature in the interest of the common good. In view of the intensification of environmental destruction, this has become an urgent question of survival.

The necessary higher stage of the materialistic worldview, the materialistic dialectic, can only win through if the social causes for the flourishing of the various forms of idealism in capitalistic class society have been eliminated. Idealism, with its limitation to subjective individual knowledge, its positivist fragmentation of science, its pragmatism, and various idealistic theoretical constructs, is nothing more than today's ideological superstructure of the profit economy.

It expresses the desire of the ruling classes to maintain traditional social conditions, which is reflected in an epistemological crisis in the natural sciences and the inability to solve urgent social problems in practice.

The new image of the cosmos that needs to gain acceptance is not a one-way street of development. Nor is it a reductionist one-way hierarchy of systems, according to which the forms of matter in the microcosm create the macrocosm "from bottom to top." It becomes clear that the structure and motion of matter in the macrocosm determine the structure and forces of the subatomic particles and their ability to form structures at higher levels, as well as vice versa. A dialectical-materialistic development theory of matter out of itself through self-organization must, therefore, advance from the analysis and synthesis of the formation and development of individual forms of matter to the mutual conditionality of all essential systems and motions, to the understanding of self-organization and transformation of matter on all structural levels.

2 Self-organization in many-particle systems

Many forms of matter in our everyday world consists of atoms. 118 different chemical elements are known today, which differ in the number of nuclear building blocks and the number of electrons in their atomic shell. Chemical bonds of the same or different types of atoms are the basis for the formation of an almost infinite variety of types of molecules, from simple molecules such as hydrogen (H_2) to complex amino acids and proteins. These consist of thousands of atoms and form the basic building blocks of life. Life with its manifold chemical, biological, psychological, and social processes certainly represents the most complex form of matter known to us today.

Even in inanimate matter, extremely diverse and complex structures occur, that arise from the collective behavior of many-particle systems through self-organization. A many-particle system of atoms can form different states of aggregation from solid, liquid, or gaseous, to plasma. In them, atoms form long range ordered, short range ordered, or irregular structures. Different orders of the particles are generally referred to as phases.

The type of order and the resulting collective forms of movement are determined by the struggle and unity of attraction by building atomic bonds and repulsion by thermal motion. Many-particle systems produce qualitatively different properties in each phase, which are to some extend independent of their composition. The mechanical properties of the solid phase are determined by volume elasticity, shear elasticity, and plasticity, the liquid phase by viscosity and vanishing compressibility, and the gaseous phase by high compressibility and low viscosity. This system behavior of the phases depends only slightly on the chemical composition of their individual atoms but strongly on the collective modes of motion. The transition between these states occurs abruptly in phase transitions and is associated with the appearance of new physical properties and laws.

Randomness and necessity interlock and condition each other in the formation of an infinite variety of structures. Even the relatively simple molecules of water not only form the aggregate states solid, liquid, and gas but also in snowflakes they crystallize in an infinite variety of ice crystals, all of which have hexagonal symmetry, although each differs individually from the other. Their structure formation is determined by the laws of self-organization, which universally determine the motion and structure of many-particle systems.

To deny them and to dismiss them like Karl Popper as "singular historical propositions" expresses a tremendous ignorance of entire research directions in materials science and the dynamics of complex systems. This applies even more to self-organization in nonequilibrium systems, in which balancing processes through the transport of matter and energy lawfully lead to completely new forms of structure formation.

https://doi.org/10.1515/9783110644203-002

2.1 The mode of existence of atoms in phases

The formation of different phases is a fundamental property of all many-particle systems. Let us consider a snowy winter landscape with a lake whose surface is frozen over. The complexity of nature that comes to light at such a sight is mainly due to the interaction of the different components: snow-covered forests, smooth water and ice surfaces, clouds, a torrential mountain stream in turbulent flow, or dripping icicles on a branch (Figure 2).

Figure 2: Coexistence of different phases of water. Left: Snow-melt in the Rocky Mountains with coexistence of solid (snow–ice), liquid (lake, drops, and clouds) and gaseous (humidity) phases. Middle: Melting icicle with phase boundaries between solid–liquid, solid–gas, and liquid–gas. Right: Ice crystal in the form of a flower. Picture left: Scott Bauer, center: Serge Melki, right: Annick Monnier.

Essential for this richness of structures of matter are the thermodynamic laws of phase formation and transformation:
- The existence of different phases of chemically homogeneous systems, such as water, which has different mechanical, thermal, and optical properties in solid, liquid, and gaseous form.
- The coexistence of these phases under suitable conditions with the formation of sharply defined phase boundaries in which one phase discontinuously merges into another. This not only increases the variety of structures: Phase boundaries are dynamic structures: water constantly evaporates from the liquid, forms clouds and rains or snows down again. Weather conditions, course of day, and season determine phase equilibrium and phase transitions via temperature, pressure, and flow conditions.
- The existence of phases of chemically inhomogeneous systems consisting of different elements and molecules. They cannot only mix. Rather, they differentiate into a variety of different phases with different chemical composition, states of aggregation, and physical–chemical properties.

Even a simple many-particle system consisting of only one type of molecule, such as water molecules, can form an amazing variety of different structures.

The states of aggregation formed are determined by the contradictory internal forces of attraction by bonding and repulsion by temperature motion. The development of contradictory forces depends in the simplest case only on the state variables pressure and temperature. While in the solid phase, the bond forms the main aspect, and atomic motions are limited to oscillations and hopping processes, in the gaseous phase, the motion due to temperature forms the main aspect, and the bonding is limited to repulsions of the atoms; see Figure 3.

An example: Phases and structures of water

The properties of the various phases of water have not only decisively shaped the earth's surface, but are also an essential basis for the emergence of life about 3.6 billion years ago in the primeval oceans. Water in the liquid phase enables a high mobility of the molecules dissolved in it. As a polar molecule, it has the ability to dissolve various substances, counteracts their aggregation and promotes the transport process. These properties are fundamental for the ability of complex organic molecules to reproduce themselves and form a state of living cells in which they enter into a metabolic process with their environment. In addition, the vanishingly small volume compressibility of the liquid state ensures sufficient stability.

The disappearance of shear stiffness, while maintaining volume compressibility, are two decisive macroscopic properties that distinguish the characteristic laws of motion of liquids from those of the solid phase: The macroscopic collective motions of atoms and molecules in the liquid phase are determined by the emergence of new collective laws, those of hydrodynamics, which play no role in the solid state.

The formation of ice, the solid state of water, makes it possible to walk on the surface of lakes and oceans. The basis for the formation of ice on surfaces is the density anomaly of water: the solid state near the freezing point has a lower density than the liquid state. This allows ice to float on water. In addition, there is a property of all solids – shear stiffness.

It arises as a collective property of the solid state of the system of water molecules. In general, it is the basis for the elastic and plastic behavior of solids.

The regular arrangement of atoms with a relative stability of their bonds results in the structure being relatively stable under strain, compression, and shear. As long as the external force does not exceed a critical limit, only a reversible elastic deformation occurs, in which the atoms of the crystal lattice return to their original position after elimination of the external force. Even above this threshold, the stability of the solid state is not lost, but a plastic deformation occurs: The crystal is irreversibly deformed by the formation and motion of crystal defects, the dislocations.

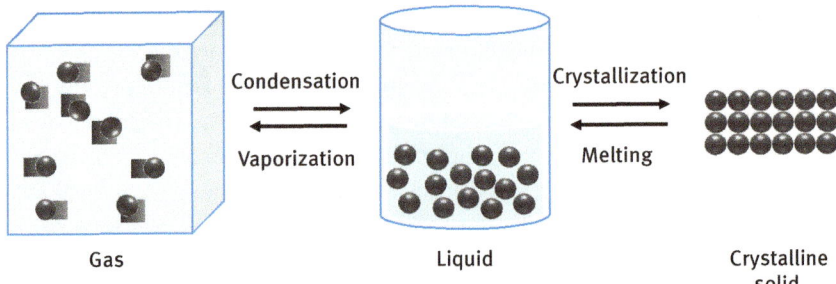

Gas — Condensation / Vaporization — Liquid — Crystallization / Melting — Crystalline solid

Figure 3: Schematic representation of the three phases: gaseous, liquid, and solid and their mutual transformations.

In the stability range of a phase, for example in the solid state with temperatures far away from the melting point, an increase in temperature will only cause a slight increase in the microscopic kinetic energy of the molecules. It is mainly the oscillations of the molecules and atoms that increase. A further process is a gradual increase of the leaps of atoms between different places in the crystal with increasing temperature. Their characteristic bond state, however, remains stable. In contrast, a slight temperature increase close to the melting point will suffice to abruptly change the bond state of the atoms in the solid. The organizational state of the same molecules suddenly loses its characteristic properties for solids. It is a general law that many particle systems at phase transitions abruptly change their properties, quantitative changes turn into qualitative changes.

Ice occurs in different crystal structures. Today, over twelve different ice phases are known, which differ considerably in their microscopic structure, and form at different external temperatures and pressures below freezing [Chaplin 2008]. Including the thousands of different non-equilibrium crystal structures of snowflakes (see also Figure 11), the number of possible crystal forms increases beyond all limits, while preserving the basic elements of crystal symmetry.

The concrete forms depend on fine details of temperature, pressure, and freezing speed, and the symmetry is determined by the electron structure of the water molecules.

The decisive difference between water vapor and water, that is, between a liquid and a gas, is the development of volume compressibility in the gaseous state, while liquids are almost uncompressible. The volume of a gas can be drastically changed by applying external pressure. We use this property of gases, which is reflected in Boyle–Marriott's Law, for example in engines and cooling systems. During the sudden expansion of a gas by pressure reduction, the molecules work against their very weak binding forces in the gaseous state and thereby reduce their thermal motion: the system cools down. The thermal energy of the disordered motion of molecules is converted into the ordered mechanical energy of the machine.

So, water can exist in many qualitatively different states or phases. They differ not only fundamentally in their macroscopic properties but also in their microscopic structure: they are determined by the organization of conflicting tendencies: on the one hand, there is the tendency of molecular bonding forces to attach the water molecules to each other and to build up solid structures. On the other hand, the opposite tendency of microscopic movements, oscillations and rotations to overcome and undermine this attachment.

The conflict and unity of these opposites can, for example, change gradually with temperature. However, the resulting collective properties, their dynamic structure, do not change gradually but form the various phases as systems that stabilize themselves and suddenly transform from one into another at certain critical points (melting temperature, evaporation temperature, etc.) (Figure 4).

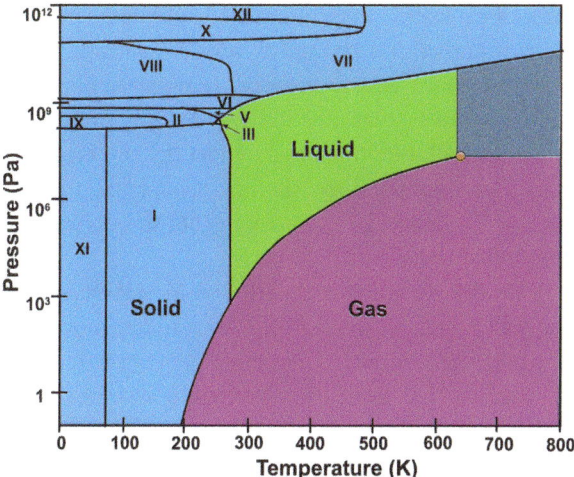

Figure 4: Phase diagram of water. The diagram shows the different states of aggregation that occur as stable states at different temperatures and pressures. Lines indicate the phase boundaries. The evaporation curve, which separates the liquid from the gaseous phase, ends at a critical point (red). Beyond the critical point, the sharp phase boundary between liquid and gaseous phase disappears. Different crystal structures occur in the solid phase, which are marked here with the Roman numbers I-XII. Today, more than twelve different ice structures are known. Water shows an anomaly in the freezing pressure curve. At the phase boundary, ice becomes liquid under increased pressure. This anomaly allows, for example, a skater to glide on ice surfaces because under the pressure of the blade the ice surface becomes liquid [Chaplin 2008].

Randomness and necessity: thermodynamics of phases and phase transitions

Stability or instability of phases can theoretically be understood with a very powerful tool: thermodynamics and its microscopic foundation in statistical mechanics. Phase stability occurs when its free energy is at a global minimum compared to other phases.

Such a minimum of the free energy of a many-particle system, and thus the energy that is actually available for a change of state, is determined by the conflict and unity of two factors:

1. The total energy of the system, which is determined by the kinetic energy of all microscopic movements of the particles and the potential energy of their interactions.
2. The distribution of the total energy among the energy of all microscopic movements of the system components. It must take into account the random, statistical character of the microscopic disordered motions that lead to energy fluctuations and are subject to the laws of probability.

This dialectic of randomness and necessity is taken into account in the free energy of a system by the entropy contribution.

It was a brilliant achievement of Ludwig Boltzmann to identify entropy in the thermal equilibrium of a closed system as the relative number of microscopic possible realizations of the motion patterns of a system. In such systems, the probability (P) that a particle in the thermal bath of all the others has the energy (E), is subject to the Boltzmann distribution. At a fixed temperature, the microscopic distribution of the energy at thermodynamic equilibrium will be just such that the entropy assumes a maximum and the free energy a minimum.

If a cup of coffee is stirred to produce a state of orderly liquid movement and thus smaller entropy, the system will return to the state without rotary movement after a short time and will redistribute orderly macroscopic kinetic energy to disorderly microscopic movements. When it is left to its own devices, it changes to the state of maximum entropy. The fact that a system in equilibrium assumes the most probable state of its microscopic movements and orders is an important statistical law. It is the basis for the success of thermodynamics in the description of phase stability, phase transformations, and energy conversion processes.

A strict prerequisite for the validity of such probability statements on energy distribution is that the system under consideration is isolated, has a uniform temperature, and only exchanges energy fluctuations with its environment. As soon as transport processes of energy and matter occur, the validity is limited or even canceled completely because ordering processes can occur. Even the concept of temperature must be used critically in such systems, since nonequilibrium occupation of microscopic degrees of freedom occurs, which deviates strongly from a thermal (Boltzmann) distribution in equilibrium.

The thesis of the heat death of the universe, which came into fashion in the nineteenth century under the influence of Schopenhauer's pessimistic philosophy [Simonyi 2001, p. 370], is not justified by anything. In order formation in open systems, the probability laws of equilibrium must be replaced by those of nonequilibrium.

Spontaneous formation of order in equilibrium

Even at equilibrium of systems, interactions can spontaneously lead to the formation of an ordered phase that minimizes free energy. Compared to the disordered phase, the entropy of the system is spontaneously reduced. Examples are the ordering of atoms by crystallization, the ordering of magnetic moments by formation of a ferromagnetic phase, the ordering of electrical dipoles in ferroelectrics, or the ordering of electron movements in superconductors. Here the probability distribution of microscopic movements is modified by the interactions between the building blocks. The higher order determines which energy fluctuations can occur in the microscopic movements at all, and thus within which limits entropy in equilibrium assumes a relative maximum relative to the respective order.

A high phase stability occurs when the internal energy of the interactions is reduced as much as possible by the order that is formed, and, simultanously there are a large number of microscopic modes of motion that stabilize this phase. The emerging order changes the microscopic random motions in such a way that those motions that contribute to breaking out of the existing phase are damped or suppressed. Other microscopic random motions, on the other hand, stabilize an existing phase and lower its free energy.

For example, atomic vacancies occurring in a solid due to the thermally induced jumping of atoms can increase the stability of the solid phase. This kind of disorder lowers the free energy via the entropy contribution of the vacancies and occurs lawfully at all temperatures above absolute zero. As the temperature increases and the melting point approaches, the number of vacancies and their mobility increase. The quantitative increase leads to a change of this kind of random motion from a system-stabilizing to a system-destabilizing character. This is shown in the sudden increase of the correlated hopping of several atoms as violent fluctuations of the atomic movements directly at the melting point.

At higher densities, vacancies can form groups that can initiate a local melting transition. At the critical temperature of a system, the type of collective behavior of the atoms changes, whereupon a discontinuous change in the properties of the phase occurs. In other words: the states in the solid and liquid phase of a system that can be achieved by random thermal motions differ discontinuously, and this is associated with a self-stabilization of the respective phases.

The Landau theory of phase transitions

These important findings were generalized by the Soviet physicist Lev Landau in the theory of phase transitions beginning in the 1940s [Landau & Lifshitz 1958]. He looked at those collective properties, orders, and movements that characterize a phase and summarized them under the term of order parameter. These can be very different quantities, the sudden volume change at the solid–liquid or liquid–gas phase transition or the shear elasticity at the solid–liquid phase transition, or the shear viscosity at the liquid–gas phase transition.

Landau also extended the concept of the order parameter to electronic orders, such as the alignment of electron spins in the ferromagnetic phase and of microscopic electrical dipoles in the ferroelectric phase.

Landau's theory of phase transition is based on the grouping of the collective modes characterizing a stable phase in the order parameter. The free energy of the system as a function of the order parameter reflects the transition from stable to unstable system behavior: imagine a sphere in the resulting mountainous "landscape" of free energy (Figure 5). Deflections of the sphere represent the random microscopic movements of the system. As long as a given phase is stable, the sphere can be deflected almost arbitrarily, it will always return to its stability valley. As long as phase stability prevails, fluctuations are not able to induce the solid–liquid, or liquid–gaseous, transition. Only when the free-energy landscape fundamentally changes, that is, an existing state becomes unstable, do the fluctuations drive the phase transition. A more detailed analysis shows that the phase transition is driven by correlated fluctuations: The transition from thermally activated jumping of single atoms to correlated hopping of many atoms drives the local nucleation of the melt at the melting point of solids.

This transition is an expression of the struggle of antagonistic motions of the atoms in both phases.

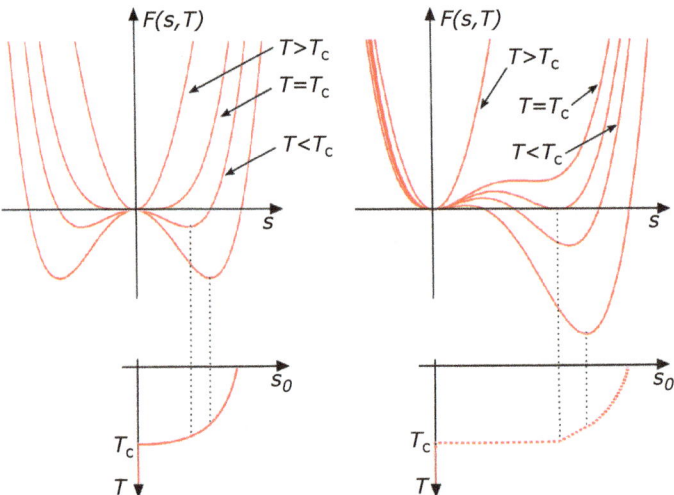

Figure 5: Landau theory for continuous (left) and discontinuous phase transitions (right). For this purpose, the free energy of a system near the critical temperature T_c of a phase transition is represented as a function of those collective modes (order parameters) which represent the new phase. For $T<T_c$ it occurs only as random fluctuation within the minimum of F, the mean value s_0 disappears. For $T>T_c$ it develops as a stable phase.

Phase transitions with and without phase coexistence

In general, phase transitions are determined by the fact that quantitative gradual changes of opposing internal forces turn into sudden qualitative changes of structure and properties. The jump marks the breaking-off of gradual changes and the transition to discontinuous changes.

There are two different basic types: phase transitions with and without phase coexistence. The transition from magnetically disordered to magnetically ordered states by aligning the microscopic moments of the electron spins is an example of phase transitions without phase coexistence. Examples of phase transitions with phase coexistence are the melting or condensing of atoms in the solid–liquid transition. The decisive difference is the necessity to supply latent heat for the phase transition in order to discontinuously change one order into another. The latent heat, therefore, causes a jump in entropy.

Ice and water can coexist at a temperature of 0 °C at standard pressure. Thus, a phase boundary exists. When cooling starts from the liquid state, supercooling of the liquid must be achieved in order to initiate solidification. In the case of water, temperatures below 0 °C are necessary for ice to form. Ice then does not form everywhere at the same time but begins to crystallize in nuclei (Figure 6). Impurities, surfaces, or container walls often represent places of preferred nucleus formation. Nuclei arise as correlated fluctuation, which is destabilizing for the old phase and stabilizing for the new phase, which is now stable. After nucleation, these can continue to grow if heat continues to be extracted, that is, the system is cooled.

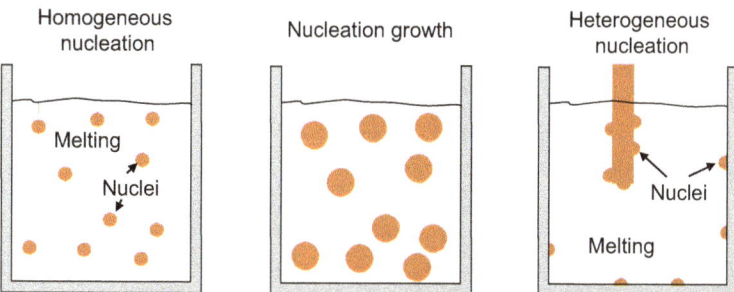

Figure 6: Schematic representation of a phase transformation with phase boundary through nucleation and growth. Left: Homogeneous nucleation using the example of a supercooled melt. Middle: With supercooling, the growth of the nuclei continues until the entire melt is transformed into the solid phase. Right: Heterogeneous nucleation. Sometimes, nuclei of the new phase develop preferably at interfaces to other materials, such as the vessel wall or a seed crystal.

However, the ice and water mixture does not cool down further at first; instead, the extracted energy serves to convert the liquid into the solid phase. Only when the entire liquid has been converted can the temperature drop further. During crystallization, latent heat is released resulting from changes in the structure and movement of the two phases. Nucleation and growth, an inhomogeneous transformation, are the decisive characteristics of phase transitions with latent heat.

Statistical fluctuations also play a decisive role in phase transitions in systems without latent heat and thus without phase coexistence. An example is the transformation of water into water vapor at the critical point or the transformation of magnetic orders. In the vicinity of the critical point such fluctuations, which already approximately represent the new cooperative state in the initial phase, increase very strongly. Correlated fluctuations occur, which means that certain fluctuating modes of movement are coordinated with each other. The critical point is characterized by fluctuations at all length scales; they become scale invariant up to scales of the system size. No matter on which length scales, from the molecule to the macroscopic system, these fluctuations are observed, for example with a microscope of various magnifications, they always look similar statistically.

One can observe this phenomenon of self-similarity very well at the critical point of water, where density fluctuations on all length scales lead to the phenomenon of what is called "critical opalescence," visible as the turbidity of the liquid.

Not only the phase stability, but also the phase transition, is determined by the dialectic of randomness and necessity of microscopic atomic bonds and motions as an expression of the struggle of conflicting inner forces.

2.2 Structure formation close to equilibrium: domains, topological defects, and structures

The concept of order has already been introduced in the previous section, in conjunction with Landau's theory of phase transitions. It is a generalization that characterizes not only static arrangements but also characteristic collective forms of movement. Crystalline solids are characterized by a static periodic long-range spatial ordering of the atomic positions, while liquids only have a dynamic short-range ordering of the nearest neighboring atoms, which constantly dissolves and forms again. Ferromagnetic materials have a long-range ordering of their microscopic magnetic moments, which are aligned, while paramagnets are characterized by an average isotropic orientation of the microscopic moments.

The ordering of magnetic moments is connected with the ordering of microscopic movements. The transition from the thermal light of an incandescent bulb to laser light occurs by ordering of the wave motion of photons. The concept of order can, therefore, refer to different microscopic degrees of freedom such as positions, movements, orientations of atoms, electrons, or light. It is common to all orders that, besides the global transformation of orders in phase transitions, there are also different types of local spatial or temporal deviations from a perfect order, which are subject to their own laws. These include domains and topological structures or defects (see also Section 9.1).

Crystal and magnetic domains

Domains occur when there are different possibilities of ordering locally in a system while preserving the global character of the order. Some examples are shown in Figure 7. The formation of a polycrystalline structure during solidification from the melt is the rule for the formation of solid bodies, unless special measures are taken for single crystal growth.

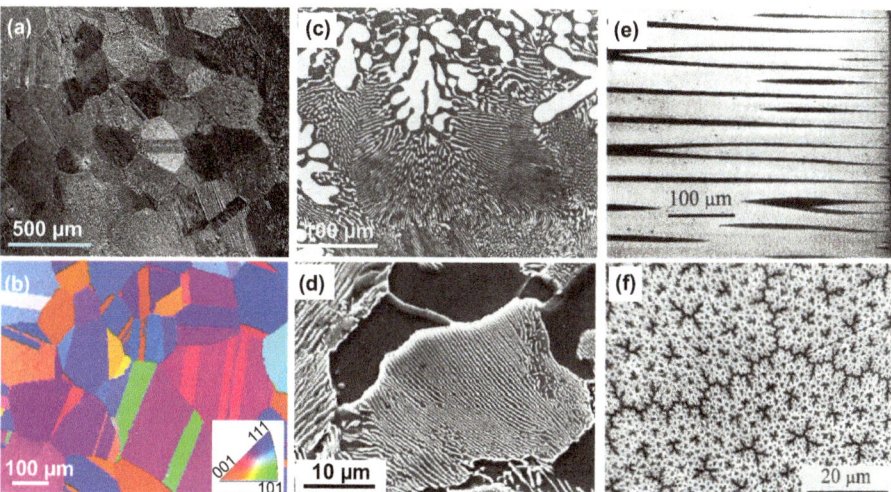

Figure 7: Structure formation through the formation of domains. (a) Polycrystalline structure of a copper layer. The different crystal orientation of the orientation domains/grains is made visible in an orientation map (b). (c) Lamellar structure of an alloy of copper and silver, which is formed during cooling from the melt by segregation of the two elements. (d) Microstructure of steel from carbon-rich perlite and carbon-poor ferrite. (e, f) Magnetic domain structures in a ferromagnetic ordered material. Dark and bright structures represent opposite directions of magnetization. Images (a+b) M. Lenius, M. Tiegel & C. Volkert, (c+d) C. Borchers, (e+f) Hubert & Schäfer, Springer 1998.

It consists of domains which each have the exact crystalline structure characteristic of the material, but have a different orientation of the crystal lattice locally. Polycrystalline domains are formed from the melt by the simultaneous formation of nuclei of the solid phase with different crystal orientations, which then grow together in the further course of the phase transformation. The result is a polycrystalline structure (Figures 7a and b). The orientation distribution of the grains was made visible by electron diffraction.

During the solidification of mixtures of several elements, self-organized chemical domain formation often occurs due to phase separation. Figure 7c shows a typical lamellar structure of an eutectic made of copper and silver. In this example, it consists of copper-and-silver-rich lamellae which arrange themselves regularly. Figure 7d shows the microstructure of carbon-rich steel. This so called "perlite" consists of regular lamellae of carbon-rich cementite (Fe_3C) and carbon-poor ferrite. Such structures determine the strength of the steel to a large degree.

Magnetic domains are another example of self-organized structure formation. They are typical structures in a material with magnetic order. Although the microscopic magnetic moments are uniformly aligned within a domain, they differ from domain to domain (Figures 7e and f). Domains generally arise when the order characterizing a system does not occur globally but locally. Orders can also be described by symmetries. For example, crystalline solids have a certain symmetry of discrete lattice positions which is lower than that of the disordered atomic positions in a melt, liquid, or gas: there are fewer possible positions. The symmetry breaking from higher to lower symmetry exists uniformly within each domain. However, their orientation changes from domain to domain.

Topological crystal defects: grain boundaries and dislocations

Grain boundaries and dislocations are examples of planar and linear topological defects of the crystalline solid state. Dislocations occur in the crystal lattice due to inserted half-planes ending at a dislocation line (Figure 9 right). In this "dislocation nucleus," the atoms are strongly displaced from the correct positions of the perfect crystal. Grain boundaries mediate the transition between crystallites of different crystal orientation, while domain boundaries mediate the transition between domains of different magnetization orientation. Some grain boundaries consist of a periodic arrangement of dislocations (Figure 8). Both types of defects cannot end inside the solid, but only on surfaces. Alternatively, geometrically closed objects such as dislocation rings or the surface of a grain are formed.

Topology is the science of the spatial connection of structures. The term topological defect expresses that the relationship of atomic bonds in the crystal lattice change discontinuously. They cannot be eliminated by a continuous local change of the atomic positions. They are topologically stable because their abolition requires a change in the overall context of the crystal lattice on a macroscopic scale.

A vacancy can move through a crystal through thermally activated hopping processes. Its annihilation is only possible if a surface atom moves to the vacancy.

Topological defects are thus local distortions of order that are inextricably linked to the macroscopic geometry changes of the system. They are also called topological structures in the following, depending on the context. Both terms are

Curved crystal lattice   **Grain boundary**

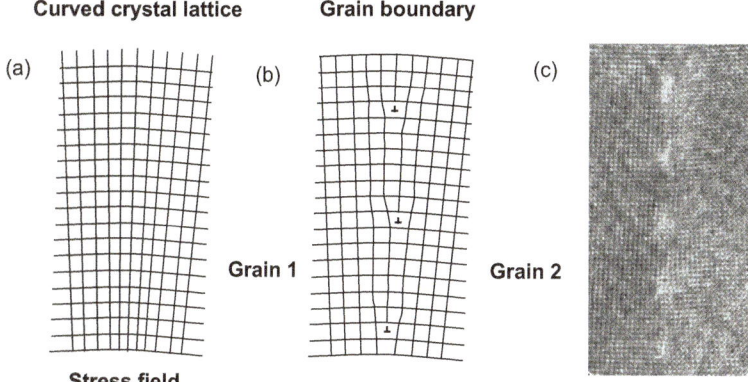

Figure 8: Grain boundaries as topological defects of the crystal lattice. (a) Continuous crystal deflections generate large and long-range stress fields. Their energy is broken down by concentration into local lattice interferences, (b) which form a grain boundary. Orientation changes between two crystal domains are thus localized in grain boundaries. In the case of a small-angle grain boundary, they consist of a chain of dislocations (see also Figure 9). (c) Image of a small-angle grain boundary composed of a periodic array of dislocations in a metal oxide, recorded by a transmission electron microscope.

Vacancies   **Dislocation**

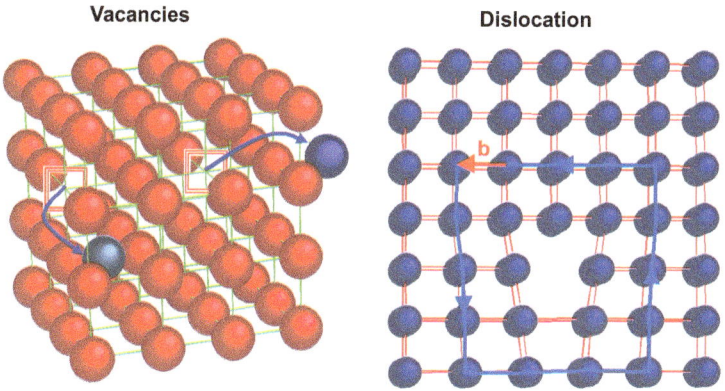

Figure 9: Vacancies and dislocations as topological defects in a crystalline solid. Left: Vacancies are caused by the thermally activated migration of single atoms to the surface or by hopping into an interstitial site. Right: A dislocation results from an inserted half-plane in the crystal lattice. The end atoms of the half-plane form a dislocation line (here perpendicular to the image plane), which can only end at crystal surfaces or must close to form a dislocation ring. The topological invariant of a dislocation is the Burgers vector b (red arrow), the closing error from a lattice circuit (blue line), which surrounds the dislocation line.

equivalent, but the term "topological defect" emphasizes the disturbance of the order of the underlying phase, while the term "topological structure" emphasizes the positive moment of the development of qualitatively new structural levels of matter.

This also applies to the annihilation of a dislocation, which is only possible by moving an atomic half-plane from the surface through the crystal to the dislocation point. Topological defects therefore have characteristic invariants. They are sharply defined quantities that can also be regarded as "generalized charges."

Topological defects are therefore local disturbances of the perfect solid. However, due to a change in the macroscopic geometry of the crystal lattice, they are the source of a macroscopic field and thus qualitatively new properties. Due to them, a non-local disturbance of ordering occurs [Thouless 1998, Kleemann 2006]. A dislocation line with a core diameter of less than one nanometre (nm) produces a displacement field of the atoms with a range of many 100 nm. Grain boundaries also have a width of less than 1 nm, but their displacement fields extend over ranges of many 100 nm. Topological defects are therefore local structures with sharply defined properties (charges) that are associated with long-range fields as expressions of qualitatively new properties.

Point defects and vacancies

Even vacancies as point defects with a diameter of approx. 1 angstrom (Å = 0.1 nm) have a long-range displacement field of the atomic positions, which slowly decrease inversely proportionally to the distance. The missing atoms of the vacancies sit in interstitial spaces or migrate to the surface of the crystal. With their local lattice interference and long-range distortion fields, they have a certain analogy to the "elementary particles" with corresponding local charges and long-range electrical fields. The formation energies of vacancies in crystalline solids are not very high, at a few electron volts (eV).

Therefore, at sufficient temperatures, vacancies can also be formed by thermal excitations, and their concentration increases with temperature. Typically, in a solid at a temperature corresponding to half the melting point, each atom changes its position every 10 seconds by occupying a vacancy and simultaneously forming a new one. The formation of vacancies increases entropy and thus reduces the free energy of a crystal, making them unavoidable in solids. At not too high concentrations, they are part of the stabilizing random motion modes of their phase.

The internal structure of topological structures is conditioned by conflicting forces. The removal of an atom during vacancy formation would create a tiny cavity in the crystal surrounded by atoms that have lost a binding partner. This increases the energy of these atoms and they tend to relax their positions to minimize the free energy. However, this creates a long-range displacement field with strain energy. The conflict and unity of locally changed binding energy in the vacancy and elastic energy in the long-range stress field determines the entire structure of the vacancy. The structure of grain boundaries is also determined by competition and interaction of two disturbances of the perfect crystal: the local refraction of bonds between atoms (topology of the defect) and the long range displacement field both contribute to the rotation of the crystal lattice (see Figure 8). The minimization of the energies of both perturbations, therefore, clearly determines the structure of the grain boundary for a given misalignment angle of the crystallites. The production of single crystals with a uniform crystal orientation and the associated absence of grain boundaries is a high art of materials production.

For example, salt crystals can be grown from a saline solution by introducing a seed crystal that determines the crystal orientation of the growing NaCl atoms. In semiconductor technology, the production of almost perfect single crystals has now been mastered with high precision.

Topological charges

It was mentioned above that topological defects have characteristic invariants – "topological charges." A dislocation is characterized by a characteristic shift of the lattice planes due to the inserted half-plane, which is called the Burgers vector (see Figure 9).

It is determined by the crystal structure of the material in which the dislocation is located and has a fixed value for a given dislocation type. The length of the Burger vector is the topological charge of the dislocation [Thouless 1998] and determines the strength of the long-range stress field. A grain boundary is clearly defined by the misorientation angle of the two crystal domains and by the direction of the grain boundary plane. However, only the misorientation angle of a grain boundary represents an actual topological invariant, since the orientation of a grain boundary plane can change locally without a macroscopic change of crystal orientations. The topological charge of a point defect is determined by the free atomic volume of the vacancy or the foreign atom. Other topological structures such as "skyrmions" (vortices) and "hedgehogs" (hedgehog structures) with corresponding invariants occur in liquid crystals. Due to the dependence of the topological charge on the geometry of the lattice, they are never only a local property of the defect, but always also a global property of the lattice.

Creation and movement of topological structures/defects

Topological defects can arise from a phase transition in which new orders form spatially inhomogeneously. An example is grain boundaries in the solidification of a solid from a melt (see Figure 10). In addition, topological defects can also form as a reaction to an external field. This results in vacancies or dislocations with compensating stress fields. With the formation of topological defects, the energy of the externally applied stress field is reduced.

Dislocations play a decisive role in the strength of a solid. When a solid is deformed by external forces, plastic irreversible deformation occurs above the limit of

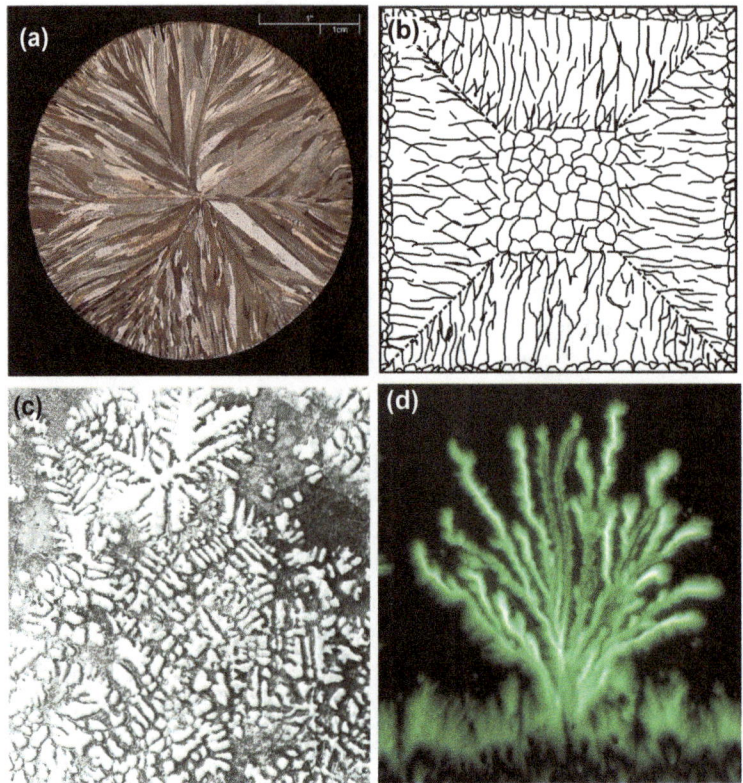

Figure 10: Self-organized structure formation during phase transformations far from equilibrium. (a) Grain structure of a solidified copper disc. (b) Schematic grain structure of the resulting polycrystalline structure with differently oriented crystallites. (c) Dendrite formation during solidification of a melt and (d) during penetration of a magnetic phase into a superconductor. Images: (a) Heinrich Pniok, (c) C. Borchers, (d) T. Johansen.

elastic reversible deformation, due to the formation and movement of dislocations. The plastic deformation does not depend on the atomic bonding forces of the solid but solely on the mobility of the dislocations. The forces required for a displacement of dislocations are orders of magnitude smaller than a hypothetical deformation by dissolution of atomic bonds. Although the mechanical deformation of metals has been investigated for centuries, the mechanism of plastic deformation was only clarified in 1933 [Orowan 1934, Polanyi 1934, Taylor 1934].

Topological defects therefore form a new structural level in the underlying many-particle system with qualitatively new properties. The structure of dislocations or vacancies in a crystal is not bound to the structure of the atoms forming it. It depends solely on the order/symmetry of the crystal lattice and the binding forces. The observation of movements of the topological defects through the crystal is fascinating: the surrounding atoms involved in a dislocation or vacancy do not move with the movement of the defect through the crystal. The atoms forming the defect are exchanged during the movement of the defect. Topological defects therefore form stable collective excitations of the lattice, which move through the crystal lattice detached from the mechanical motion of the atoms which are forming these objects. In this sense, topological defects are subject to a "metabolism."

Their existence is bound to a well-defined ordering of the underlying many-particle system, for example, the crystal lattice. In a phase transition into a disordered phase, for example the liquid state, the existence of topological defects is lost: point defects, dislocations, and grain boundaries exist in crystalline solids, but lose their existence in the liquid or gaseous phase.

There are also topological defects that lose their existence in transformations between different solid phases, such as defects due to disorder in occupation of lattice sites in the transition from ordered to disordered alloys.

The general mechanisms for the formation of topological defects in equilibrium and non-equilibrium are of fundamental importance for the general understanding of the structure formation of matter such as "elementary particles", neutron stars and galaxy nuclei as different stages of topological defects in quantum æther.

2.3 Self-organization in transformation and transport processes

Ultimately, all systems in nature interact with other systems sooner or later. This causes differences, gradients, which cause transformations or transport processes. Conversion processes between different phases of a system are caused by changes in state variables such as temperature, pressure, or density. The speed of the changes, for example the temperature change, determines the degree of supercooling or superheating and thus the driving force of the transformation. During the transformation, new structures and forms of matter can form, such as textures in the solidification of melts or bubbles during the boiling of liquids.

Transport processes of charges, atoms, heat, magnetic moments, etc., are an expression of a contradiction in the electric field, composition, temperature, or magnetic order. They generate new structures such as flow patterns or convection cells.

Influence of the velocity of phase transitions on structure formation

Phase transitions are only possible in a nonequilibrium state since they require the negation of a state and the transition to a new state.

Solidification of a melt by lowering the temperature requires internal or external heat sinks to be connected to the system. The idealization of the transformation at the thermodynamic equilibrium frequently used in theoretical physics is based on the assumption that a uniform temperature can be assigned at least approximately to the system to be described, that is, that the microscopic movements and fluctuations of the atoms can be described by local mean values of the states and simple distribution functions. As long as the deviation from equilibrium is not large, such as when the temperature is only weakly inhomogeneous, it is assumed that the description via location-dependent mean values is still valid. Temperature gradients, cooling velocities, material and field gradients play an important role in the nonequilibrium since they represent the driving forces for the formation of heat flows, inhomogeneities, material flows, and electrical transport processes. These are an important basic condition for the large variety of pattern formation and self-organization processes in nonequilibrium.

During the solidification of melts, the microstructure of the solid that forms depends on the degree of supercooling below the solidification temperature and thus on the cooling rate. The supercooling determines the number and the cooling rate, the growth of the nuclei of the solid phase. The number of nuclei increases substantially with supercooling. Figure 10a shows a copper disc whose polycrystalline structure is determined by the degree of supercooling on the outer wall of the vessel, on which crystal nuclei first form and then grow inwards.

Pattern formation during rapid phase transformations

If a critical threshold of supercooling is exceeded, a new process comes into play with the growth of the nuclei. Tree-like, branching growth, structures called "dendrites," can develop (Figure 10c). The solidification front shoots into the melt at high speed and with continuous branching and enlargement of its surface. It forms an instability of the transformation, which is driven by the fact that as much as possible of the heat of transformation at the solidification front can flow off into the supercooled melt and therefore the surface of the spreading phase boundary is maximally enlarged. The dialectic of heat generation in the transformation zone and heat discharge across the phase boundary determines the pattern formation of the dendrites during solidification. Such dendrite structures occur at pronounced nonequilibrium in many types of phase transformations, for example also in superconductors (Figure 10d). The penetrating magnetic phase (bright) increases its phase boundary with the magnetic-field-free region (dark) to a maximum, in order to dissipate the resulting heat of transformation optimally.

The nonequilibrium during solidification also determines e.g. the shape of ice crystals in snowflakes. All ice crystals crystallize in a hexagonal crystal structure. However, each ice crystal has an individual structure of growth dendrites, which form after the nucleation of the ice crystal (mostly heterogeneous on dust grains) (Figure 11). It is obvious that the resulting diversity of thousands of forms cannot be attributed to the properties of individual water molecules, which are all identical but depend on collective processes of their formation. Dendritic pattern formation also

Figure 11: Selection of ice crystals in snowflakes. They all have a hexagonal crystal lattice with six-fold symmetry. Due to the different nucleation and dendrite growth, thousands of different forms are formed. Images: Alexey Kljatov.

occurs at nonequilibrium in various other processes, such as electrical discharge and lightning in thunderstorms.

In atomic many-particle systems, extremely high cooling rates of the melt can even lead to the crystalline equilibrium structure not being established and the disordered structure of the liquid phase being virtually frozen. The result is an amorphous solid, in which the atoms retain the frozen order of the liquid and the crystalline order is suppressed.

Hydrodynamic structure formation by convection

Transport processes in gases and liquids are the response to a spatial difference in temperature, composition, or density. They create hydrodynamic forms by self-organization of the atomic or molecular motion into an ordered convection motion (Figure 12). The driving force for convection is a temperature gradient between the hot and cold sides of the liquid or gas that must exceed a critical threshold. Below this critical threshold of temperature difference, heat is transported from the hot to the cold side by a diffuse heat flow. It is a transport process of microscopic disordered movements of atoms. If the critical threshold is exceeded, a mass transfer of hot gas or liquid occurs in addition to the heat transfer.

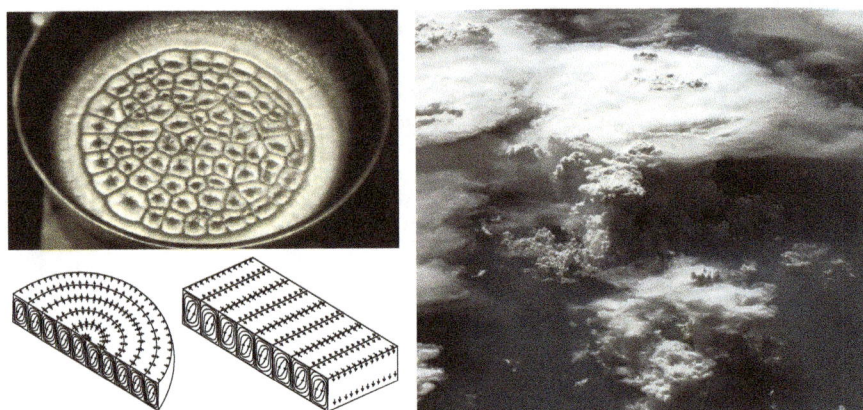

Figure 12: Structure formation by convection. Left: A viscous liquid is heated from below in a pan. Depending on the temperature difference between the bottom and the top and the container geometry, honeycomb, concentric, or linear arrangements of convection rolls are formed (schematic drawing below). Right: Aerial view of the formation of thunderclouds over Brazil by convection currents of the air. Images: Upper left: University of Nebraska-Lincoln, Right: NASA.

Billions of atoms or molecules form the cooperative process of a directed convective motion. Molecules absorb heat on the hot side and release it on the cold side. Once they have released heat, they return to the hot side and form a circulation process. Disordered thermal movement of the molecules is transformed into ordered convection movement.

Self-organized hydrodynamic structures of atoms and molecules under nonequilibrium conditions produce a huge variety of forms. A developed form of convection is the formation of Bradley cells, which form in air currents of the atmosphere under the influence of the earth's rotation. They also produce vortex-shaped low-pressure and high-pressure areas. Their formation, movement and collision are essential factors that determine the weather in different regions of the Earth. Vertical temperature gradients and the associated air currents can also lead to various forms of cyclones, from small tornadoes to huge hurricanes.

Filament formation through self-organization in plasmas

At temperatures at which the thermal kinetic energy of the atoms reaches the binding energy of the outermost shell electrons, electrons are released by impact processes between the atoms. The plasma state is consisting of an ionized gas of electrically negatively charged electrons and positively charged atomic cores. The term plasma is due to Irving Langmuir [Langmuir 1928], who investigated motion processes in ionized gases. The plasma state is another aggregate state and occurs naturally in lightning, and flames, and technically in fluorescent lamps or plasma monitors.

The majority of the atoms in the part of the universe known to us today are present in the form of diluted plasmas, so that, strictly speaking, it is the most widespread phase on the structural level of atoms.

Plasmas exhibit the formation of magnetic filaments by self-amplification as a special law of structure formation in nonequilibrium. Under the influence of a magnetic field, circular currents of the charged particles are generated, which in turn generate a magnetic field. If a self-amplification situation occurs in which the magnetic field and circular currents gradually amplify each other, a plasma instability can evolve which leads to the formation of filaments and spiral-shaped structures.

This effect, named after Birkeland, can be investigated on a laboratory scale and is also the cause of the polar auroras (Figure 13). It occurs at the poles of the Earth because the fast-moving plasma particles of the solar wind is directed by the earth's magnetic field to the magnetic North or South pole. There, the ions collide with the upper layers of the earth's atmosphere and excite luminous phenomena. Plasma instabilities are also important for astrophysical processes such as the formation of jets and spiral arms in galaxies [Alfvén 1990].

Figure 13: In the beautiful play of light of the aurora borealis, luminous filaments appear. Right: Filaments arise as instability in plasmas, which is caused by self-organization due to the Birkeland effect – a randomly occurring current of electrically charged particles in the plasma generates a concentric magnetic field, which constricts the current density distribution to a filament via the Lorentz force. This is a self-reinforcing nonequilibrium process of the plasma. Image: Y. Pashnin.

Chemical pattern formation by coupling of different reactions

When different elements or molecules are mixed, chemical transformations can occur spontaneously. The driving force is the conversion of chemical bonds with different free enthalpy. The free enthalpy is based on the free energy, but in addition to the energy of chemical bonds and the entropy, it also takes into account the work of the reacting molecular system against the external pressure by volume changes. In the reaction, the bonding of existing molecules is dissolved, and/or new molecules are formed. Such processes are strongly dependent on environmental conditions such as temperature, solvent, pressure, illumination, etc.

Often, reactions that lead to a reduction of free enthalpy do not occur because they require intermediate states that represent a high energy barrier for conversion. However, suitable intermediates or end products can also have a positive effect on the reaction. Such catalytic or autocatalytic processes result in a nonlinear amplification of the reaction. If the end product is subject to backreactions, even cyclic chemical reactions occur.

Around 1950, the Soviet chemist B. P. Belousov discovered a reaction in the course of which not only a temporal but also a spatial pattern formation evolves. Despite careful documentation, his results initially met with great skepticism, so that he was only able to publish the results in a medical journal years later [Belousov 1959]. A. M. Zhabotinsky then extended the investigations until 1964 [Zhabotinsky 1964]. The Belousov–Zhabotinsky reaction is today one of the best studied pattern-forming reactions.

Coupled chemical reactions can also lead to spatial and temporal pattern formation. Figure 14 shows the formation of standing and moving waves by coupling the reaction rate of the oxidation of carbon monoxide to the surface structure (and thus the effectiveness) of a platinum catalyst.

This reaction takes place every day in the catalytic converters of passenger cars. However, it was not until the 1990s that it was possible to elucidate that the exact processes under certain conditions are associated with spatial and temporal structure formation [Jakubith 1990]. It is essential for the oscillatory pattern of the surface structure of the platinum catalyst and thus its effectiveness depends on the oxygen content at the surface.

When under supply of oxygen carbon monoxide (CO) oxidizes to carbon dioxide (CO_2), the surface becomes depleted of oxygen, so that a structural transformation of platinum takes place that stops the catalytic reaction.

The consequence of this is that the oxygen concentration on the surface gradually increases again and the platinum jumps back into the structure with catalysis capability. However, the speeds of the two opposite steps are very different, resulting in a temporal and spatial wave propagation.

Dialectical laws of self-organization in nonequilibrium

In the formation of dynamic structures, disordered microscopic movements are transformed into ordered forms of movement. The ordering of movements in transformations

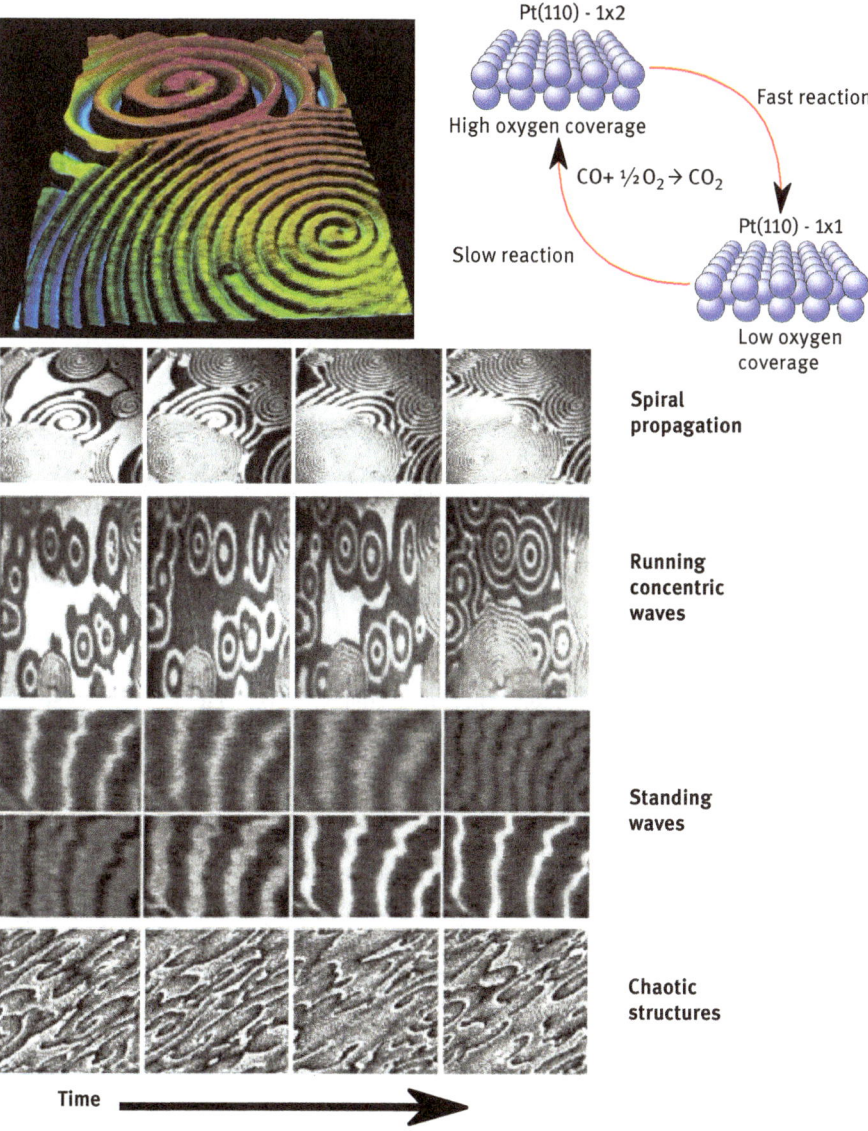

Figure 14: Self-organization of oscillating chemical patterns during the conversion of carbon monoxide (CO) to carbon dioxide (CO_2) on a catalytically active platinum (Pt) surface. Top left: False-color representation of the spiral chemical wave propagating on the Pt surface. Top right: Underlying mechanisms, where the speed of the oxidation reaction from CO to CO_2 depends on the Pt surface structure, which in turn depends on the concentration of oxygen. This results in a coupling of the structural transformation of the Pt surface and its efficiency of conversion with the concentration of the reaction partners. Bottom: Overview of the different running and stationary wave patterns depending on the environmental conditions of the reaction. Images: Bottom and top bottom left, H. H. Rotermund, FHI Berlin; Top right [Meissen 2001].

and transport processes is an expression of the struggle of opposing driving forces and development of a higher level of cooperative action. The ordering at nonequilibrium requires the exceeding of a critical threshold in the driving force. The individual thermal random motions must be subordinated to the ordered motions by the nonlinear coupling of different collective motions.

For example, the convection current starts when a critical temperature difference is exceeded. In the struggle between different collective modes of motion, the one with the greatest heat transfer prevails. Microscopic disordered heat movement combines with the buoyancy movement of hot and thus less dense molecules. A stable convective transport of the molecules is only possible through the formation of convection cells.

The formation of filaments by plasma instability begins when a critical electrical current density is exceeded. It combines the incipient order of the movement of the ions with the formation of the magnetic field. Both mutually amplify each other in the self-constriction in the formed filament.

Self-organization is thus a qualitative leap in which disordered movements of microscopic degrees of freedom turn into ordered movements and structure formation of some degrees of freedom of the system. The concept of entropy as an expression of the system, to distribute its energy equally to all microscopic forms of movement, becomes meaningless for such self-organized nonequilibrium structures, because it is precisely the abolition of the probable distribution in equilibrium and the transition to an ordering of motion driven by a contradiction that is characteristic. Whether extended concepts of entropy far from equilibrium are meaningful is completely open for debate [Mahulikar & Herwig 2004, Martyushev & Seleznev 2006]. Various principles for the formation of stationary states of self-organized structures have been discussed, such as the principle of maximum production of entropy with minimum energy dissipation [Glansdorff & Prigogine 1971]. However, the existence of an universal principle of minimal energy dissipation for stationary states is questioned [Nicolis 1999].

The Second Law of thermodynamics reflects the equilibrium situation of closed systems, where microscopic random movements are distributed over all microscopic degrees of freedom, and entropy strives for a maximum. Any attempt to apply this law to open systems or even to self-organization far from equilibrium is scientific nonsense, because it is precisely the abolition of equal energy distribution. Self-organization is pattern formation by ordering of microscopic movements overcoming entropy driven random movement. Such ordering processes arise through the conflict and unity of opposing driving forces and not through abstract principles. For example, it has even been attempted to base the Second Law of thermodynamics on the "directionality of time" or vice versa to attribute the "directionality of time" to the irreversible devaluation of ordered forms of motion into the disordered form of heat [Weizsäcker 1985, Zeh 1989]. Thus the Second Law would be highly stylized into a natural law underlying all development. Even the physicist and philosopher C. F. von Weizsäcker, who certainly appreciates the positive findings of the laws of self-organization, cannot escape the theory of general decline by heat death: *"Heat death would not be, provided that the temperature is sufficiently low, a porridge, but a collection of complicated skeletons"* [Weizsäcker 1985, p. 178]. *"The most probable content today of the assumption of heat death in the future is that the energy sources of the stars finally dry up and the spiral nebulae in space disperse completely"* [Weizsäcker 1958, p. 148]. This ideologically shaped conclusion is in fundamental contradiction to the observation of an increase in the complexity and diversity of matter structures in self-organization processes.

Every materialistic theory of the development of forms of matter in the cosmos must focus on phases of nonequilibrium by coupling processes of different structural levels of matter in the microcosm and macrocosm. Any equilibrium that is established over time at a certain structural level of matter will sooner or later be overcome by development processes at other structural levels. In the contradiction between the trend towards equilibrium in an isolated system and its negation by interactions and developments in other structural levels, non-equilibrium is the main feature. This produces a rich variety of matter structures, the complexity of which is likely to gradually increase in the cosmos. Chapter 8 returns to this question.

2.4 Structure of atoms and their bond

The internal structure of atoms with a nucleus that has more than 99.9% of the mass of an atom and a very light, but crucial for the chemical, optical, and electronic properties, electron shell is essential for the formation of chemical bonds and thus the attachment of atoms to molecules as well as evolution into different bonding states in solids, liquids or gases.

The lightest atom, the hydrogen atom, has a nucleus of one proton. Its positive electrical charge is compensated by the negative charge of the electron shell which consists of exactly one electron. The next heavier stable atom is the helium atom. Its nucleus consists of two protons and two neutrons. The two positive electrical charges of the proton require an electron shell consisting of two electrons for charge neutrality. In fact, all known elements, be they oxygen and nitrogen in the air or metals, can be traced back to different numbers of protons and neutrons in the atomic nucleus and the number of electrons in the electron shell necessary for electrical neutrality.

The formation of the various chemical elements is determined by the law of the transformation of the quantity of the number of particles into different qualities of physical and chemical properties, as vizualized in Figure 15.

Each atom of the same species has exactly identical properties, although they occur in countless numbers. A certain number of basic building blocks, for example, six neutrons, six protons, and 6 electrons that form the carbon atom, always produce exactly the same atomic structure and the associated physical and chemical properties, at least under conditions that are not too different from those on Earth. These 18 particles therefore always organize themselves in the same way, in a stable ground state

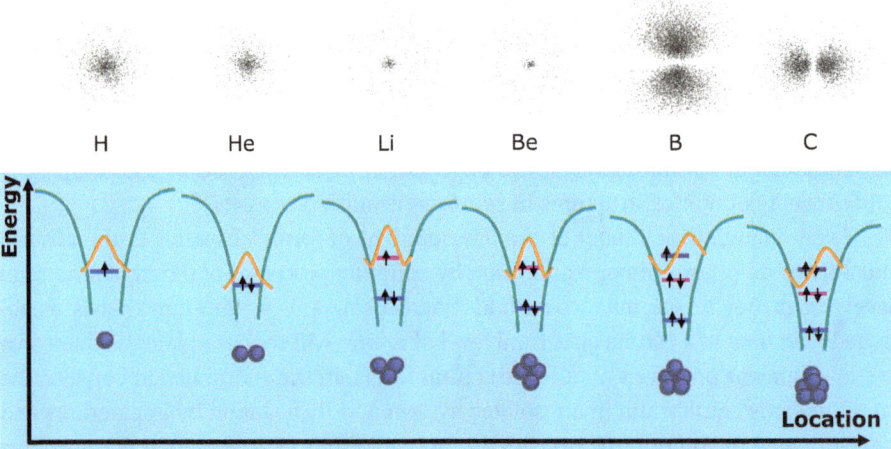

Figure 15: Schematic representation of the transition from quantity to new quality in the electromagnetic structure of nuclei and electron shells of light atoms. The representation starts with the lightest element, hydrogen (H), with a proton (blue sphere) and an electron (arrow as an expression of its own rotation) and ends with carbon (C), with six protons and electrons. Top: Electron density distribution of the shell electron with lowest binding energy. Below: Energy levels of the electrons in the attracting electrostatic potential of the protons and oscillation state of the matter wave of the highest occupied state. Neutrons in atomic nuclei are not shown for simplicity.

of this system, namely a heavy atomic nucleus with six protons and six neutrons and an electron shell of six electrons, with a certain shell and intrinsic angular momentum structure. Oxygen also has a characteristic structure with an atomic nucleus of eight protons, eight neutrons, and an electron shell of eight electrons.

The dynamic structure of the electron shell is determined by the interaction and competition of four essential factors:

1. The repulsive electrostatic forces between the electrically negative electrons and the attracting electrostatic forces between the electrons and the positive atomic nucleus.
2. The spatial movement of electrons and the associated matter waves. Electrons are not simple structure-less small "billiard balls," which move in an "empty space," but rather complex objects, whose movement is connected with the occurrence of wave phenomena (see Chapter 4).
3. The angular momentum structure of the electron shell with the orbital momentum and the intrinsic momentum of the electrons, which determine the spatial distribution.
4. What is called the "Fermi pressure," caused by the destructive interference of incompatible matter wave states, when two or more electrons are in the same motion and angular momentum states.

Molecules are forms of matter in which the atomic building blocks form relatively stable bonds. They are formed when atoms share one or more outer electrons of their shell. The simplest molecule is the hydrogen molecule, H_2. It consists of two hydrogen atoms (Figure 16).

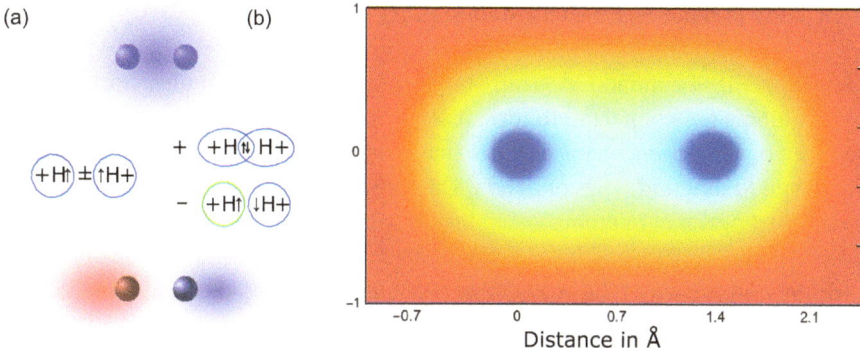

Figure 16: Formation of chemical bonds by common electrons using the example of the hydrogen molecule, H_2, consisting of two protons (H^+) and two electrons with an intrinsic angular momentum (arrow). (a) Schematic representation of the two possible common electron states, consisting of two protons (p) and two electrons with angular momentum (arrow). The binding state (top) is accompanied by an increase in the electron density between the protons, with the electrons creating an antiparallel intrinsic angular momentum. The antibinding state (bottom) is associated with a reduction in the electron density between the protons. The two electrons then have an identical state of the intrinsic angular momentum. This increases the mutual repulsion by the "Fermi pressure." (b) Quantum-mechanical calculation of the electron density distribution of the two electrons in the binding state after [Foulkes 2001].

The repulsion of the positive nuclear charges of the two protons counteracts the negative charge of the two electrons, which are preferably arranged between the atomic nuclei. Compared to two separate hydrogen atoms, the total energy of the hydrogen molecule is reduced by 4.7 eV, due to the newly formed common matter-wave state of the electron shell.

The bond in the molecule must not be imagined as a static, solid connection of the atoms involved, but rather as a dynamic structure in which both the electrons and the two atomic nuclei are in constant motion.

The resulting hydrogen molecule is nevertheless an extremely precisely defined material structure; for example, the equilibrium bond distance between the two atomic nuclei is precisely determined at 0.074 nm at standard pressure. However, at normal ambient temperatures, the two protons carry out a constant oscillation movement around this distance, in which the electron arrangement must also constantly adapt. Only at very low temperatures do these oscillations gradually die out; even at the lowest achievable temperature of −273 °C, at absolute zero, however, a state of oscillation remains, the so-called zero-point oscillation. It is an expression of the fact that the molecule is not isolated from the zero point electromagnetic field fluctuations of its environment.

Dialectic unity of bonding type and atomic structure

The development of common electron states is the principle that also works when billions of atoms are connected to form a solid. If many atoms form a solid, a self-consistency of the arrangement of the atoms arises with the reorganization of the electron shell to different types of bonds. This can result in a fascinating almost perfect lattice arrangement of atoms in crystalline solids.

In simple metals such as copper or gold, each atom releases exactly one electron from the outermost shell to the newly formed sea of electrons in the solid. The characteristic feature of metals is that the electrons released can move almost freely in the resulting crystal (Figure 17, left).

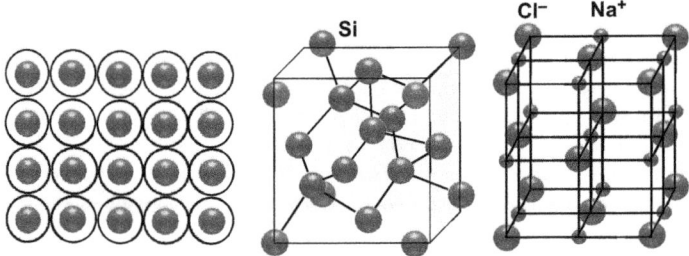

Figure 17: Schematic representation of the possible bond types in a solid body. Left: In the metal, the outermost electrons of the electron shell form a common sea of electrons. Middle: In semiconductors, such as silicon, the common electron distribution of the atoms forms a directional structure, known as a covalent bond. Right: In insulators such as common salt (NaCl), the shell electrons pass completely from one type of atom to the other. In the case of common salt, a lattice of positively charged Na ions and negatively charged Cl ions is formed, which is stabilized by the electrostatic forces. All three types of bonds are only extreme cases, and intermediate forms usually occur.

This means that a large number of atoms together share a huge number of electrons. As a result, these outer electrons of the shell can no longer be assigned to one or a few atoms, but only to the solid body as a whole. Therefore, metals have very good electrical conductivity. Metals also have no preferred binding directions, and the resulting lattice structure is therefore characterized by forming a densest ball packing. The model of "free electrons" in metals, however, is only a rough approximation of the actual state as a Fermi liquid (Chapter 3).

Another class of solids are the ion crystals. These include common salt, which consists of sodium and chlorine. Due to the different binding energy of the electrons in sodium and chlorine, one electron per sodium atom is transferred to the chlorine. The result is a stable and regular arrangement of single negatively charged chlorine ions and single positively charged sodium ions. The outermost electrons of the shell are not divided equally among many atoms but are transferred from one type of atom to the other. This also has the consequence, for example, that the electrical conductivity in ion crystals is extremely poor: they are insulators. The arrangement of atoms in a lattice follows the requirement of balancing the electrical charge.

There are many more variants of the formation of solid bodies from atoms and an infinite variety of lattice arrangements connected with them. An important example is silicon, which is of great importance as a semiconductor in microelectronics or photovoltaics. The covalent bond between silicon atoms is determined by the formation of electron pairs between the atoms. Silicon atoms move into a tetrahedron arrangement.

The lattice structure of crystalline solids is thus determined by the unity of attracting and repulsive partially shielded electrostatic forces which lowers the total energy of the system of negatively charged electrons and positively charged atomic nuclei as far as possible.

Qualitative changes of the bonds in the liquid and gaseous state

Even in a solid state, atoms are not at all immobile and are subject to lattice oscillations as well as changes of location by jumping. On average, at room temperature in metals, all atoms change their locations within less than one hour. In the liquid phase, the atoms change of location, which is orders of magnitude faster, turns into a collapse of the lattice structure: a dynamic, near-ordered state is formed in which the next neighboring atom forms a bond for only fractions of a second, which then dissolves again.

Another qualitative change occurs in the gas phase: In the gaseous state, the density of the molecules is lower, their mobility increases, and the formation of bonds is greatly reduced. The interaction between the atoms in the gas phase is characterized by their collisions with each other.

In addition to the covalent bond between hydrogen and oxygen atoms in the H_2O molecule, a hydrogen bridge bond is formed between adjacent H_2O molecules. The oxygen atom of an H_2O molecule is capable of electrically polarizing the electron shell of a hydrogen atom of the adjacent H_2O molecule. The mutual polarization leads to fluctuating electrical dipoles which attract each other. Hydrogen bridge formation is also associated with an exchange of protons between adjacent water molecules. When the temperature rises above the melting point of ice, the proton exchange becomes so rapid that the ice structure collapses and the liquid phase is formed. The formation of hydrogen bonds is the cause of special properties of water, such as its low density in the solid state, its relatively high melting and boiling points compared to other similar molecules, and the anomalous increase in the density of water during melting.

This density anomaly of the water is the reason that the ice surface melts due to the contact pressure during skating and thus the skate blades glide particularly well.

2.5 Phases and their "Zoo of Excitations"

Phases are characterized by a certain kind of order. Crystalline solids by the long range ordering of the atomic positions on periodic lattice sites, the ferromagnetic or antiferromagnetic phase by the parallel or antiparallel alignment of microscopic magnetic moments of the electrons, and the ferroelectric phase by the alignment of microscopic electrical dipoles. In addition, each phase produces characteristic excitations as movements of atoms, electrons, and magnetic or electrical dipoles. The exact structure of the ordering and the possible excitation and movement forms a dialectical unit.

Imagine a fictitious observer in the inner world of a solid body, for whom the visible forms of matter would consist only of the excitations and topological defects of the solid body. For him, the existence of the ideal crystalline structure of the atoms at low temperatures would not be perceptible; these would form the so-called vacuum. For the fictitious interior observer, the "zoo of excitations" would be the visible universe. An external observer studying the solid would, of course, in the investigation of the "zoo of excitations" establish the connection to the microscopic structural level of the many-particle system, that is, the solid with its atoms in a certain phase.

The Unity of Order and Collective Excitations in Materials

A crystalline solid body is characterized by a strictly periodic lattice arrangement of atoms with well-defined symmetries. Nevertheless, they are not a rigid structure, because at finite temperatures all atoms and electrons are in constant motion. But on average all atoms of the lattice have a sharply defined lattice distance, which is proven by the occurrence of sharp diffraction reflexes in X-ray diffraction. Each type of order has a zoo of different individual and collective excitations of atoms and electrons of different energy:

- Among the excitations with the lowest energy are the lattice oscillations of the atoms around their equilibrium position. They have a quantized amplitude of the oscillation and therefore require a minimum energy of their excitation. Such quantized lattice excitations, therefore, occur as particles, as phonons. The propagation of the atomic oscillations takes place in waves. The relationship between their oscillation frequency and the wavelength, also known as dispersion, is determined by the lattice structure and the inter-atomic forces. Their oscillation frequencies range from less than 1 Hz to 100 THz (10^{14} Hz) and their wavelength from meters to less than 1 nm (10^{-9} m). Any change in the crystal structure changes the excitation spectrum of the phonons, their dispersion.

– All electrons in materials, no matter if inner bound states of atoms or outer shell electrons forming chemical bonds, are in constant motion. In metals or doped semiconductors, the outermost shell electrons move almost freely through the entire solid and form a "sea of electrons." Even in the ground state, the state of lowest energy that exists at a temperature of absolute zero, sea electrons have velocities in the range of 1,000 to 10,000 m/s. However, their individual directions of motion are random and average out collectively. Excitations of this lowest energy state are identical with the collective reorganization of the random motion to a directed effective motion. This is caused, for example, by an applied electric field and leads to electric transport processes with electron drift velocities of nm/s to μm/s. An excitation type of higher energy is electron-hole pair excitations, either by thermal random motion or by the absorption of light.

– In addition to the excitations of the lattice and the electrons, an entire zoo of other collective forms of movement is formed, depending on the order of the material: polarons consist of an electron which is surrounded by a polarization cloud of the ion shells of the lattice. When it moves, it carries this lattice distortion with it, which increases the electron mass many times over. Plasmons are collective electron oscillations with energies of some eV. Magnons are collective oscillations of the ordered magnetic moments of the electrons of a ferromagnetic solid. Polaritons are the coupled oscillations of photons and the lattice of atoms. The list of these "new particles" could be extended even further and depends on the ordering of the solid and the interactions between the different forms of microscopic motion of the electrons and the atoms.

In addition to the unstable excitations, topological defects are formed. Their structure as a point defect, line defect, or interface is associated with a macroscopic topological change of the material and thus characteristic invariants as "topological charges." They are created by shielding external fields or in the phase transition during materials production.

The structure of topological defects also depends on the order of their phase. Thus, single vacancies in metals or double vacancies from cations or anions in ion crystals are formed to avoid electrical charging of the material, which would greatly increase the energy of the system. With higher temperature/energy the concentration of vacancies increases. Neighboring atoms hop into the vacancy, which then migrates to the neighboring site. When crystalline solids are heated to just below their melting point, the number of thermally excited vacancies, i.e. the number of particles in the "vacuum state" increases drastically, as does the number of atoms hopping from one lattice site to the next. Despite the increasing diffusion movement of the atoms, the symmetry of the lattice positions remains exactly the same up to the melting point.

The excitations of a phase are not arbitrary. They are dependent on the exact structure, order, and mutual influence of the forms of motion and are subject to laws which are shown in their "dispersion": the relationship between energy (frequency) and momentum (wavenumber) of the particle. Unstable thermal excitations of phonons, polarons, magnons, polaritons, etc., die out at low temperatures. Such collective excitations of the solid therefore do not have a topological invariant corresponding to a conservation quantity.

In contrast, topologically stable defects such as point defects, dislocations, or grain boundaries do not die out at low temperatures but remain as a stable component of the system. Their topological charges are a conservation parameter. In the image of the "solid-state universe" with a fictitious "interior observer" they would form the stable forms of matter.

The change of the excitation spectrum during phase transitions

At the melting point of a material in its solid state, the entire cooperative behavior of atoms and electrons changes. In liquids and gases, the formation of relatively solid bonds between atoms in the solid changes into fluctuating dipole forces of dynamically deformed electron shells of atoms or molecules. The structure of the electron shell of gases even approaches that of free atoms or molecules. The resulting collective excitations change fundamentally:

- High-frequency phonons of the crystal become vibrational excitations of the atoms in the molecules and rotational excitations of the molecules in gases and liquids.
- Low-frequency phonons in the form of sound waves persist as longitudinal waves (density waves), while transverse waves disappear, because the shear elasticity of the solid is canceled at the phase transition.
- Collective electronic excitations of solid materials become more or less individual excitations of electrons in individual molecules and atoms. Magnons of the ferromagnetic order become individual excitations of magnetic moments in disordered phases.
- The existence of topologically stable defects is bound to a long-range ordering of either the atomic positions in the crystal or other quantities (magnetic and electrical dipoles). The concentration and movement of point defects such as vacancies first increase when approaching the melting point. Above the melting point theses topological structures disappear and turn into density fluctuations. These density fluctuations characterize the dynamically short-range ordered structure of liquids.

- In the plasma state, the excitation spectrum continues to change. In contrast to the plasma oscillations of the "sea of electrons" with respect to the lattice of the atomic cores in metallic solids, collective oscillations of the electrically positively and negatively charged particles form, called "plasma waves." This results in a series of qualitatively new properties in the electrical conductivity of ions and electrons, as well as in their interaction with light.

The change of the "zoo of excitations" in many-particle systems at phase transitions has its material origin in the change of the structural and electronic order in the phase transition. All excitations characteristic of a phase change abruptly – the "zoo of particles" is modified. Such phase transitions are also discussed in high-energy and particle physics as "changes in the vacuum state." In contrast to particle physics, in the physics of condensed matter, nobody would have the idea of separating the changes in the collective excitations occurring in the phase transition from the structural changes occurring in the underlying material structures.

Stability and excitations in different phases – "laws of nature"

What laws of nature would our fictitious "interior observer" perceive in the solid state, and how would they change during a phase transition? For the "interior observer," the laws of nature are the laws of excitation, movement, and interaction of the excitations with each other. Irrespective of his knowledge, these laws of nature are determined microscopically by the atomic and electronic structure of the material. Conversely, we have seen that the excitation spectrum, the type of collective motion, also determines the structure of the underlying material.

The structure and motion of a macroscopically large number of atoms are mutually dependent. Natural laws as laws of motion of a certain type of matter do not exist separately from the underlying material structure out of them they are formed.

Material systems with their forms of motion are self-stabilizing, that is, within the limits of stability neither the alteration of individual building blocks nor the increase in low-energy excitations are capable of breaking out of collective laws of motion of the system. The laws of motion of excitations of a given phase thus appear as "laws of nature". The lawfulness of Ohm's law for electrical conductivity is an expression of the possible collective modes of motion of the sea of electrons and the scattering of electrons among themselves and to the phonons. Hooke's law for the elasticity of a solid material is nothing more than an expression of the crystalline order with relatively solid electronic bonds of the atoms. The laws of plastic deformation are nothing more than the result of the occurrence of dislocations as topological defects and their movement and interaction with each other.

The forms of matter and the related laws of nature of the different phases are mutually inderdependent and are not the result of "fundamental laws" above nature.

Any attempt to detach the laws from the structure and organizational form of matter and to place them above matter in the sense of a "world formula" would be ridiculed in the field of solid-state and materials physics. Natural laws have the character of collective laws of organization and motion of material systems. They can neither be attributed one-sidedly to the individual properties of the microscopic components (atoms), according to the intentions of "reductionism," nor can it be the task of natural science to understand natural laws as purely empirical laws according to the intentions of "positivism" and to deny their real objective existence. It characterizes the dialectic of the self-organization of matter, that it understands the formation of laws of motion of systems of matter as the unity of struggle between structure formation through the interactions and abolition of this order by the various collective low-energy and individual high-energy excitations of a system.

3 Quantum gases and liquids

Quantum gases and quantum fluids are qualitatively new states of matter in many-particle systems in which the motion of the individual particles is strongly influenced by the interference of their matter waves. For example, quantum gases and quantum fluids are formed in a dense sea of electrons in a metal or semiconductor. At low temperatures, they condense into the superconducting state – a state with vanishing electrical resistance. At low temperatures, even a gas of helium atoms initially changes into a liquid state and at very low temperatures into a superfluid state, which is characterized by disappearing viscosity, representing friction.

In such systems, an important new property of the constitutants comes into play: the occurrence of matter waves during the motion of atomic and subatomic particles and the ordering of particle motion due to the interference of matter waves. The classical states of plasma, gas, liquid, or solid are determined by the conflict and unity between repulsive thermal motion and attracting binding forces between the particles. In contrast, the properties of quantum gases and quantum fluids are determined by the conflict and unity of disordered thermal motion and co-ordinated ordered motion through the interaction of their matter waves [Legett 2006]. This presupposes that the formation of chemical bonds between the particles forming them plays a subordinate role.

Frictionless superconducting or superfluid states were already discovered in metals or liquid helium at the beginning of the 20th century. The theoretical prediction was made for the first time by Einstein and Bose in 1924. Steps towards a fundamental understanding of such states, however, did not come until the second half of the 20th century with advances in the theory of many-particle quantum systems. In the 1930s and 1950s, scientists in the Soviet Union made outstanding experimental and theoretical contributions to this end. In the last 20 years alone, the Nobel Prize in Physics has been awarded seven times for discoveries in the field of quantum fluids. Quantum gases and liquids of various particles form an important state of matter in the cosmos, for example inside "dead" stars such as neutron stars, or galactic nuclei. In the author's opinion, quantum fluids are also a model system for understanding the essential characteristics of quantum æther, among other things because of their frictionless forms of motion and the formation of unstable and stable field-like and particle-like excitations.

3.1 Superfluidity

In 1995, after an exciting race, first the working group of Eric Cornell and Carl Wieman in California and shortly afterwards, that of Wolfgang Ketterle in Boston

https://doi.org/10.1515/9783110644203-003

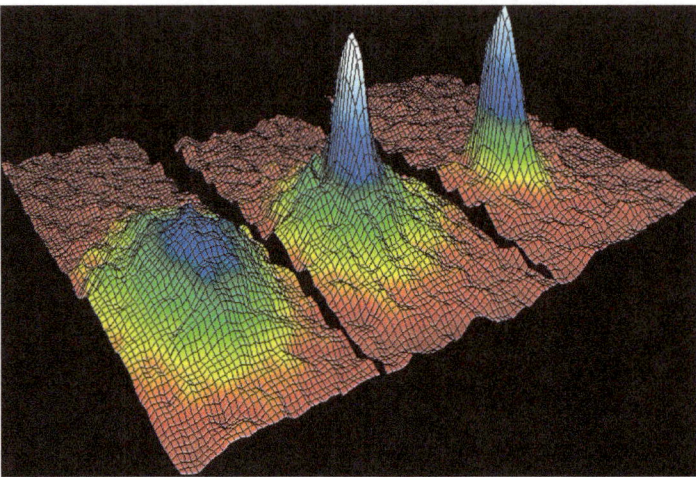

Figure 18: Visualization of the Bose–Einstein condensation of a gas of rubidium atoms by the optical imaging of its density distribution. The temperature is lowered from left to right. In the middle picture, the transition into the condensate is visible as a concentrated tip of the condensed atoms above a "thermal background" of noncondensed atoms. On the right, the "thermal hill" has also disappeared, almost all the atoms are in the superfluid. The sharp maximum in density indicates that many of the atoms are in a single quantum state with uniform velocity (in this case $v = 0$), while the noncondensed atoms have a thermal velocity distribution [Anderson 1995]. Picture: NIST/JILA/CU-Boulder.

succeeded in transforming a gas from alkaline-earth atoms into a superfluid state [Ketterle 1999, Anderson et al. 1995]; see Figure 18. For this purpose, extremely low temperatures in the range of less than one microkelvin had to be implemented by means of revolutionary new cooling methods.

This qualitatively new superfluid state of extremely cold atoms is called a Bose–Einstein condensate because its formation at low temperatures was predicted in 1924 by the Indian physicist Satyendra Nath Bose and Albert Einstein on the basis of quantum theory. Bose is also the eponym of the particles in this quantum fluid, the bosons, as particles without or with integer intrinsic angular momentum (spin) in units of angular momentum quantum (see Chapter 4). Bosons include atoms in which the individual angular momentums of the nuclei and their electron shells saturate down to a total angular momentum of zero, such as ^{4}He, ^{23}Sodium, ^{87}Rubidium or "elementary particles" such as photons. Their theory says that an ideal gas of bosons at low temperatures change into a state in which macroscopically, many particles adopt the same matter-wave state (Figure 19). The collective of particles forms a common matter wave. The uniformity of the wavelength and wavefront of the matter waves of different particles is called coherence. In the superfluid state, the coherence length becomes macroscopically large compared to the particle distance. For the borderline case of very low temperatures ($T \rightarrow 0$), the coherence length becomes identical with the system size of the many-particle system.

This means that their motions must be coordinated by interference of the waves. The result is an ordered atomic motion in contrast to the disordered thermal velocity distribution in normal liquids. Consequently, entropy in the condensed state is drastically reduced.

The transition from a gas of bosons to a quantum gas or a quantum fluid depends on the temperature and density of the many-particle system, as vizualized in Figure 19. At high temperatures, the high kinetic energy of the particles is associated with a very small wavelength of their matter waves. If this wavelength is smaller than the particle distance, the effects of the matter waves are negligible and a classical gas is formed. Chance and necessity in classical gas are expressed in such a way that on average all gas particles have the same kinetic energy.

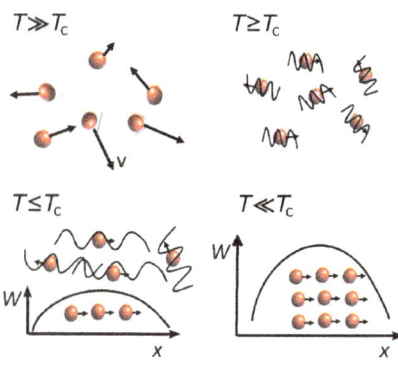

Figure 19: Bose–Einstein condensation of particles with matter waves to a superfluid as a function of temperature, T (from top left to bottom right). The condensate is formed by cooling below a transition temperature, T_c, by overlapping the matter waves and forming a uniform, macroscopic coherent wave state with amplitude, W. The arrows indicate the velocity, v, of the individual particles.

However, the disordered thermal motion is accompanied by a statistical fluctuation of their kinetic energy, the probability of which depends exponentially on the energy deviation. Thus, they are subjected to what is called a Maxwell velocity distribution. As soon as a significant overlap of the matter waves occurs due to cooling or increasing density, their velocity distribution turns into the Bose–Einstein energy distribution that is characterized by a lower fluctuation of their kinetic energy. As a result, quantum gases have a lower heat storage capacity than classical gases. When further cooling occurs, the transition of the quantum gas to the quantum fluid state, the Bose–Einstein condensate, occurs abruptly at a critical temperature. The condensate is formed when the wavelength of its matter waves become similar to the mean distance between the atoms.

A normal fluid is viscous. An initiated flow comes to a standstill due to the dissipative transformation of macroscopically ordered movement of the flow into microscopically disordered thermal movement. Due to the orderly motion of the particles in the superfluid state, no dissipation of energy by internal collisions occurs. Their viscosity vanishes and a superfluid, frictionless form of motion is created.

Properties of superfluid helium

The helium isotope with mass 4, ^{4}He, occurs in minute concentrations in the earth's atmosphere with a proportion of about 10^{-5}. It was the discoverer of superconductivity, H. Kamerlingh Onnes, who succeeded in liquefying ^{4}He for the first time in 1908. This happens at a temperature of 4.2 K at the standard pressure of one atmosphere. The Soviet physicist P. L. Kapitza succeeded in 1938 for the first time in reaching the superfluid state of ^{4}He at a temperature of about 2.2 K; see Figure 20 for the phase diagram. In 1978, Kapitza was awarded the Nobel Prize in Physics for this work.

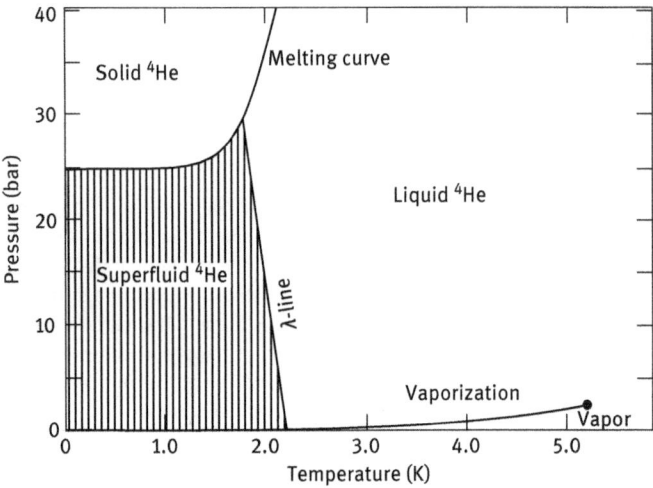

Figure 20: Areas of existence of different states of aggregation (phases) of the helium isotope ^{4}He. Under standard pressure (1 bar), the superfluid state occurs below 2.2 K in ^{4}He. Depending on temperature and pressure, a solid, normal liquid, or gaseous state may be present [Swenson 1950].

At first, it was disputed whether a macroscopic matter wave in the sense of a Bose–Einstein condensation really develops, or whether simply the internal friction of the liquid, the viscosity, disappears at low temperatures because the disorderly atomic movement, which causes the friction force, comes to a standstill. Superfluidity is shown by the fact that a transport motion of the liquid does not decay, but continues for very long periods of time, without a demonstrable weakening. As will become clear in the following, however, experiments on rotating fluids have clearly shown that superfluid helium not only has a vanishing friction, but also a quantum state with macroscopic coherence of its matter wave [Legett 2006].

The macroscopic coherence of the matter wave is proven by the investigation of rotary flows of the superfluid. When a container of superfluid helium is rotated, the fluid inside the container remains stationary because the frictional forces between the superfluid and the container walls disappear. These properties were first discovered by Kapitza in the late 1930s. He found out by his experiments that there is no uniform rotation of the liquid in the superfluid state (as shown in Figure 21, top left), despite a rotary motion of the container containing it.

A uniform rotation of the superfluid would have the consequence that the velocity of the flow must be greater outside than inside, in contradiction to the state of macroscopic coherence with uniform velocity of all condensed particles. Since 1999, quantized vortex filaments have been detected in the superfluid states of Bose–Einstein condensates [Ketterle et al. 1999, Shaeer 2001]; see Figure 21, right.

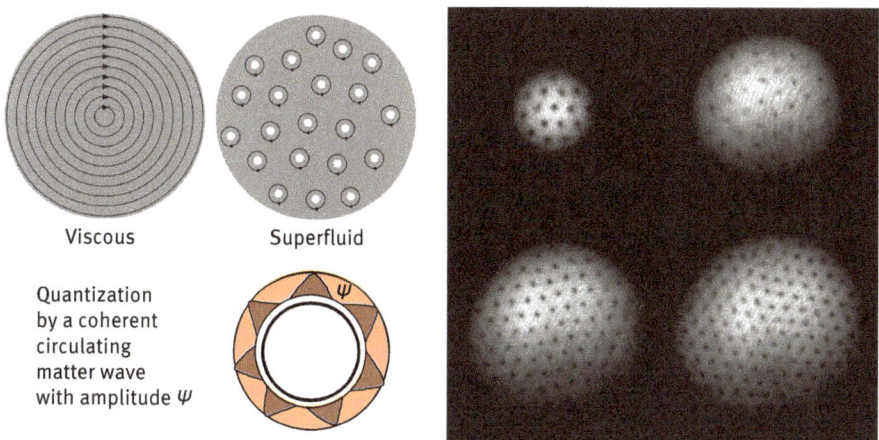

Figure 21: Formation of quantized vortices as "particle excitations" in the superfluid. Left: A macroscopic rotary motion of the viscous normal liquid state is incompatible with the organization of the atoms in a uniform matter wave in the superfluid state. Microscopic vortices with quantized rotary speed are formed. Bottom left: Quantization condition by circulation of coherent matter waves. Right: Experimental observation of quantized vortices in a Bose–Einstein condensate. Pictures above: W. Ketterle, right from J. R. Abo-Shaeer et al. Science, 292 (2001) No. 5516, reprinted with permission of the AAAS.

Exactly in the center of the vortex filaments, a peculiarity occurs: The phase position of the matter wave is undefined there, it should have all phases at the same time, which is not possible. The superfluid state inevitably collapses in the center of the vortex and a "normal fluid core" of the vortex is formed. Thus, in a core of the vortex the fluid adopts a different phase. Vortex filaments form a stable quantized excitation, a new type of topological defect in the superfluid condensate. Its topological invariant (charge) is the quantized circulation.

As with other topological structures, the cause of the formation of the vortex filament is shielding from a macroscopic field: the macroscopic rotary motion of the condensate. "Rotary movements" of the superfluid caused from outside cannot penetrate the volume of the condensate. This rotary motion is shielded off because it is incompatible with the macroscopic quantum state.

With a homogeneous rotary movement of the condensate with uniform angular velocity, the velocity of the condensed particles would increase continuously with increasing radius. However, this is incompatible with a uniform matter wave of the particles, which must also be phase-coherent for each radius. Superfluids and macroscopic rotary states are mutually exclusive – which is why they can be used, for example, for extremely sensitive detection of rotary motions.

These experiments allow a far-reaching conclusion to be drawn for the mechanisms of the formation of quantized particles by self-organization in quantum fluids: When a macroscopic field is applied, the homogeneous ground state of a superfluid system changes into an in-homogeneous two-phase state with the continuum of the superfluid and particle-like, stable, and quantized vortices as excitations. This is the state with lower free energy of the whole system.

Fountain effect: condensate and quasiparticle excitations

In quantum fluids at finite temperatures, a coexistence of superfluid and normal-fluid components occurs. Such a coexistence of two different liquid states was postulated by the Hungarian physicist Tisza on the basis of the experiments by Kapitza in 1938: The fountain effect is observed when superfluid helium flows through a capillary (Figure 22). It proves the existence of two liquids with qualitatively different forms of motion. While the superfluid moves frictionlessly through the porous sealing material, the movement of the normal-fluid helium is inhibited by frictional forces. The resulting overpressure in the experiment is made visible by a fountain of liquid helium through a capillary opening.

Figure 22: Exceptional properties of superfluids. Left: Superfluid creep against gravity: A thin film of superfluid covers the entire inner wall of the container to minimize the potential energy in the Earth's gravitational field. Middle and right: Fountain effect in superfluid helium. Middle picture: [Allen & Jones 1938].

Lev Landau, however, criticized Tisza's too simple idea of the "coexistence" of two liquids, since the superfluid component would then experience a frictional force due to scattering with the atoms of the normal-fluid component. In the 1940s he developed the revolutionary idea that the collective movements of the superfluid, its quasiparticle excitations, form the normal-fluid component [Landau & Lifshitz 1992]. The importance of this theory can hardly be overestimated, since here for the first time the metaphysical idea of a mechanical juxtaposition of condensate (which Landau imagined to be similar to a frictionless cosmic æther) and uncondensed particles was overcome.

The low-energy excitations of superfluid helium are not individual helium atoms "broken out" of the condensate, but collective oscillations and motions of the condensate similar to a quantized sound wave. Such low-energy excitations are called quasiparticle excitations and, unlike quantized vortex filaments, are unstable.

According to Landau's theory, the superfluid state consists of condensate and collective excitations, which can be considered approximately like two liquids that have penetrated each other. Besides the vanishing friction, the condensate is characterized by the disappearance of the rotation on a macroscopic scale. The quasiparticle excitations form the normal-liquid component. They have to subordinate themselves to the state of motion of the condensate – they are carried along, so to speak, by the "superfluid background." They are also solely responsible for the heat transport of the superfluid and the entropy, while the thermal resistance of the superfluid disappears. The number of quasiparticle excitations of the superfluid per unit volume is a function of temperature, and disappears at absolute zero.

In the framework of the Landau theory, the fountain effect is caused by the hydrostatic pressure of the condensate excitations. Local heating at the place of heating increases the density of quasiparticle excitations. The conversion from superfluid to normal-fluid components inevitably generates a superfluid flow to the heater through the porous seal, while the back-flow of the quasiparticle excitations generated is slowed by their viscosity. The resulting overpressure creates the fountain.

Bose–Einstein condensation in interacting systems

In the interpretation of superfluid helium as Bose–Einstein condensate with macroscopic matter waves, however, there were initially major theoretical difficulties related to the fact that the inter-atomic interactions between the helium atoms are not negligible. Inter-atomic interactions of helium atoms can constantly scatter atoms from one matter wave state into another, raising the question of how they can condense into a uniform, macroscopically coherent, wave state. This was solved theoretically in 1947 with the Bogoliubov theory of Bose–Einstein condensation of interacting gases or liquids developed in the Soviet Union (see for example [Pines & Nozieres 1966]).

It was shown that the interaction between the atoms is not an obstacle, but in a certain sense even a necessary condition for them to be able to organize themselves into a uniform matter wave state, because it contributes to the tuning of the matter waves of individual helium atoms to each other and to producing a coherent matter wave state (see Figure 24 above).

The theory of Bose–Einstein condensation in ideal and interacting systems leads to a central organizational principle of quantum fluids: The theoretical principle of the indistinguishability of particles in their matter-wave states. Due to the interference of the matter waves of the particles, the quantum fluid state cannot be broken down into the matter wave states of the individual particles. In many-particle quantum theory, therefore, the many-particle matter wave is constructed from the single-particle waves in such a way that, firstly, the superposition and interference of all matter waves of all individual atoms are taken into account. Secondly, the indistinguishability of the particles is taken into account theoretically by "exchange of the particles in the single-particle wave states." This peculiarity of coherent matter wave states of many-particle systems is also called the entanglement of states. It is objective, real and present (see also Chapter 4).

Atoms lose their individual movements in the superfluid condensate

In recent years, experiments have been carried out at the Max Planck Institute for Quantum Optics in Munich on Bose–Einstein condensates for the controlled alteration of the entanglement of atomic matter waves [Mandel 2003]. To this end, a collective of about 10,000 rubidium atoms is first converted into a superfluid state and then into an "insulator" by means of a spatially modulated light field ("optical lattice"). The isolation of the individual atoms in the potential wall of the optical lattice results in vanishing of the mass flow currents in the system. Thus, their common matter wave decays into that of the individual atoms. Changes to the nuclear spins by microwave pulses and changes to the optical lattice then produce an entangled state in a controlled manner, which is proven by interference experiments. These experiments may open the way for the development of new computer technology called quantum computers. Today, entangled states of interfering superfluid beams of atoms, so called atom lasers, can also be demonstrated (see Figure 23).

Due to the idealistic world view, however, the entanglement by interference of matter waves is interpreted as a "ghostly action at a distance." The experimental quantum physicist Anton Zeilinger writes: *"The only way to avoid this problem is not to see the probability wave as a real wave that actually propagates in space. It is only a tool to calculate probability (...)."* [Zeilinger 2003, p. 191] The real and objectively existing coherent state of tens of thousands of atoms in the superfluid state, with which even an atom laser can be realized, is supposed therefore not to be created by real coherent matter waves, but an object of calculations? His experiments on

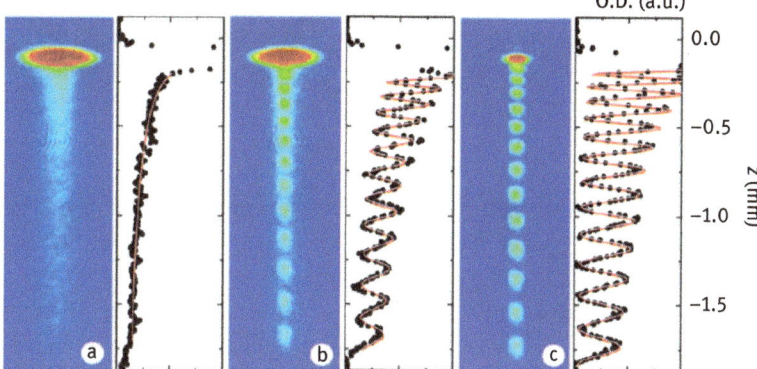

Figure 23: Realization of an atom laser by Bose–Einstein condensation and decoupling of atom beams from the condensate. The color code of the images is a measure of the density of the condensed atoms. (a) For temperatures $T > T_c$ the beam has no interference capability. After the Bose–Einstein condensation commences, (b) $T < T_c$ and (c) $T \ll T_c$, outputted beams of atoms show interference and thus the existence of a macroscopically coherent matter wave of the condensed atoms. Image from: I. Bloch et al. Nature 403 (2000) 16, reprinted with permission of Macmillan Publishers Ltd.

entangled two-particle systems (see Chapter 4) obviously do not prevent Anton Zeilinger from explaining the real interference effects of matter waves as mathematical fictions.

In fact, in the quantum fluid, the individual atoms lose their individual identity and no longer appear as individual objects, although they continue to exist as real atoms. The latter can be seen by comparing the condensation energies for quantum fluids of a few meV with the self-energy of the extremely massive and stable objects in the range of GeV. Their "decomposition" would require completely different energies. Just as the elasticity of a solid body is not the property of individual atoms, so is the coherence and entanglement of the matter waves of a quantum fluid not the property of individual particles. The characteristic new quality of this state of atoms is that by ordering of their matter waves the condensed atoms become an indissoluble part of the ordered collective movement of the condensate.

3.2 Fermi gases and liquids

Other many-particle systems also form new phases due to the interference of their matter waves. Besides the just-discussed quantum gases and liquids made up of bosons, today the Fermi gases and liquids made up of fermions are known. In contrast to bosons as particles with integer angular momentum (in units of the angular momentum quantum), fermions are particles with a half-integer intrinsic angular momentum. Fermions include, for example, not only electrons, but also the components of nuclei, neutrons, and protons. Fermi liquids made up of electrons occur in many solids with moving electrons. Fermi liquids made up of components of nuclei are of crucial importance for understanding the structure of burnt-out stars. The term fermion refers to the Italian physicist Enrico Fermi, who in 1925 made a significant contribution to the understanding of kinetic energy in collectives of particles with half-integer spin.

The distribution of the kinetic energy of fermions was explained by P. Dirac in 1926 by the nature of the matter-wave state of particles with spin. The energy distribution of fermions is named the Fermi–Dirac distribution after them.

The difference between Bose and Fermi gases (see Figure 24) is due to the fact that the condensed atoms differ in their intrinsic angular momentum. The quantization of the angular momentum in half-integer or integer units is associated with different properties of their matter waves. When the particles revolve, matter waves of the bosons must return to their initial state after one revolution, those of the fermions after two revolutions. Therefore, the intrinsic angular momentum occurs in integer

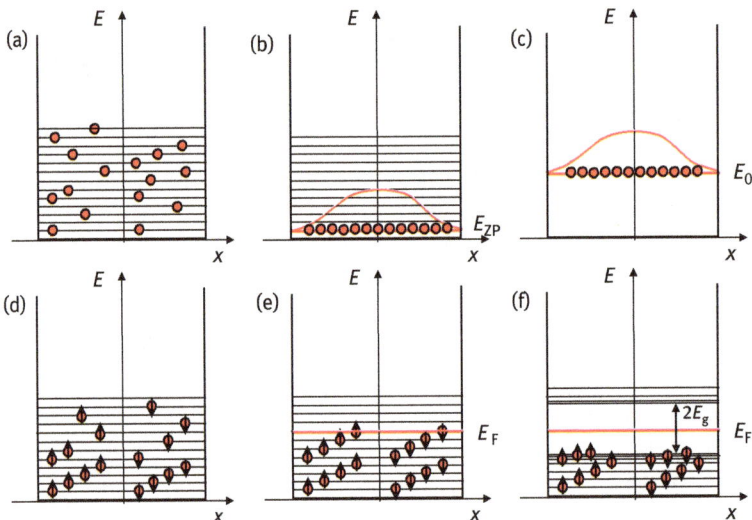

Figure 24: Comparison of the condensation of gases made up of bosons (top) and fermions (bottom) in an energy diagram: horizontal lines are the energies of the single-particle states, arrows are the directions of spin of the fermions. (a) High-temperature state of bosons as a quantum gas with disordered energy distribution; (b) Bose–Einstein condensate ($T = 0K$) of an ideal boson gas with zero-point energy E_{ZP}. (c) Bose–Einstein condensate ($T = 0K$) of a real gas of interacting bosons with energy E_0, determined by the interactions of the particles. (d) High-temperature state and (e) Ground state ($T = 0K$) of a quantum gas of fermions. Each single-particle state is occupied by a maximum of 2 fermions of different spin orientation; the highest occupied state at $T = 0K$ is called Fermi energy E_F. (f) Superconducting state of a Fermi liquid for $T = 0K$. An energy gap of $2E_g$ is formed in the excitation spectrum.

(bosons) or half-integer (fermions) multiples of a minimum angular momentum quantum, i.e. Planck's constant h/2 π. Fermions in identical states repel each other not only by the electrostatic interaction between their charges, but also by the Fermi pressure due to the interference properties of their matter waves.

The repulsion by the "Fermi pressure" is a general property of Fermi gases and fluids and is the basis for the "Pauli principle." A matter wave state with identical energy can be occupied by at most two fermions of different spin orientation. In a many-particle system, successive states must, therefore, be occupied at ever higher energies. The energy distribution at the thermal equilibrium of a Fermi gas is therefore fundamentally different from the distribution of the values of the energy of the particles in a classical gas.

This becomes clear in Figure 24e: In the ground state of a gas of fermions at temperature $T \rightarrow 0$ each state is occupied by two fermions with an antiparallel spin direction. The highest occupied state has the "Fermi energy" E_F. However, this picture of the ground state of an ideal Fermi gas is a rough approximation because it neglects the interactions between the particles that occur in addition to the interference of their matter-wave states. A system of interacting fermions forms a Fermi liquid. At low temperatures, the fermion liquid collapses into a superconducting state (Figure 24f), in which some of the fermions with energy in the range of the Fermi energy condense into a quantum state with macroscopic matter wave, thereby forming an energy gap in the excitation spectrum.

Fermi liquids made up of electrons

Electrons have a mass that is more than a thousand times smaller than that of atoms, and thus a de Broglie wavelength of their matter waves that is greater by the same factor. At typical densities of electrons, as they are present in metals and semiconductors, the effects of the formation of Fermi gases, Fermi liquids and their condensation into states with macroscopic matter waves caused by the interference of matter waves, therefore, occur at much higher temperatures compared to the Bose–Einstein condensation of atoms.

In a metal, the electron density is on the order of 10^{21} per cm^3. Since electrons are particles with electric charge, an interaction between them occurs not only via their matter waves but also through the electrostatic repulsion between their charges of the same sign. Due to the strong interactions, many-particle systems of electrons in metals and doped semiconductors form a Fermi liquid. Today, the entire modern semiconductor technology and computer industry is based on the understanding and technical application of Fermi liquids.

The Fermi liquid is a quantum fluid state of the electrons, which is decisively determined by the cooperative behavior of the matter waves and the interacting electrons. The conditions for this are a sufficiently high electron density and a sufficient degree of delocalization of the electron states. The Fermi liquid, like all other quantum fluids, is characterized by the fact that its state can no longer be understood as independent particles and individual matter waves. Every motion of a single electron inevitably leads to a reaction of the surrounding electrons, which in turn affects the motion of the selected electron.

The Soviet physicist Lev Landau contributed significantly to the theoretical understanding of Fermi liquids in the 1940s and 1950s. He developed the surprising and brilliant theory that the collective excitations of this liquid at finite temperatures again behave like a gas of noninteracting effective particles, what are called "quasiparticles." Such a gas of quasiparticle excitations of a Fermi liquid behaves in turn exactly like an ideal Fermi gas (electron gas).

The approximation of independent electrons and the band model

The consequences for the field and our world view of Landau's revolutionary work have hardly been realized in physics to this day. Instead of rigorously turning the development of individual particle properties from the collective movements of a quantum fluid into the starting point for electron theory for solid bodies, Landau's brilliant theory of replacing the liquid with the gas of its excitations is made absolute as a concept of "independent electrons." The method of reducing the complex many-particle motion to a single-particle motion of particles with "effective properties," which take into account the interactions with all other particles, is very successful, on the one hand.

But the widespread ideological suppression of the origin of the formation of single-particle excitations from the collective motion of quantum fluid is, however, a fatal influence of the positivist and pragmatic ideologies on modern physics. The collective behavior of the electronic many-particle system in a solid is not only determined by the interactions between the electrons and the interference of their matter waves, but also by the influence on the states by the electrostatic potential of the atomic nuclei.

In 1928, Felix Bloch considered the problem of the behavior of the matter wave of a single electron in a solid in the structure of periodically arranged atomic cores. Bloch came to the conclusion that the possible stationary energy states of the electrons are concentrated in bands separated by "energy gaps" with forbidden states (Figure 25). Possible matter waves of the electrons that are stable over time are precisely those that are not scattered or diffracted by the periodic lattice of the atomic cores. This excludes matter wave states which have the same periodicity as the atomic lattice, as well as higher multiples (harmonics) of this periodicity.

The stationary matter wave states of the electrons are therefore determined by the unity of the matter wave structure of the moving electron and the electrostatic forces on the electron by means of the arrangement of the atomic cores. The possible states that electrons take up in the lattice of the atomic cores can be separated into stationary

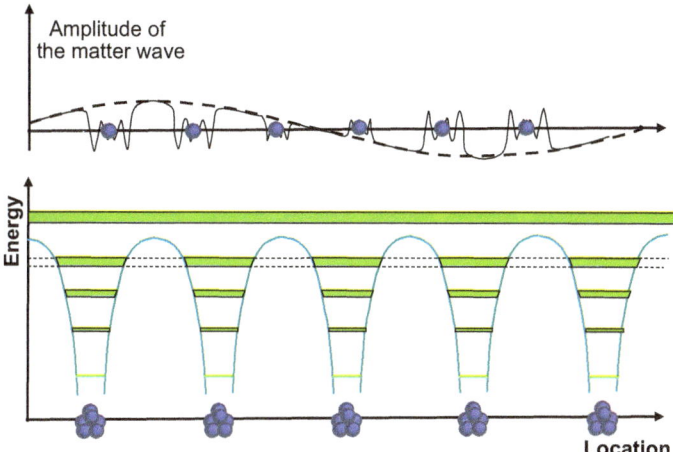

Figure 25: Electron states in the periodic electrical potential of the atomic cores of a crystalline solid. Above: The matter wave of freely moving electrons (broken line) experiences a modulation in the periodic potential and a Bloch state is formed (solid line). Matter waves with wavelengths corresponding to an integer multiple of the lattice period are diffracted at the periodic potential of the atomic cores and therefore do not propagate in the crystal. There is a band gap in the excitation spectrum of the electrons. Below: Energy bands of the electrons in a solid body, taking into account the bound core electrons and the moving conduction electrons. The overlap of the matter waves of electrons of lowest energy is small and they are localized at the atomic core. The matter waves of the higher-energy electrons are delocalized, forming Bloch states with energy bands.

Bloch waves, in which the matter waves correspond to the propagation through the lattice structure (Bloch waves) and unstable electron states, whose matter waves are diffracted at the lattice. Bloch wave states are stable and form a permitted electron band. In contrast, states in the energy gaps are unstable, i.e. electrons in such states rapidly decay into stationary states within the permitted energy bands. Transitions between different energy states of the electron occur by emission or absorption of photons.

This band model is successful. It can explain why certain materials occur as insulators, semiconductors, or metals. Different materials differ in their electron density and the position of their bands. Therefore, they differ in the highest energy state that can be occupied by the electrons.

Metallic behavior occurs if the Fermi energy happens to lie within a band. Electrical conductivity requires the occurrence of low-energy excitations of the electrons, and this is only possible if permitted states lie above the Fermi energy. If there is an energy gap just above the Fermi energy, then there can be no redistribution of the electron movements and thus no macroscopic current flow: the material is then an insulator. The success of the band model is based on the fact that it correctly reflects essential aspects of the self-organization of the electron movement, in dialectical unity with the atomic structure of the respective material. This band model is based on the framework of describing the highly entangled collective motion of electrons in the liquid as independent particles.

Quasiparticle excitations as a qualitatively new collective form of motion of Fermi liquids

The great success of a one-particle concept of independent electron states and electron excitations, without interaction with all surrounding electrons, has its basis in the laws of the collective motion of electrons in a Fermi liquid. Lev Landau's decisive discovery in the development of the theory of Fermi liquids in 1957 was that the macroscopic properties of a Fermi liquid are completely determined by its excitation spectrum. At low energies, these consist of electronic quasiparticles, and at high energies of plasmons. Both are nothing more than different collective forms of electron motion. In general, quantized collective motions in interacting many-particle systems are called quasiparticles.

In his fascinating theoretical work, Landau was able to show that the electron quasiparticles, as excitations of the electron liquid, behave in exactly the same way as individual electrons in an ideal gas of quasi-electrons, that is, collective electron movements that behave like electrons that do not interact.

The low-energy excitations of a Fermi liquid form an ideal gas of quasi-electrons to a surprisingly good approximation. However, this only applies to excitation energies that are not too high and moderate temperatures, at which the density of the quasiparticles is so low that the interaction between them can be neglected. The fact that the excitations of the Fermi liquid form a gas of independent quasiparticles is the basis for the great success of the single-particle model of electrons in solids. The entire interactions of the electrons with each other are taken into account in the quasiparticle by modified properties. Compared to the "free electrons," quasiparticles have a renormalized mass as a result of the collective movements of the underlying electron liquid, and a renormalized charge.

In most metals and semiconductors, electron quasiparticles have a smaller effective mass than free electrons at low excitation energies and a reduced negative electrical charge due to electrical shielding. The reduced effective mass can be demonstrated by the fact that the acceleration of electron quasiparticles in an electric or magnetic field is greater than that of free electrons. Their inertial force and thus their inertial mass are reduced. At higher excitation energies, the quasiparticle excitations correspond more and more to the free electrons that constitute the many-particle system.

However, there are also materials in which the effective mass of the electron quasiparticles is many times greater than that of the free electrons. Such systems are called "heavy fermion systems" with effective masses of electrons up to a thousand times the mass of free electrons. The size of the effective mass of the quasiparticles is determined both by the band structure of a solid and by the interactions of the electrons with each other and with other quasiparticles (e.g., phonons).

The real existence of hole quasiparticles in Fermi liquids

Quasiparticles can have both particle and hole character, the latter having a positive electric charge despite being composed of electrons. On the one hand, hole quasiparticles are created dynamically from collective electron movements, which can produce unoccupied electron states. In metals, the hole quasiparticle therefore appears dynamically as a "correlation hole" when electron quasiparticles are excited. In semiconductors such as silicon, the density of the electrons can also be adjusted by dopants. Depending on the type of dopant, this results in electron doping or hole doping. In electron doping, a previously unoccupied band is partially filled with electrons (Figure 26).

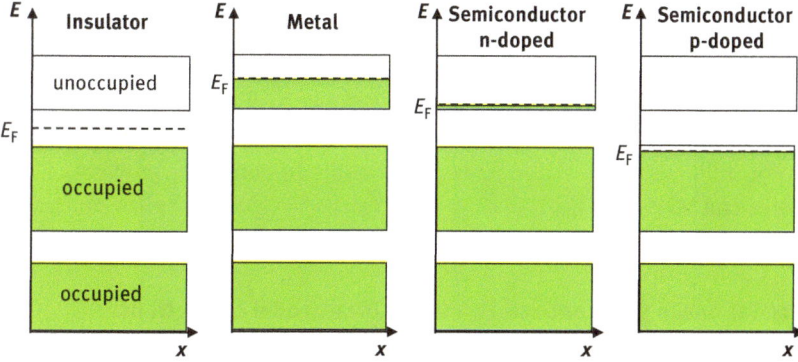

Figure 26: The difference in the electrical properties of solids is due to a different position of the highest energy state occupied by electrons (Fermi energy, E_F). In an insulator, a band is completely occupied. The energy gap is so large that there are only a negligible number of electrons in the conduction band due to thermal excitation. In a metal, the Fermi energy lies in the middle of a band and low-energy excitations of electrons, such as in a transport process, are easily possible. Like insulators, semiconductors have an energy gap, but it is usually smaller, so that the temperature movement of the electrons is sufficient to excite a certain number across the energy gap.

During hole doping, the occupied states of the highest energy of a fully occupied electron band are emptied. The resulting "missing electrons," the holes, behave like particles with positive charge. Their effective mass is usually greater than that of the free electrons. Thus, hole quasiparticles are also a collective effect of the motion in the electron liquid, which manifests itself as a new particle.

Just as with the electron quasiparticles, the question is also raised with the hole quasiparticles whether they really exist, or whether they are only an elegant, simple description of the complex motion of missing electrons, as the pragmatic-positivistic world view would suggest. In fact, the hole is a real existing, qualitatively new, quasiparticle. This is proven by the Hall effect: If a semiconductor through which current flows is exposed to a magnetic field oriented perpendicularly to the direction of the current, electrical charge carriers are deflected due to the Lorentz force. If electrons were moving and motion of the holes would only occur as an indirect consequence of the electron motion, a negative Hall constant would have to result. But it is positive and is thus called p-conduction. Here the holes manifest themselves as real positively charged particles which are exposed to a Lorentz force in a magnetic field. For this reason, the p-conduction is an independent type of conductivity, which differs from the n-conduction, the conduction of negative charge carriers. Before the development of the quantum theory of Fermi liquid, the positive Hall constant found in a number of materials was a baffling phenomenon.

The theoretical understanding of electron gases and liquids has also enormously stimulated the development of applications in microelectronics and energy technology. Electron-doped and hole-doped semiconductors form the essential basis of microelectronics and the enormous technical progress based on it. By a suitable combination of electron-doped and hole-doped semiconductors, electronic components such as diodes, transistors, and more complex components up to the integration of millions of semiconductor components in computer processors and memory chips can be realized.

Electron excitations in semiconductors and the Dirac model of electrons

The importance of Landau's theory of Fermi liquids cannot be overestimated as it not only provides a quantitatively correct theory of the many-particle system of electrons in many areas but also reverses the perspective and, for the first time in the recent history of physics, describes particle properties as collective, quantized excitations of an underlying "continuum." The role of the "elementary particle" here is reversed – away from a fundamental "primordial entity" that cannot be fathomed any further, to a well-defined cooperative motion in a system of matter. Many scientists have an inkling of the far-reaching consequences of this development for their entire world view, as shown by the way in which the term "vacuum" is used. For example, D. S. Chemla and J. Shah wrote in their review article "Many-particle and correlation effects in semiconductors":

> More than five decades ago, Landau suggested that strongly interacting real particles in a real vacuum could be mapped on "quasiparticles." These quasiparticles are "clothed" with part of the interactions and represent relatively long-lived excitations of the many-particle system,

which develops in a "new vacuum" containing the "rest" of the many-particle system and the part of the Coulomb interaction not contained in the quasiparticle. The quasiparticles are complex objects (Cooper pairs in a superconductor, excitons in a semiconductor, etc.) and the new vacuum may be structured (it may have an antiferromagnetic order in copper oxides) and dynamic (it may contain phonons and magnons). [Chemla & Shah, 2001]

Indeed, the excitation spectrum of the quasiparticles in the semiconductor has a number of similarities with the Dirac model of the free electron in a "vacuum": It replaces the "empty space," in which electrons move, by a condensed background of a completely occupied sea of electrons with lowered energy.

The Dirac Sea of the electron corresponds to the uppermost fully occupied band of the undoped semiconductor. In the ground state, the presence of the quantum fluid (of the sea of electrons) is not noticeable for an "inner observer." This is why pragmatism speaks of a "vacuum state" in which the electrons are located. If an excitation takes place across the band gap, i.e., an electron of negative energy passes into the range of positive energy, a particle is born. It leaves a hole state in the Dirac Sea, which corresponds to the generation of an electron–positron pair.

The positron as a "hole state" with a positive charge and as an antiparticle of the electron, is represented in Dirac's model as a collective excitation in the sea of electrons with negative energy. In both cases, in the semiconductor and in the Dirac sea, the excitation of a particle–antiparticle pair is possible by the absorption of photons with sufficient energy. However, there are also significant differences between an electron liquid in a solid body and the Dirac sea, which are discussed in Chapter 6.

3.3 Superconductivity of electrons

Fermi liquids have the remarkable property that at low temperatures they enter a new condensed phase of electrons, the superconducting state. This state was first observed experimentally as early as 1911 by H. Kamerlingh Onnes in Holland during the cooling of mercury. He reported on the "disappearance of the electrical resistance." Like in the superfluid state of the atoms, in the superconducting phase, the electrons organize themselves into a collective state with macroscopically coherent matter waves. In connection with this, the resistance for a continuous electric current becomes immeasurably small. For T→0 and in the absence of external electromagnetic fields, it must be assumed that the electrical resistance actually "disappears." In the course of the 1930s, it was discovered that the electrons in almost all metals change to the superconducting state when they cool down to sufficiently low temperatures. It is a general property of Fermi liquids. However, a microscopic understanding of this phenomenon only began to be developed in the 1950s by Bardeen, Cooper, and Schriefer, and independently of them by Bogoliubov in the Soviet Union.

Condensation of electrons and Cooper pairing

The model, also called BCS theory after Bardeen, Cooper, and Schriefer, has been extensively tested to date and is at least in simple superconducting metals in excellent agreement with experimental observations. In fact, the theory should be called BCS-Bogoliubov theory in appreciation of the Soviet contributions.

The theory says that, in the presence of a weak attracting interaction between the electrons, and sufficiently low temperatures, the Fermi liquid becomes unstable towards the formation of a new many-particle state, superconductivity. This state has two aspects: On the one hand, two electrons in pairs form a new composite particle, called a "Cooper pair." This allows them to change from fermions with a half-integer angular momentum (spin) to bosons with an integer spin. The pairing as bosons enables them to overcome the Fermi pressure and to override the Pauli principle as the organizational principle of the Fermi liquid. The Cooper pairs condense into a macroscopic quantum state with uniform energy and matter wave, similar to the Bose–Einstein condensate.

There is a major difference from Bose–Einstein condensation: the diameter of a Cooper pair, called coherence length ξ_0, is much larger than the mean electron distance (see Figure 27). This means that the pairs penetrate each other, and the superconductivity cannot be regarded as condensation of completed pairs (a "molecule" of two electrons). Also, condensed Cooper pairs are not static entities, but a dynamic state in which a constant exchange of electrons takes place in and between the pairs.

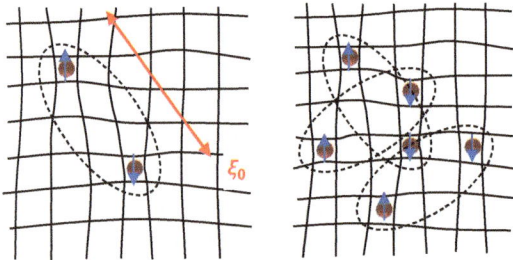

Figure 27: Formation of Cooper pairs and condensation in the superconducting phase. Left: An attractive force between two electrons is created by the distortion of the lattice of the atomic cores (black lines) induced by them. A bound pair of electrons with opposite intrinsic angular momentum is called a Cooper pair (broken ellipse). It has the size ξ_0. Right: The superconducting state results from the condensation of many Cooper pairs into a macroscopically coherent matter-wave state. The mean size of the Cooper pairs exceeds the mean electrical distance so that the pairs penetrate each other, and their individual motions indissolubly merge into the collective motion of the condensate.

The pairing of the electrons is triggered by a small attracting force between the electrons. One cause is the electron-phonon coupling: The movement of negatively charged electrons induces a vibrational movement of the positively charged atomic cores, a phonon, which exerts an attracting force on a second electron.

Cooper pairs are therefore examples of a new composite quasiparticle, called the "polaron," consisting of electrons and phonons. Above a certain level of electron-phonon interaction, the Fermi liquid becomes unstable with respect to the superconducting state. In fact, superconductivity is observed in almost all metals for sufficiently low temperatures. However, the superconducting state cannot be separated from the Fermi liquid; rather, it is its ground state at sufficiently low temperatures, unless other interactions oppose it.

The normal conductive state of the Fermi liquid only occurs above a critical temperature at which the individual thermal movements of the electron quasiparticles begin to dominate the coherence of the matter wave; the long-range order of the electron waves in the superconducting state changes into a dynamic short-range order in the normal conductive state of the Fermi liquid.

The occurrence of an energy gap in the excitation spectrum

At the transition to the superconducting state at the superconducting transition temperature T_c, an energy gap opens in the excitation spectrum of the electrons (Figure 28). It corresponds to the condensation energy gained per electron during the

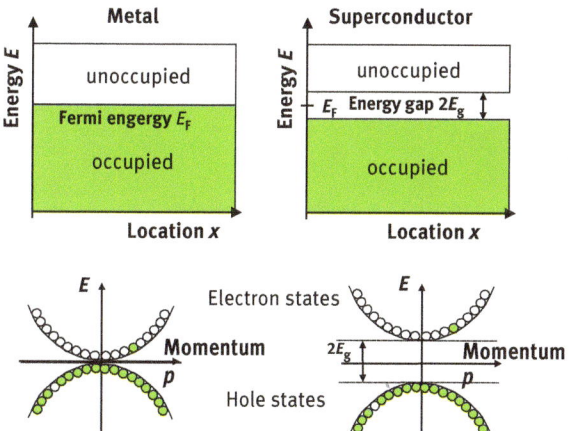

Figure 28: Transition from Fermi liquid (normal conducting metal, left) to superconductor (right). The electrons near to the Fermi energy (E_F) form a new collective state in the superconductor, in which a uniform macroscopic matter wave is formed. It has an energy gap E_g in the excitation spectrum. A minimum energy is necessary to produce a collective excitation (change of the collective motion of the electrons) and thus a quasiparticle. The lower part of the graph shows the excitation energy of electrons or holes as a function of their momentum (dispersion) in metals and superconductors. It shows one electron-hole pair excitation for each case.

transition. The condensation energy is determined by the binding energy of the Cooper pairs and, depending on the superconducting material, is between 1 and 20 meV.

Although the excitation spectrum in the superconducting state changes only for electrons with energies in the Fermi energy range (Figure 24f), condensation in the macroscopic quantum state of superconductivity affects all electrons. It changes the overall collective behavior of the electron liquid.

The superconducting condensate of these electrons has the remarkable property that a minimum energy of magnitude of the energy gap E_g is necessary for collective excitation. For the scattering of electrons from the coherent motion in the condensate, a minimum energy must be overcome. The energy gap therefore directly reflects the stability of the loss-free state of motion of the condensed electrons. The occurrence of an electrical resistance requires low-energy excitations by scattering processes of the electrons. However, this is suppressed in the superconductor.

The cause of the energy gap in the superconductor is fundamentally different from that in the semiconductor: While the energy gap in the semiconductor is a direct expression of the incompatibility of traveling electron waves with wavelengths of the size of the lattice spacings of the periodic lattice of the atoms, the superconducting energy gap is an expression of the subordination of the disordered electron motion to that of the ordered superconducting condensate. At finite temperatures $T > 0K$, however, thermal disordered motions are always present. They are quasiparticle excitations as disordered collective modes of motion of the condensate. A superconductor therefore also consists of "two different electron fluids": the superconducting condensate and the collective of fluctuating quasiparticle excitations. The quasiparticle excitations are unstable. As with the Bose–Einstein condensate, however, one cannot speak of two "fluids existing side by side," but it is one and the same electron system which forms the condensate and the collective motions as excitations.

Topological defects: quantized vortex tubes

At higher energies, in addition to the unstable quasiparticle excitations of the condensate, topologically stable excitations, the vortex filaments, also occur. Movements of electrical charges in the superconducting electron condensate inevitably lead to magnetic fields. Therefore, quantized vortices in the superconductor are associated with the quantization of the magnetic flux. Quantized vortex filaments in a superconducting electron condensate, therefore, form quantized flux filaments (see Figure 29).

For a better understanding, let us analyse the behavior of superconductors in a magnetic field. A magnetic field applied externally to a superconductor is shielded from the interior by a continuous superconducting electron current flowing at the surface.

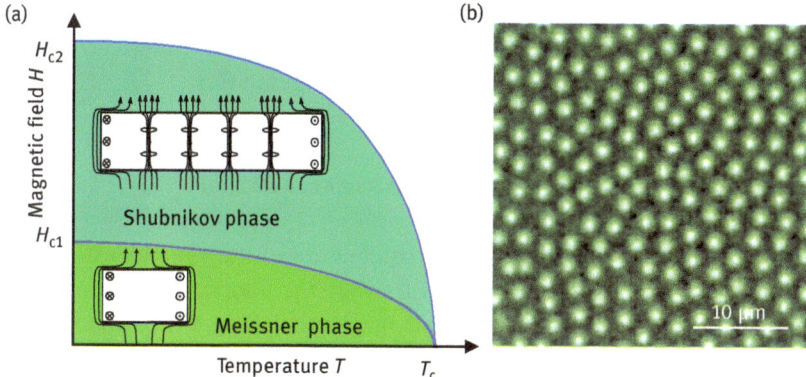

Figure 29: Left: Magnetic phase diagram of superconductors. Below a first critical field H_{c1}, the Meissner phase occurs in which the magnetic field is completely shielded. In the region of the first and second critical fields H_{c1} and H_{c2}, quantized vortex filaments are formed, and the magnetic field is displaced in magnetic flux filaments connected to the current vortices. Schematically, both types of flux shielding are represented by a cuboid superconductor with superconducting shielding currents. Right: Magneto-optical image of flux tubes in the superconductor $NbSe_2$. Bright points represent local magnetic fields, in the dark areas the magnetic field disappears. Image right: T. H. Johansen.

It is triggered at the moment when the magnetic field is applied to the superconductor. Just as no macroscopic rotation of the condensate can occur in superfluids, the same applies to superconductors for the superconducting current. A macroscopically coherent matter-wave of electrons requires a uniform speed of all electrons, which is incompatible with a rigid rotary flow. The superconducting condensate, therefore, produces a surface current that shields the interior of the superconductor: it remains field-free. This is called the "Meissner–Ochsenfeld effect."

However, the shielding of the interior by superconducting continuous currents at the surface is only possible up to a critical magnetic field H_{c1}. If it is exceeded, the macroscopic shielding of the entire superconductor collapses. The magnetic field penetrates into the interior of the superconductor and is enclosed there in the magnetic flux filaments mentioned above, by means of quantized vortex filaments, comparable to those in superfluids. According to Ampere's law, a macroscopic magnetic field in the superconductor would be associated with macroscopic currents that are incompatible with the coherence of the matter wave. The electric eddy current constricts the macroscopic magnetic field into the interior of the vortex filament.

Since the matter wave of the superconducting eddy current has to coincide in phase in the course of a revolution, both the eddy current and the total magnetic flux of the vortex filament are quantized. A quantized magnetic flux tube has been formed, an "elementary particle" of the magnetic field in the superconductor.

However, it does not have spherical but linear geometry. Like a dislocation line in a solid, flux tubes can only end at the surfaces of superconductors or they can form a closed ring. Their topological charge is the quantized circulation.

Particle formation by partial shielding of the electromagnetic field

Flux quanta form topological defects in the condensate of the superconductor by partial shielding of the electromagnetic field. They are the solution of the contradiction of the incompatibility of the macroscopic magnetic field and the macroscopic phase coherence of the matter wave by a process of self-organization of "particles."

The structure of the vortex filament is determined by two different length scales: The magnetic penetration depth is an electromagnetic shielding length that determines the inclusion of the magnetic field. It expresses a stiffness of the superconducting condensate against spatial variation of its current density and is a few 100 nm for typical superconductors.

The coherence length is determined by the stiffness of the condensate against a spatial variation of its condensation energy. It is the length scale of the formation of a normally conducting core in the center of the vortex filament, and is determined by the phase stiffness of the matter wave of the superfluid (see also appendix Chapter 9, Section 9.4). The condensed component of the superconducting electrons disappears in the core. Its formation represents the response of the condensate to the continuously increasing eddy current density towards the center: When the kinetic energy of the eddy current exceeds the condensation energy of the superconductor, the local phase transition to the noncondensed phase sets in: Thus, the system finds a way to limit the eddy current close to the center. In the center of the vortex filament, an undefined phase of the macroscopic matter wave and thus a "singularity" would also arise: For a coherently rotating wave, the phase in the center of the rotary motion is not defined.

Breaking of a "gauge symmetry" and mass formation of the photon

Partial electromagnetic shielding in superconductors does not only affect static magnetic fields. In addition, alternating electromagnetic fields, photons, are also shielded from the superconductor at the length scale of the magnetic penetration depth, as long as their energy does not exceed the condensation energy of the superconductor. This has the consequence that photons with energies below the energy gap of the superconductor effectively acquire a mass. In the superconductor, they are transformed into a spatially strongly damped plasma wave. The photon mass depends on the density of the electron condensate and is typically one hundred thousandth of an electron mass. The mass formation of the photon is identical with the breaking of what is called a "gauge symmetry." This relationship is enormously important for understanding the self-organization of "elementary particles" in the superfluid model of quantum æther, and will therefore be discussed in detail, using the example of superconductors.

In electrodynamics, the electrical and magnetic fields can be derived from what is called the "potential field" A. While the magnetic field corresponds to a rotation of the potential field, the electrical field is given by its change over time. Until the 1950s, the potential field A was regarded as a purely calculated quantity in electrodynamics, since it did not appear to be physically perceivable. However, Bohm and Ahanorov showed in a remarkable experiment that the potential field (also called vector potential) causes a shift in the phase position of matter waves. Conversely, any local displacement of the phase of a matter wave is identical to the occurrence of a quantum of the electromagnetic field, a photon.

This identity of leaps in the phase of the matter wave of electric charges with the emission or absorption of photons is the basis of the quantum electrodynamics of interacting charges and electromagnetic fields, and is called the gauge symmetry $U(1)$. This stands for a unitary transformation of a one-dimensional quantity, the phase of the matter wave. Gauge invariance, invariance among global $U(1)$ transformations, means that a global change in the phase of the matter wave of an electron has no physical effect on its interaction with an electromagnetic field. Only local phase changes of the matter wave occur in the electromagnetic interaction and are identical to a photon.

But what happens in the superconductor? The global gauge symmetry $U(1)$ is broken in the superconductor by the electron condensate with macroscopically phase-coherent matter waves, because all condensed electrons have a uniform phase of their matter waves. A local phase jump of an electron by absorption or emission of a photon would catapult the electron out of the superconducting condensate.

The phase rigidity of the electrons in the superconducting state, i.e., the breaking of the gauge symmetry for the phase of the matter waves, is the cause of the partial shielding of the photons and the occurrence of a photon mass. The occurrence of a superconducting condensate with uniform phase, the partial shielding of the electromagnetic field by formation of a photon mass and the formation of quantized vortex filaments by partial shielding, i.e., by electromagnetic structure formation, form an inseparable unity.

In the center of a flux filament, however, such a relationship between magnetic field, eddy current, and phase is not possible. The shielding superconducting eddy current would have to become infinitely large in the center, and the system reacts to this intensification of the contradiction with a local phase transformation to a normally conducting nucleus in which the superconducting electron condensate is destroyed. This means that the formation of topological defects in the condensate is intimately connected with the partial or full shielding of a macroscopic field: For photons of energy below the bandgap of the superconductor E_g, the background medium becomes "opaque" due to the formation of the superconducting state; the photons acquire a "mass," and cannot propagate in the superconductor.

Self-energy and rest mass of the quantized vortex filaments

The self-organization of superfluid flow and magnetic field in the magnetic vortex tubes leads to "identical particles." All vortex filaments carry the same magnetic flux, have the same circulation, and identical self-energy. Self-energy is the energy difference between a superconducting condensate with and without a flux filament. The self-energy corresponds to a rest mass and has has different components [Suhl 1965]: the condensation energy of the suppression of superconductivity in the nucleus of the flux quantum forms the largest part. A second part is contributed by the kinetic energy of the ring current of the vortex, and a third part by the energy of the magnetic field of the flux tube generated by the eddy current. Although the energy of the eddy current and the magnetic field add to the energy of the "particle," from the point of view of the superconductor system they deduct from the energy, since they lower the total energy of the system compared to a state with a macroscopic magnetic field without formation of a vortex filament.

Vortex filaments resist an accelerating force as a result of their rest mass. Depending on the superconductor material, it amounts to approximately 4,000 electron masses per unit length of the vortex filament. In addition to the inertial forces during the acceleration of vortex filaments, electromagnetic and hydrodynamic forces also act on them: Due to electromagnetic forces, magnetic vortex filaments repel each other, and can therefore form an ordered vortex filament lattice. Hydrodynamic forces occur due to macroscopic superconducting currents, and cause the vortex filaments to float along. Whether magnetic vortex filaments in superconductors also have a gravitational mass, i.e., are the source of a gravitational field, is unclear. Hydrodynamic forces on vortex filaments are, however, to a certain degree analogous to the occurrence of a gravitational field due to spatial variation of the Higgs condensate in the quantum æther.

Superconductors as the origin of the "Higgs model" for the formation of the mass of "elementary particles"

In fact, what is called the "Higgs mechanism" for the formation of the masses of "elementary particles" in a phase transition developed from the theory of superconductivity of electrons in solids [Higgs 1964]. In 1964, Peter Higgs suggested that below what is called the "electroweak phase transition," a superfluid condensate occurs in the quantum æther, which, by breaking a "gauge symmetry," gives a mass to the W and Z bosons as exchange quanta of the weak nuclear force.

According to his theory, the partial shielding of the weak nuclear force (more precisely the electroweak force, see Chapter 6) by the Higgs condensate is the cause of the formation of mass of all elementary particles. At the electroweak phase transition, the gauge symmetry (SU(2) x U(1)) of the electroweak interaction is broken. SU(2) stands here for "special unitary" transformations which act on a two-dimensional quantity, the fields of the weak interaction. The electroweak phase transition in the quantum æther is, however, a high-energy effect and occurs at temperatures of about 200 GeV (about 2,300 trillion kelvin).

In contrast to the topological defects of the vortex filaments in superconductors, however, the Higgs mechanism assumes that the "elementary particles" already exist as massless particles above the temperature of the phase transition, which gives this model a rather artificial character. However, it was able to predict successfully the existence of an excitation in the Higgs condensate, the "Higgs" boson (see Chapter 6).

At the end of this section let us again imagine an "interior observer" who lives in the "world of superconductors" and whose visible matter consists of "quasiparticle matter." He would not notice anything of the superconducting background. His universe would consist of the "zoo of excitations" of quasiparticles, which range from plasmons (as low-energy excitations) and electron-hole quasiparticle pairs to topological defects.

The reader as the "outside observer" of the superconductor, however, would never seriously come up with the idea of defining away the superconducting material with all its components including the superconducting condensate simply as "vacuum," as "nothing." Nor would an "outside observer" in determining the energy of the particles, ever seriously get the idea of separating them from the energy contributions of the variation of the superfluid condensate. We will discuss in Chapter 4 that all the difficulties of infinite energies in describing particles in quantum field theories are due to the fact that they are to be understood as an effective theory, that does not explicitly address the nature of the deeper structural level of matter forming the particles.

3.4 Superfluids with spin – Helium-3

Even the relatively simple quantum fluids discussed so far show a fascinating variety of properties and excitation states that illustrate the physical mechanisms of the development of unstable and stable particles as self-organized complex dynamic structures in this new form of matter. In addition, there are much more complex quantum fluids in which the internal degrees of freedom of the condensed particles, such as their own rotation (spin), also play a role. Examples are the superfluid phases of the helium-3 isotope with three nuclear components (^{3}He) and the superconducting state of the electrons in the compound $SrRuO_3$. We limit ourselves here to the representation of some properties of ^{3}He, which will play an important role as a model system for the qualitative description of properties of the quantum æther in later chapters.

In addition to the helium variant ^{4}He (a helium isotope with 2 protons and 2 neutrons in the atomic nucleus and 2 electrons in the shell) already discussed in Chapter 3.1, there is also the isotope ^{3}He (with 2 protons and 1 neutron as nucleus and 2 electrons in the shell). This lighter ^{3}He isotope is extremely rare in the Earth's atmosphere. It was not until 1933 that it was identified by M. L. E. Oliphant. Sufficient amounts of the ^{3}He isotope for low-temperature experiments could only be obtained by nuclear reactions.

The superfluid states of ^{3}He were discovered in 1971 by David Lee, Douglas Osterhoff, and Robert Richardson. They were awarded the Nobel Prize in Physics for this discovery in 1996. The condensation into the superfluid phases takes place at temperatures of 2.6 mK and 1.8 mK. Two different superfluid phases ^{3}He-A and ^{3}He-B occur (see Figure 30).

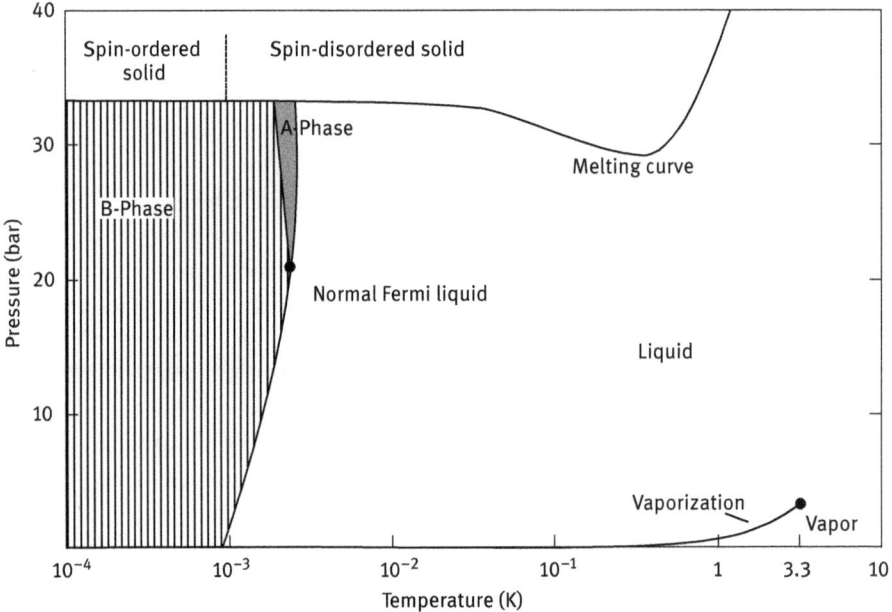

Figure 30: Areas of existence of different states of aggregation (phases) of the helium isotope ^{3}He. Two different superfluid states occur (^{3}He-A at temperatures below T_c = 2.6 mK and ^{3}He-B at temperatures below T_c = 1.8 mK). Furthermore, normal-liquid, solid, and gaseous phases may be present. According to [Vollhardt & Wölfle 1990].

Pair formation and the A and B phases of superfluid Helium-3

The ^{3}He isotope is a Fermion and, with an odd number of nucleus components, has a total spin of ½ angular momentum quanta. To condense into a macroscopic matter wave state, ^{3}He atoms must form pairs, like the electrons in a superconductor. In this case, the attractive interaction causing pair formation is supplied by the magnetic dipole interaction of the helium spins, which is reinforced by an orbital motion of the two helium atoms in the pair around each other. In order to understand the occurrence of two superfluid phases, the relationship between rotating and orbiting motions of the helium atoms must be considered. It can be seen that in both phases, as can be expected from the magnetic dipole coupling, the intrinsic angular momenta of the two helium atoms align in parallel, so that their total spin is $s = 1$, and thus each pair has a uniform spin axis. In the superfluid A-phase, the total spin s of a pair couples strictly parallel to the direction of the total orbital angular momentum, l, and both align macroscopically uniformly (Figure 31b). This results in a macroscopic orbital moment, L, from the sum of all microscopic orbital moments l and the phase thus acquires a spontaneous magnetic moment and a "ferromagnetic" character.

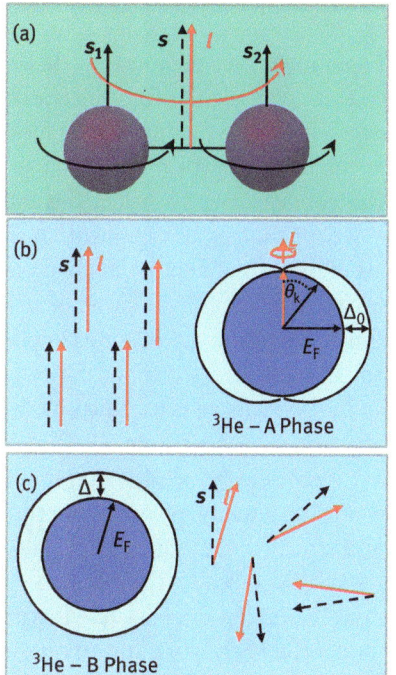

Figure 31: (a) Illustration of the coupling, spin, and orbital motion of a pair of ^{3}He atoms. The individual spins of the two helium atoms are strictly parallel due to the magnetic dipole coupling, so that the total spin is $s = s_1 + s_2$. The orbital angular momentum is parallel in the ^{3}He-A phase and at an angle to s in the ^{3}He-B phase (b and c). (b) In the superfluid ^{3}He-A phase, the orbital moment of all helium pairs is uniformly aligned and forms a macroscopic angular momentum L. This results in an anisotropic energy gap in the excitation spectrum, which is at a zero in the direction of the macroscopic orbital momentum, while perpendicularly to this, it is at a maximum Δ_0. In the superfluid ^{3}He-B phase, on the other hand, an isotropic energy gap Δ that is constant in all spatial directions is formed. While the spin and orbital moments of each individual helium pair are rigidly coupled, each pair has a random direction of the total moment, which therefore averages out. E_F denotes the Fermi energy.

Therefore, in the superfluid A phase, the energy gap for the excitation of quasiparticles from the condensate is anisotropic. Parallel to L, the energy gap disappears, and the amplitude of the macroscopic matter wave has a node. Perpendicular to the L-direction, the energy gap is at a maximum. The anisotropy of the energy gap has drastic effects on the quasiparticle excitations of the condensate, which do not have to supply a minimum energy parallel to the L-direction.

In the superfluid B-phase, however, the orbital and the spin angular momentum are slightly tilted in relation to each other and continue to be rigidly coupled via the spin-orbital interaction. However, no preferred direction of the total angular momentum of the condensed system is established: The angular momenta of the individual He pairs are averaged out, because they do not have a common orientation.

The B-phase, therefore, has an isotropic energy gap in the excitation spectrum, although it also has intrinsic motion patterns of the spin and the orbital moment for each individual pair. The study of two superfluid phases in 3He, the gapless A-phase with massless excitations and the B-phase with energy gap and massive excitations, provides mechanistic insights into understanding phases of the quantum aether involved in the electroweak phase transition.

Phase rigidity and flexural energy for angular momenta

In general, a condensate with internal orbital and spin moments, such as the ^{3}He superfluid, has a macroscopic coherent matter-wave state with significantly increased complexity. The matter-wave of the condensate possesses a phase rigidity against local phase shifts (such as due to a rigid rotation of the condensate), which induce superfluid eddy currents as a response, as in ^{4}He. The condensate also has a flexural strength in relation to the local deflection of the axis of the angular momentum or spin. Thus, local splays, twists, bends, and rotations of the axis of angular momentum or spin axis generate further orbital and spin components of superfluid currents as a response [Kleinert 2013]. In fact, the intrinsic rotation of the helium pairs, for example, contributes directly to a superfluid transport current. The law of the disappearance of rotation for a superfluid due to the phase rigidity of the condensate is modified by the possibility of superflows in conjunction with l and s textures in ^{3}He: while in ^{4}He, a rotation occurs only through formation of quantized vortices with a destruction of the coherent matter-wave state in the core of the vortex, in ^{3}He, a rotary motion of the condensate also occurs through l textures without local destruction of the superfluid state.

In order to describe the ordering of the matter-wave theoretically, the influence of the three possible values of the total spin ($s = 1$) and the orbital moment ($l = 1$) with respect to a fixed spatial axis (the projection of l and s can each be ±1 and 0) on the matter-wave is characterized by a nine-component order parameter (three times three components). The local deflection of the angular momentum axis, as well as the possibility of local deflection between spin and angular momentum axis represent qualitatively new excitation possibilities in superfluid ^{3}He. This results in a series of fields which have a certain similarity to the electromagnetic, weak, and gravitational fields of the quantum æther. The theoretical description of condensates with nine intrinsic motions is very similar to the theory of quarks and gluons in high-energy physics.

A zoo of topological defects/structures

The drastically increased variety of possible excitations due to the intrinsic motion of the superfluid gives rise to a whole zoo of topological defects. In addition to linear vortex structures, planar and spherical structures can also occur as monopoles or dipoles. It would exceed the scope of this book to deal with all excitations and topologically stable defect structures of superfluid phases with intrinsic motion patterns (see for example [Vollhardt & Wölfle 1990, Volovik 2003]). Some topologically stable structures are presented in Figure 32. The fundamental law of the formation of topological defects by partial shielding of macroscopic fields also determines the self-organization of the defects in superfluid ^{3}He:

- Vortex structures are created by partial shielding of external macroscopic rotary movements. In addition to vortices with one quantum of circulation in the isotropic ^{3}He-B phase, vortex structures with a half-integer and a double

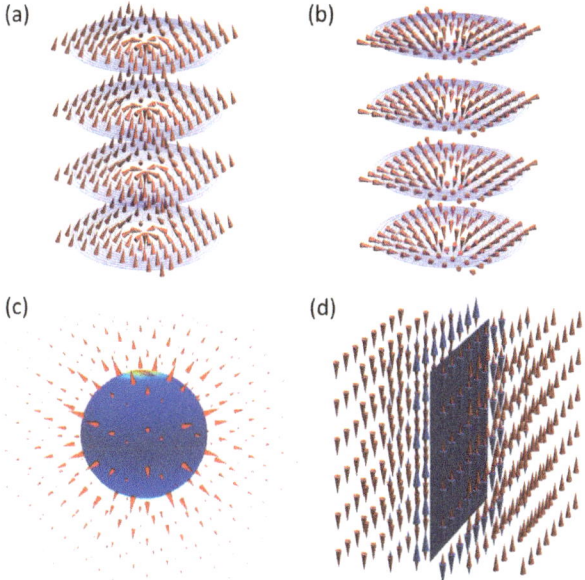

(a) (b) (c) (d)

Figure 32: Illustration of typical stable topological structures in the superfluid states of ^{3}He by means of the orientation of the orbital angular momentum axis L (red cones) and the superfluid current density (blue lines or arrows). (a) Anderson–Toulouse skyrmion. It is a vortex tube with 2 quanta of circulation. (b) Mermin–Ho vortex tube with one quantum of circulation. Both occur in the ^{3}He-A phase. (c) Spherical defect in the ^{3}He-B phase. The color coding on the sphere indicates the amplitude of superfluid current density. In the interior, the superfluid ^{3}He-B phase is suppressed, while in the immediate vicinity a "hedgehog structure" is induced in the orbital angular momentum. Such a topological defect has particle properties. (d) Planar texture with domain wall in the L field. In the vicinity of the domain wall, superfluid continuous currents occur parallel to the wall (blue arrows).

quantum of circulation may also occur in ^{3}He-A. The latter form a vortex pair which is bound to one another by the common circulation field of the superfluid, similarly to quarks in a nucleon [Volovik 2000].

– Planar structures and textures are formed in ^{3}He-A either at phase boundaries between different superfluid phases or to reduce magnetic leakage fields generated by the ferromagnetic condensate on surfaces, minimizing the free energy of the system.
– Monopoles and dipoles as spherical defects can form in ^{3}He-A only under special conditions at the edge of the container of the superfluid [Kleinert 2013]. Far from container walls, in the free superfluid, the flexure energy is too high for the formation of a "hedgehog structure" of the L axis due to the anisotropy of ^{3}He-A to form a stable topological spherical defect.
– In contrast, in isotropic ^{3}He-B, the formation of spherical topological structures is possible (see Figure 32c), if the induced change of the dipole energy and spin-orbit coupling energy is taken into account [Volovik 2003, p. 174]. At the defect, the ^{3}He-A phase is induced with a "hedgehog-shaped" structure of the L axis, which, however, decreases rapidly with distance.

Such spherical defects can contribute much to the understanding of the formation and structure of "elementary particles" and are discussed in Chapter 6 in conjunction with the modeling of "elementary particles."

Dialectics of self-organization in quantum systems

Are there fundamental differences between classical and quantum systems from the point of view of the development of qualitatively new matter structures and laws of motion by collective self-organization of an interacting many-particle system? This question must be largely answered in the negative. Both the self-organization in classical many-particle systems (Chapter 2) and in quantum systems is determined by thermodynamic laws. The resulting stable phases are determined by the dialectical unity of interactions and forms of motion, whereby the individual motion of the constituent particles must subordinate itself to the collective state of the system. In quantum systems the resulting phases of quantum gases, quantum fluids and superfluids also possess well-defined collective ground-state properties, such as ordering of their forms of motion, sharp transitions between different orders at phase transitions, and well-defined condensation energies.

When considering the mutual relationship between ground state and collective forms of excitation, not only can the excitations be derived from the type of collective ground state, but conversely, as Lev Landau showed by means of Fermi fluid, the excitation spectrum also determines the collective properties of the ground-state of the respective phase. Classical as well as quantum-physical multiparticle systems produce, besides the unstable collective excitations, also stable excitations as topological defects with sharply defined "particle-like" properties, which can be generalized to topological charges. In both systems, they are usually created by shielding macroscopic fields. This creates a new structural level of matter with qualitatively new laws that cannot be traced back to the individual properties and movements of the particles of the deeper structural level. For example, quantized vortex filaments in a superfluid can develop ordered structures of a periodic lattice with shear elasticity – a property that the underlying superfluid lacks. In 2016, physicists David Thouless, Duncan Haldane and Michael Kosterlitz were awarded the Nobel Prize in Physics for their theoretical description of collective properties of topological defects. Vortex filaments even have an inertial and rest mass, which has nothing to do with the mass of the atoms or electrons forming the superfluid, but is caused solely by the necessary condensation energy for the formation of a normal-fluid nucleus.

The main differences in the properties of quantum fluids resulting from self-organization compared to classical systems are due to the frictionless continuous currents. The underlying ordering and coordination of particle motion prevent energy dissipation by converting macroscopic motion into intrinsic microscopic forms of motion. It modifies the laws of electrodynamics (breaking of the gauge symmetry) and thus reveals that these laws are neither fundamental, nor apply independently of the underlying material structure and its forms of motion.

Quantum fluids as model systems for the understanding of the quantum æther

Therefore, it is not a superficial analogy, but an essential step towards a deeper understanding, when quantum fluids as model systems are examined today with regard to the self-organization processes of particles and interactions in the quantum æther, as exemplified by G. E. Volovik [Volovik 2003]. Particularly with regard to the large variety of possible excitations and the associated "intrinsic symmetries" of the condensate, this approach offers many interesting insights into mechanisms that play a role in particle and high-energy physics.

In fact, phase transitions in the quantum æther, such as the electroweak phase transition, and the transition into the quark–gluon plasma, can be compared with phase transitions in quantum fluids, with the difference that they occur on much smaller energy scales in the range of a few 10 meV, instead of GeV in the field of "particle physics."

The mechanisms of the formation of energy gaps in the excitation spectrum of quantum fluids at low energies today help to understand the cause of the energy gap for the excitation of electron–positron pairs in the quantum æther. The associated "Higgs" field as a superfluid æther condensate was predicted directly from the analogy to the theory of superconductivity, and the Nobel Prize in Physics in 2014 was awarded for analogy to the likely discovery of the Higgs boson. Furthermore, progress in understanding the mechanisms of the formation of quarks with fractional elementary charges in the building blocks of atomic nuclei results from the study of the formation of "electrons" with a fraction of an elementary charge from their collective dance with magnetic flux quanta in quantum fluids, as they have been realized in thin semiconductor layers in the laboratory, the "Hall fluids" [Anderson 1997]. The dialectical view of the formation of "elementary particles" from the collective self-organization of deeper structural levels is supported in an article by Nobel Laureate winner B. Laughlin and his colleague D. Pines, comparing solid-state and particle physics. In their remarkable article "The World Formula," they vehemently oppose the futile attempt, shaped by the reductionist world view, to trace the material world back to the final "elementary particles" and a world formula:

> The low-energy excited quantum states of these systems are particles in exactly the same sense that the electron in the vacuum of quantum electrodynamics is a particle (...) Rather than a Theory of Everything we appear to face a hierarchy of Theories of Things, each emerging from its parent and evolving into its children as the energy scale is lowered. The end of reductionism is, however, not the end of science, or even the end of theoretical physics.
>
> [Laughlin and Pines 2000, p. 29]

Quantum fluids are therefore a crucial model system for understanding the formation and transformation of particles in the quantum æther, and the associated interactions.

4 Matter waves and quanta as expression of the effect of deeper structural levels of matter

In the everyday world surrounding us, the movement of bodies along well-defined trajectories in time and space is a fact of experience confirmed millions of times over. The laws of classical mechanics were abstracted from the observation of the movement of masses. According to them, a body continues its state of motion until an acting force changes its state of motion. This principle of inertial motion is a central component of the revolution in natural science triggered by Galileo Galilei. It had to assert itself against the principle "A movement needs a mover," which was attributed to the philosopher Aristotle, and elevated to a dogma in medieval scholasticism. The law of inertial motion of bodies on a trajectory is based on the fact that in the macroscopic world no influence of the "vacuum" on particle motion can be detected.

However, the inertial law of particle motion through a space conceived as "empty" is not a fundamental law of nature but itself only an expression of the development of stable macroscopic forms of motion from complex microscopic movements of particles. In the microcosm, the influence of randomly fluctuating and unstable particle excitations of the zero-point fields on the motion of atomic and subatomic particles cannot be neglected.

The impact of deeper structure levels of matter on motion, i.e. a material continuum out of which particles are evolving and to which they are connected [Dickhut 1987], takes the form of matter waves. Each state of motion of particles is linked to a characteristic matter wave, which affects the motion of the particle. The laws of motion of classical mechanics, therefore, change into the laws of quantum mechanics. This deals with the motion processes allowing for the matter-waves, by means of probability laws.

Instead of studying the fluctuations of speed and location of a moving particle as an expression of the objective effect of a deeper structural level of matter, however, physical idealism sticks rigidly to the concepts of classical mechanics: according to Niels Bohr, one of the founders of quantum mechanics, the fluctuations in the momentum and location of particles are the result of the uncontrollable interaction between measuring apparatus and micro-object, "which cannot be analyzed in more detail because of the indivisibility of the quantum process." According to him, the "observer" plays a central role in the description of the quanta, by means of the "measuring process." In this widespread Copenhagen interpretation of quantum theory, the subject creates reality through the "act of observation." The basic concept of the materialistic world-view of the objective existence of matter, independent of consciousness, is thereby refuted by quantum physics, it claims.

The objective nature of matter waves is also disputed far beyond the supporters of the Copenhagen interpretation. Thus the well-known string physicist Brian Green writes in his book called *the Fabric of the Cosmos*:

https://doi.org/10.1515/9783110644203-004

> No one has ever seen a probability wave directly, and according to common quantum mechanical readings, no one ever will. Instead, we use mathematical equations (developed by Schrödinger, Niels Bohr, Werner Heisenberg, Paul Dirac, and others) to find out what the probability wave should look like in a given situation. (. . .). It is still controversial whether we should say that the probability wave of an electron is the electron or that it is associated with the electron, that it is a mathematical tool for describing the movement of electrons, or that it is the embodiment of what we can know about the electron. [Green 2004, p. 113/ 114]

Such widespread subjectivistic–idealistic interpretations were criticized by the Soviet physicist D. T. Blochinzew:

> The wave function is not regarded as an objective characteristic of a quantum mechanical whole, but as an expression of the observer's knowledge, obtained from the result of the measurement. The reality of any state of the microsystem in this view becomes identical to the observer's knowledge of the microsystem, i.e. it is transformed from an objective category into a subjective one. [Blochinzew 1953]

In order to understand the occurrence of matter waves and the resulting qualitatively new quantum-mechanical laws of motion for microscopic particles, one has to deal with the objective properties of deeper structural planes of matter, which is in this book approached by the physical model of the quantum æther.

4.1 The filled "vacuum": zero-point fields

Every thermal movement of atoms and electrons is associated with the emission and absorption of electromagnetic radiation, called "thermal radiation." In 1906, Max Planck introduced the quantization of the electromagnetic field, with photons as discrete quanta of energy $E = h\nu$ and Planck's constant h as quantum of action, in order to solve the contradiction between the observed and theoretically calculated frequency curve of this radiation. It will prove to be fundamental for the transition between unstable and stable excitations in the quantum world. Only by quantizing the light field was it possible to avoid the paradox of the classical theory, according to which the intensity of thermal radiation should grow to infinity at small frequencies through energy transfer between atomic motion and radiation field in "arbitrarily small portions."

From the theoretical treatment of thermal radiation with quantized photons, Max Planck predicted for the first time in 1912: The existence of zero-point motions of atomic oscillators [Planck 1911, 1912]. The atomic motions should also be present at the absolute zero of temperature and therefore not represent a temperature motion of the atoms, he claimed. Their energy per oscillation mode is $E = \frac{1}{2}h\nu$. While Planck gave no further meaning to the zero-point motion, Walter Nernst and Albert Einstein took up this theory. In particular, the investigation of the specific heat of solids at low temperatures indicated that zero-point motion really existed. In 1916, Walter Nernst transferred the idea of zero-point energy from atomic motions to the electromagnetic field. He argued that the zero-point motion of atoms must be in statistical equilibrium with the radiation field. Therefore, there must be an electromagnetic zero-point field, with an energy of $\frac{1}{2}h\nu$ per oscillation mode. He regarded this field as a kind of minimal fluctuating motion of the "light æther," which should exist independently of the presence of atoms.

Electromagnetic zero-point field – real or virtual?

The prevailing doctrine today regards zero-point energy as a result of the "principles of quantum theory." In almost every textbook, the actually existing statistical fluctuations of location, momentum, or energy of quantum particles are regarded as a "*consequence of Heisenberg's uncertainty relationship.*" Even Stephen Hawking, who acknowledges the real existence of a zero-point field, writes:

> As we saw in (...), quantum theory means that fields can't be exactly zero even in what is called the vacuum. If they were zero, they would have both an exact value or position at zero and an exact rate of change or velocity that was also zero. This would be a violation of the uncertainty principle, which says that the position and velocity can't both be well defined. (...). Vacuum fluctuations can be interpreted in several ways that seem different but are in fact mathematically equivalent. From a positivist viewpoint, one is free to use whatever picture is most useful for the problem in question. [Hawking 2001, p. 118]

The zero-point fluctuations are thus not reflected in the mathematical uncertainty relationship of quantum theory but rather the latter causes the zero-point field. The theory thus does not approximately reflect objective reality but imposes its principles on reality.

In contrast to this ideologically idealistic position, Walter Nernst was one of the first to regard a real zero-point field as the actual material cause of the quantum effects and to interpret, for example, the structure and stability of the electron shell as a result of the resonant movements of electrons with the modes of the zero-point field [Nernst 1916]. According to Nernst, the electrons bound in the atom are in equilibrium with the zero-point radiation and thus are subject to stochastic movement. Despite this random motion, they cannot emit or absorb any radiation on average in the stationary state. In contrast to a thermal radiation field with disordered photons, Nernst assumed a higher ordering of radiation for the zero-point field; we will come back to this. He recognized that just as with temperature radiation, the energy density of the zero-point field per mode must increase with the third power of the frequency of the oscillations. Only such a radiation field has the property that for bodies moving uniformly: no radiation pressure and thus no "friction effects" due to radiation interaction with moving bodies, occur on average.

Well-known quantum physicists such as Wolfgang Pauli vehemently denied the reality of a zero-point field with the argument that its large energy content would make it noticeable in large physical effects that could hardly be overlooked. For example, photo plates should blacken spontaneously even in the absence of light or a dark current should occur in photodetectors (see for example the discussion of Pauli's arguments in de la Pena & Cetto 1996, p. 101). This raises the question of the qualitative difference between the unstable excitations of the zero-point field and the stable photons. This question is closely related to the importance of Planck's constants as critical thresholds for the emission and absorption of stable photons. Only stable photons propagate across space and transfer energy. Electromagnetic fluctuations below this threshold, on the other hand, occur in the form of unstable photons, which decay again after a short time and release their energy. In quantum physics, however, the term "virtual photons" has become established. It suggests that they are only "mathematical quantities" without real existence and opposes a real understanding of the origin of zero-point fields. Zero-point fields are now directly detectable experimentally, for example in the Casimir effect and in quantum optics experiments.

The Casimir effect: forces due to the zero-point field

Hendrik Casimir, who studied electromagnetic fluctuations in molecules and their role for the practical application of coatings on materials, dealt with the effect of the zero-point field on metal plates in a theoretical paper in 1948. He predicted an attracting force between two parallel metal plates at a close distance, which are brought into an evacuated vessel at $T \to 0$ K. The zero-point fluctuations occur as fluctuating electromagnetic waves with all possible wavelengths. However, these must have a node of their oscillation amplitude on a metal surface (Figure 33a), otherwise, they would be immediately damped in the metal by the generation of electric currents and the associated energy dissipation.

For this reason, all wavelengths of electromagnetic fluctuations that do not have an oscillation node on the metal surface are suppressed. Therefore, the number of zero-point modes in the space between the two plates is reduced, thus reducing their energy density compared to the external area (Figure 33b). Casimir predicted an attracting force between the two plates, which was experimentally demonstrated a few years after his prediction [Sparnaay 1958].

Sparnaay was able to show that the attracting force is still present even when the plates are cooled to very low temperatures, although the temperature radiation has disappeared. Very accurate measurements were made in the 1990s [Lamoreaux 1997], which gave excellent confirmation of Casimir's prediction of the inverse-square law of force (Figure 33c).

Other consequences of the zero-point field that are fully accepted in physics today are the existence of fluctuating dipole forces between molecules that lead to what are called Van der Waals binding forces. Furthermore, the natural line width in the spectral lines of the atoms is caused by the energy fluctuations of the zero-point field, which leads to a finite lifetime of excited electron states. In addition, there is the Lamb shift, a shift of the energy of transitions in the electron shells of

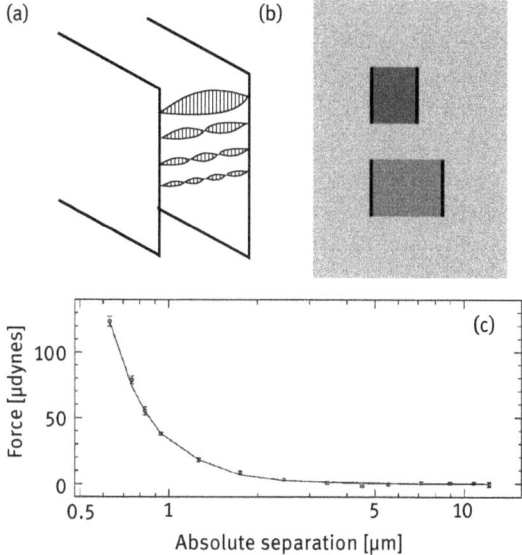

Figure 33: Casimir effect. (a) Wave trains of the electromagnetic zero-point field between two metal plates show a changed mode density since only waves with nodes on the plate surfaces can occur between the two plates. (b) The energy density of the zero-point field is lowered between the two metal plates. (c) Experimentally measured force between two metal plates as a function of their distance. Figure (c) from S. K. Lamoreaux, Phys. Rev. Lett. 78 (1997) 5, reprinted with permission of the APS.

atoms by the polarization of the zero-point field in the Coulomb potential of the atomic nuclei. The connection between the zero-point field and the existence of matter waves, to say nothing of the related material causes for the modification of the laws of classical mechanics by quantum mechanics is, however, rejected due to the influence of the positivist world-view and method.

Zero-point fields in quantum optics

Using sophisticated experimental techniques in quantum optics, it is now possible to generate and measure changes in the electromagnetic zero-point field (see Figure 34). What is called the "quantum homodyne" method is used, which allows the development of the zero-point field over time to be reconstructed via the influence of the zero-point field on the stable photons of laser light [Breitenbach 1997].

Light fields can be generated in different order states. In the thermal light field of a light bulb the photons have not only different energies and wavelengths but also arbitrary phase position of their electrical field. In contrast, coherent laser light is characterized by a uniform energy, wavelength, and phase position of the electromagnetic waves. Figure 34 shows the development over time of the differently ordered light fields by measuring their electrical field strength. The amplitude of the fluctuating electrical field strength is a measure of the number of photons; however, fluctuations also occur below the photon threshold. For very small amplitudes of laser light, the influence of the zero-point field is therefore visible as a randomly fluctuating variation of the electrical field amplitude in the coherent light wave. The zero-point field also causes a fluctuation to occur both in the phase position φ and in the instantaneous photon number N.

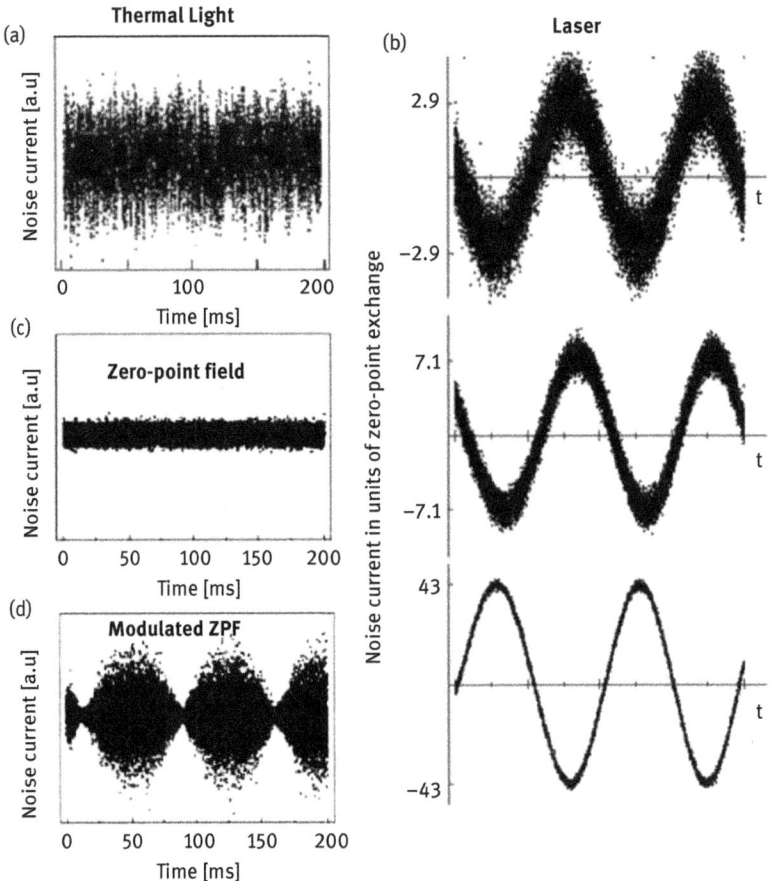

Figure 34: Reconstruction of different states of light fields including their zero-point fluctuations by means of quantum homodyne measurements of the electrical field amplitude [Breitenbach 1997]. (a) Thermal light. (b) Coherent light fields of a laser with increasing amplitude (from top to bottom) in units of the amplitude of the zero-point field. (c) Development over time of the field amplitude of the zero-point field, with equally distributed fluctuation of phase and amplitude. (d) Zero-point field with an artificially impressed amplitude modulation. Images: G. Breitenbach, S. Schiller, and J. Mlynek [Breitenbach Web site], with permission.

The product of the fluctuations $\Delta\varphi\,\Delta N$ of the photons never falls below a threshold of minimal uncertainty. The coherent state of the laser represents a state of the photon field in which the fluctuations have the smallest possible value for both of its phase and of its amplitude or photon number.

The ground-breaking quantum optical experiments of Gert Breitenbach in Konstanz made it possible to detect different states of the zero-point field of photons via the fluctuations of the electrical field amplitude. Figures 34c and d show two different zero-point field states. In the nonmodulated state, the variation in amplitude and phase is about the same. The amplitude-modulated state of the zero-point field is achieved by a nonlinear coupling of the zero-point wave with a laser wave [Breitenbach's Web site, Breitenbach 1997]. It should be noted that the development over time visible in Figure 34 corresponds to the time evolution of the probability distribution for the photon number of the Schrödinger matter wave of the light field.

This proves a connection between the zero-point fluctuations of the electromagnetic field and the matter waves. This relationship was taken up by various scientists in the 1970s in the context of work on "stochastic electrodynamics" [de la Pena & Cetto 1996]. However, this progressive development is restricted to a scientific niche by the effect of the positivist world-view.

4.2 The dialectical unity of matter wave and particles

In addition to the investigation of temperature radiation, the investigation of the photoelectric effect at the beginning of the twentieth century forms another source for the prediction of photons as quanta of the light field. In the photoelectric effect, electrons are emitted from the surfaces of solid bodies by the absorption of light quanta. From the classical wave theory of light, it would follow that the energy of the emitted electrons depends on the intensity of the light, that is, on the amplitude of the incident electromagnetic wave. In contradiction to this, in the photoelectric effect it is seen that the energy of the emitted electrons depends on the wavelength/ frequency of the incident light. Such experiments were conducted in 1902. In 1905, there followed Albert Einstein's explanation with the theory of light quanta, as mass-less particles with energy $E = h\nu$ and momentum $p = h\nu/c$, with c as the speed of light.

The energy of the photons is proportional to their frequency ν – in contrast to the classical wave concept, where the energy is proportional to the square of the amplitude of the field strength of the electromagnetic wave.

Einstein was awarded the Nobel Prize in Physics in 1921 for this work. For a long time, however, the theory of light quanta remained controversial – although no other explanation of the photoelectric effect was possible. It was not until 1922 that the Compton scattering of high-energy X-ray photons at electrons unquestionably proved the particle aspect of light. In both the photoelectric- and the Compton-effect, the energy and momentum transfer occur abruptly and thus particle-like in the interaction of the electromagnetic field with an electron. This is in contradiction to the classical theory, according to which the electron is gradually excited to oscillate in the oscillating electromagnetic light field.

The de Broglie matter waves

In 1924, the physicist de Broglie recognized that even subatomic particles with a rest mass, such as electrons, cannot simply be understood as extremely small "billiard balls in empty space."

He transferred the idea that light has both wave and particle properties from the photon to the electron. He assigned a matter wave with wavelength λ to the moving electron, which is inversely proportional to its kinetic momentum p and thus to its velocity via the de Broglie relationship $\lambda = h/p$.

As with photons, the frequency of the electron wave is determined according to $E = h\nu$ by the kinetic energy of the particle. The wave and particle aspects of the electron are linked to each other via the quantum of action. Davison and Germer succeeded in experimentally confirming the electron waves in 1925: They detected interference phenomena in electron diffraction at a crystal (Figure 35), which could only be solely by a wave associated with the electrons. The detection of wave

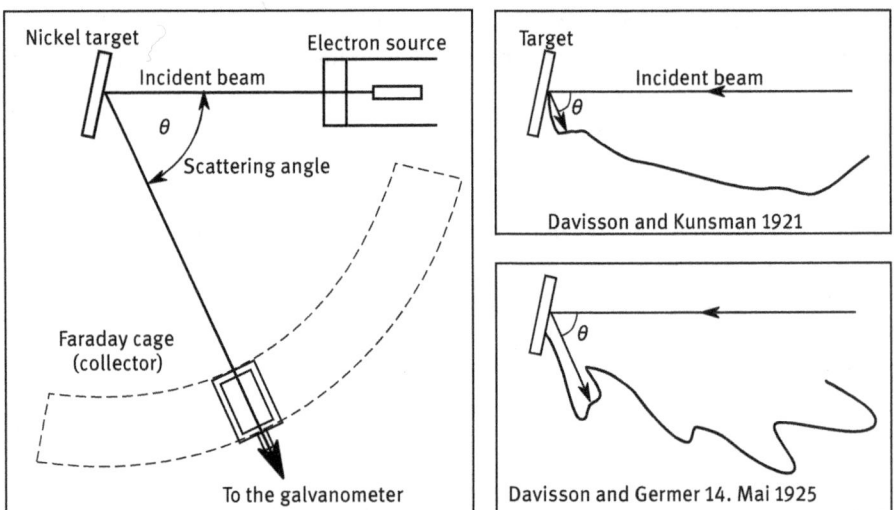

Figure 35: Experiments to prove the wave properties of electrons. Left: Experimental setup: The incident electron beam is diffracted by a nickel crystal. Right: The measured angular distribution of the electrons in various experiments in 1921 and 1925. The improved experiment of 1925 shows diffraction maxima of the electrons, which can only be explained by a wave accompanying the electron.

phenomena during electron movement was the key to Erwin Schrödinger's theoretical description by means of a wave equation in 1926.

Diffraction of single neutrons at the slit

Diffraction experiments on small slit openings with electrically neutral particles, for example, neutrons, are better suited than experiments with electrically charged electrons to investigate the complex dual wave and particle nature, due to their high precision and low susceptibility to interference from electromagnetic fields. The experiments can be carried out with such low particle fluxes that there is only one neutron in the apparatus at a time. This is of central importance for studying the nature of matter waves.

The precision experiments with monochromatic neutrons (see Figure 36) show a characteristic diffraction pattern which is in exact agreement with the Schrödinger theory. It is caused by the interference of matter waves emanating from different locations in the slit. This observation clearly confirms the wave nature of neutrons. The observed spatial distribution of the neutrons behind the slit is incompatible with a pure particle concept, in which only scattering at the slit edges would lead to a diffuse shadow. The detector behind the slit discretely detects individual neutrons, each of which occurs at a specific location. The diffraction pattern only occurs after a large number of neutrons have passed through the slit, as a spatial distribution of the incident neutrons. The experiment with single neutrons which pass through the slit one after another proves that the matter wave connected to the neutron interferes with itself at the slit, contrary to the idea that the diffraction pattern is the result of the interference of the matter waves of different neutrons with each other.

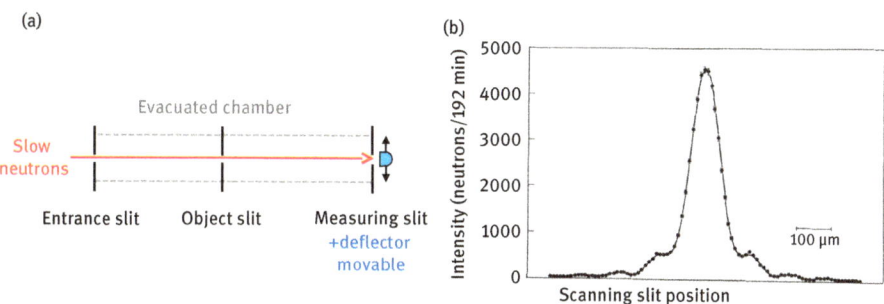

Figure 36: Diffraction of monochromatic neutrons at one slit. (a) Experimental set-up with entrance slit, object slit at which diffraction takes place, and a movable measuring slit with detector. (b) Neutron distribution at the measuring slit. The neutron flux was chosen so that only one particle is in the apparatus at a time. Images: (a) after [Rauch 2000] (b) reprinted with permission of A. Zeilinger, R. Gähler, G. C. Shull, W., Treimer, W. Mampe, Rev. Mod. Phys. 60 (1988) 1067, copyright 1988 by APS.

Interference of particles at a double slit

Interference experiments at a double slit pose a major challenge for the theoretical understanding of the interaction of particles and matter waves (see Figure 37). The experiment shows the interference of neutrons, where again the neutron flux is so small that only a single neutron is in the apparatus. The detector behind the double slit registers individual particles as a function of the location at the detector. The interference pattern is generated as a spatially oscillating frequency of the arriving neutrons. The location of each individual neutron is random. However, the probability for the impact at a certain location at the measuring slit is determined by the interference of its matter wave. Thus, the interference pattern of the oscillating neutron abundance only establishes itself as the sum of the impact of many neutrons. The distance between the interference maxima is a direct reflection of the wavelength of the matter wave of the monochromatic neutrons.

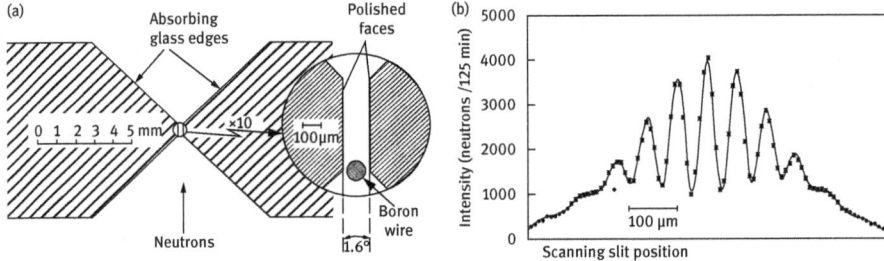

Figure 37: Neutron diffraction and interference at a double slit. (a) The experimental set-up is similar to that in Figure 36, except that the object slit is divided into a double slit by a boron wire. The slit consists of polished glass which absorbs neutrons. (b) Neutron number as a function of the location of the measuring slit. An oscillation of the particle number occurs due to the interference of the neutron waves from both slits. Again, there is only one neutron in the apparatus at a time. The theoretical curve from Schrödinger's theory of the matter waves is shown as a solid line. Images reprinted with permission of A. Zeilinger, R. Gähler, G. C. Shull, W., Treimer, and W. Mampe, Rev. Mod. Phys. 60 (1988) 1067, copyright 1988 by APS.

Similar experiments on single and double slits were also carried out with electrons, photons, even atoms, and whole molecules. They show the dual nature of these matter systems with discrete particle properties and the delocalized matter wave associated with their motion. Experimenting with individual particles shows that the matter wave of each individual particle interferes "with itself." The matter wave of a single particle must, therefore, pass through both slits in the double slit – in contrast to the particle, which of course only passes through one of the two slits. When one of the two slits are closed, the interference pattern is lost – only the diffraction phenomenon of the single slit remains.

All experiments show the objective existence of the dual nature of microscopic particles and their matter wave. It is not possible to attribute the particle property to the matter wave or vice versa. A theory that equates the particle with a "wave packet" leads to blatant contradictions with the experiment: Such wave packets would not only dissipate due to the different phase velocities of the individual wave trains but would also be literally fragmented by diffraction at a slit. Neutrons, atoms, or electrons as particles, however, in such experiments occur indivisibly.

With the kinetic energies of the particles in the eV range present in the experiment, it is impossible for the particles to decompose into sub-components as they pass through the double slit, which then reunite at the detector. This also applies to unstable particles such as neutrons. The process of neutron decay into protons and electrons is irreversible. A hypothetical process in which the neutron decomposes into an electron and a proton at the double slit, which fuse back to a neutron at the detector location after passing through the two slits, can be ruled out.

Recently, an experiment with carbon C_{60} macromolecules succeeded in proving not only the existence of matter waves of a macromolecule of 60 carbon atoms by interference [Arndt 1999] (see Figure 38). The molecules, which look like microscopic footballs, are large enough to make them glow, which makes it possible to measure their position. This makes it possible to conduct an experiment in which not only the "speed" of the molecules is determined via the interference pattern that is created and thus the wave property. By observing the photons emitted from the C_{60} molecule, it should in principle also be possible to determine along what path the molecule passed through one of the two gaps. But this would contradict the laws of motion of quantum mechanics. However, the matter-wave state of the molecule is influenced by the emission or absorption of photons. In particular, it is subject to a phase shift. Through emission of photons, therefore, the coherence of the matter wave of the C_{60} molecule and thus its interference capability is lost bit by bit [Hackermüller et. al. 2004].

The experiments in Figure 38 show that if the temperature of the C_{60} molecule is gradually increased to a value at which higher-energy photons are emitted as temperature radiation, the interference capability of the C_{60} molecules is first gradually and then completely lost. As soon as a temperature is reached at which the emitted photons have a sufficiently small wavelength to distinguish their emission location at one of the two slits by detection, the interference pattern is completely lost. Thus, it is actually not possible to simultaneously determine the momentum of quantum objects via the interference pattern and their path through the double slit.

Heisenberg's uncertainty relationships and their interpretation

Heisenberg's uncertainty relationships state that the momentum and location of a quantum object are subject to a random fluctuation whose product is at least the size of Planck's quantum of action. This central result of quantum mechanics is confirmed

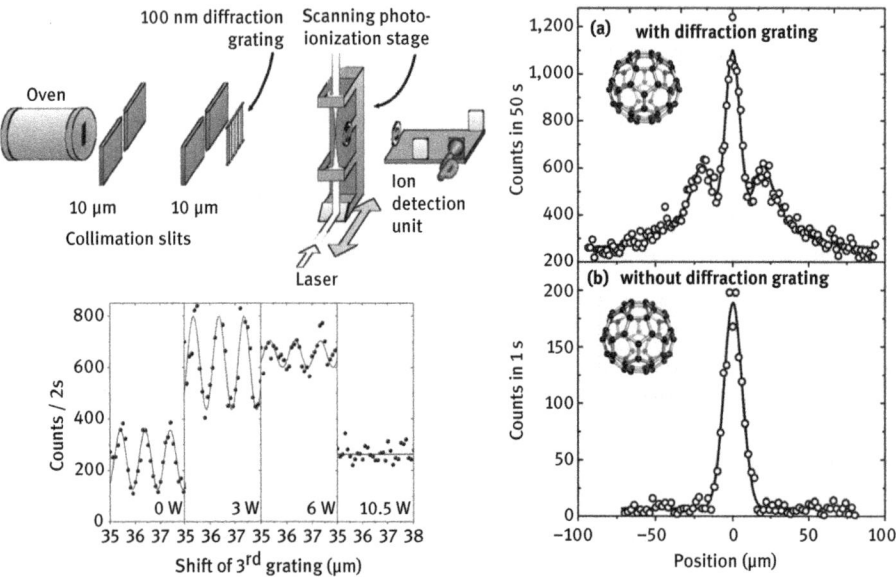

Figure 38: Detection of matter waves for a C_{60} macromolecule by interference at a grating. Top left: Experimental setup with oven for producing a beam of C_{60} molecules, a 100-nm grating, a photo-excitation unit, and detection unit. Right: Frequency of the incoming molecules as a function of the location with diffraction grating (top) and without (bottom) diffraction grating. In the top half, the first-order interference maxima can be seen clearly. Below left: If the temperature of the C_{60} molecules is increased by a laser directly after leaving the oven, so that they emit heat radiation, the interference pattern of the C_{60} molecules is gradually lost. The photon emission is associated with decoherence. Reprinted with permission of Macmillan Publishers Ltd.: Top left and right: M. Arndt et. al., Nature 401 (1999) 680, copyright (1999). Bottom left: L. Hacker-Muller et. al., Nature 427 (2004) 711, copyright (2004). C_{60} macromolecule: Sponk.

by all experiments carried out to date for the simultaneous determination of the location and the wavelength (or the momentum). An essential aspect can already be understood with classical waves: The spatial extension of each wave increases the more precise the value of its frequency and wavelength are. A monochromatic wave with an exactly defined wavelength would have an infinite extension of its wave train. As soon as a spatial localization of the wave occurs, the associated wave packet consists of the superposition of waves of different wavelengths λ or of wave number $k = 2\pi /\lambda$, which is the number of oscillation periods per length. The stronger the spatial localization, the wider the distribution of the wavelengths and frequencies of the wave packet forming it must be. From classical wave theory, it follows that the localization range Δx and the range of the wave numbers k forming the wave packet are inversely proportional to each other, or $\Delta x\, \Delta k \sim 1$. In other words, a spatially localized wave packet cannot have a single value for the velocity, and a wave with a defined velocity cannot have a precise location.

The decisive new aspect in quantum physics is that a wave property (the wavelength λ of the matter wave) directly determines a particle parameter (the momentum p of the particle) via the de Broglie relationship $p = h/\lambda$. The correlation between the localization range of a wave packet and the range of the wave numbers contained, known from classical wave theory, therefore expands to a statement about the possible certainty of the trajectory of a particle with location x and momentum p. Thus, Heisenberg's uncertainty relationships describe the relationship between the degree of minimum uncertainty Δx for the location x and the uncertainty of the momentum ($\Delta p = \hbar \Delta k$), which results in $\Delta x \, \Delta p \approx \hbar$, with $\hbar = h/(2\pi)$.

It is not too difficult to imagine that the location and momentum of a particle moving in the chaotic fluctuations of the zero-point field is subject to a range of variation, the minimum size of which is determined by Planck's quantum of action (h), as the threshold between unstable and stable excitations.

Under the influence of the positivist world view, however, a subjectivist interpretation of Heisenberg's uncertainty principle has developed. It claims that fluctuations in momentum and location are not the expression of qualitatively different processes in the microcosm, but the result of measurement by an observer. Werner Heisenberg writes in his essay "Über die Grundprinzipien der Quantenmechanik" ["On the basic principles of quantum mechanics"]:

> In our view, even for very small particles, such as electrons, it always makes immediate sense to speak of the location and velocity of a particle. The physicist, on the other hand, takes the position that these words only have a meaning if one can state how location and velocity can be determined, i.e. measured experimentally. One can very well imagine experiments that make it possible to measure, say, the location with any degree of accuracy: for example, one can in principle observe the electron under a microscope with a very high resolution; there are also measurements to determine the velocity (for example by the Doppler effect). However, it seems to be a general law of nature that we cannot simultaneously determine location and velocity with arbitrary accuracy. The more precisely we determine the location, the less precisely the speed can be determined at that moment and vice versa.
>
> [Heisenberg 1927, p. 83, translation by the author]

The scientist Anton Zeilinger, who played a decisive role in advancing the double-slit experiments with C_{60} molecules, also regards Heisenberg's uncertainty principle as a subjective determination: "... *what is always important in measurement and observation, however, is that we can ultimately speak of a sensory impression. Because without a sensory impression there is no observation. In the case of the Heisenberg gamma microscope, this sensory impression is ultimately the impression of the flash of light that appears on the observation screen. We can talk about this with each other. It is a direct experience. (. . .). Strictly speaking, we can only talk about these classic objects. Everything else is our mental constructs.*" [Zeilinger 2003, translation by the author]

In the double-slit experiments with C_{60} and in similar experiments with neutrons [Rauch 2008], however, the interference capability of the wave is destroyed, because the decoherence of the matter wave of the molecule and the emission of a photon are two sides of the same process. The influencing or even undermining of the coherence of matter waves takes place objectively and constantly everywhere in the microcosm. It does not require any consciousness of an observer. Instead of investigating the process of decoherence of matter waves in connection with the quantization of action in the interaction between particles, some quantum physicists take refuge in subjectivist ideologies.

The Schrödinger theory of the matter wave

The laws of motion of microscopic building blocks of matter in conjunction with their matter–wave states for not too large energies are very successfully reflected by the Schrödinger theory of matter waves. The spatial structure and temporal propagation of matter waves as a function of internal particle properties (mass, charge, and velocity and in Dirac's extended version, also spin) is determined by means of a differential equation for a matter–wave field. The influence of the macroscopic matter environment is included as a boundary condition of the wave field. The core of this theory is Max Born's probability interpretation, according to which the structure of the matter–wave field ψ determines probability distributions of particle properties such as location, velocity, angular momentum, etc., of particles. The probability distribution $W(x)$ of finding a particle at a certain location x is obtained from the magnitude of the square of the wave field using $W = |\psi(x)|^2$.

As further explained in the following sections, the Schrödinger theory is the statistical low-energy theory of particles interacting with the zero-point field. Although it does not contain a microscopic model of this interaction, that is, it is not a theory of the material causes of wave processes, it correctly reflects statistical laws of particles under the influence of the fluctuating field environment of the zero-point field. The matter wave cannot, in fact, be interpreted directly as the wave of a medium. Rather, it only reflects certain wave properties of existing highly dynamic processes in the quantum æther. But to speak therefore of a "pure probability wave" or even of a "knowledge function of an observer" is subjective idealism.

The Bohm theory of quantum potential

The British physicist David Bohm developed the theory of quantum potential in the 1950s [Bohm 1952, 1985], which represents a mathematically exact reformulation of the Schrödinger theory. Its strength is that it opens up new insights into the physical processes of wave and particle propagation [Dürr 2001, Passon 2005].

Bohm's theory of quantum potential is conceptually a further development of de Broglie's theory of the pilot wave [de Broglie 1960]. Just like de Broglie's theory, it assumes that both particles and waves exist continuously in space and time.

According to it, particle motion causes a matter-wave field, which is influenced by the macroscopic matter environment and affects the particle motion. The influence of Bohm's work, however, remained very limited due to the ideological bias of many physicists. While John Bell enthusiastically accepted these ideas, *"In 1952 I saw the impossible done"* (quoted from [Selleri 1990]), it was ignored by most physicists and simply hushed up in many textbooks of quantum theory.

David Bohm describes his theory as follows:

1. The wave function, ψ, is assumed to represent an objectively real field and not just a mathematical symbol.
2. We suppose that there is, beside the field, a particle represented mathematically by a set of coordinates, which are always well defined and which vary in a definite way.
3. We assume that the velocity of this particle is given by $\vec{v} = \frac{\vec{\nabla S}}{m}$ where m is the mass of the particle and S is a phase function obtained by writing the wave function $\Psi = Re^{iS/\hbar}$, with R and S real.
4. We assume that the particle is acted upon not only by the classical potential $V(x)$, but also by an additional "quantum *potential*" $U = \frac{\hbar^2 \nabla^2 R}{2mR}$.
5. Finally, we assume that the field ψ is indeed in a state of very rapid random and chaotic fluctuation, such that the values of ψ used in quantum theory are a kind of average over a characteristic interval of time τ (...).

The fluctuations of the ψ-field can be regarded as coming from a deeper sub-quantum-mechanical level, in much the same way that the fluctuations in the Brownian motion of a microscopic liquid droplet come from a deeper atomic level. [Bohm 1980, p. 98]

The quantum potential U is generated by the particle motion by means of the matter–wave field. At the same time, its stochastic fluctuations cause forces on particles and thus fluctuations in the particle trajectory. Instead of a smooth trajectory, a complicated zigzag motion is created. However, there are characteristic mean values of the velocity of particles. The particle energy and momentum are preserved.

The quantum potential has no punctiform sources, such as with the classical potential – the mass or the electrical charge. It depends both on the state of motion of the particle and on macroscopic ambient conditions, such as walls or wall openings, which are incorporated as boundary conditions. The theory of quantum potential allows a clear interpretation of the double-slit experiment with single particles: Each individual particle always passes through only one of the two slits, whereas the matter wave accompanying the particle penetrates both openings and causes a highly complex quantum potential due to interference (see Figure 39). This potential becomes only slightly weaker with increasing distance to the slits. The particle motion is therefore modified in such a way that the trajectories of the particles shown in Figure 39 occur. For a macroscopic number of particles, this results in a

(a)

(b)

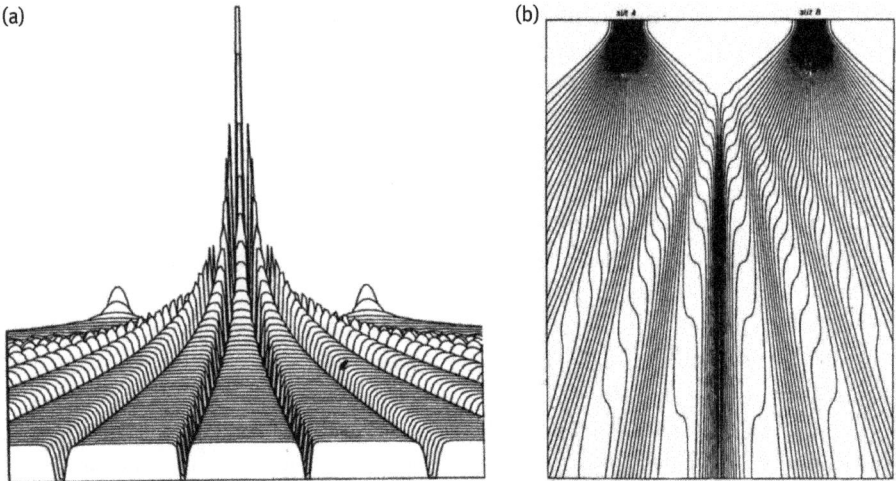

Figure 39: Interpretation of the double-slit experiment according to David Bohm by the unity of wave and particle by means of the quantum potential. (a) Representation of the spatial structure of the quantum potential for the double-slit system. In reality, it fluctuates in time so that only a momentary view can be seen. (b) Trajectories of particles under the influence of the quantum potential, which in total generate the interference pattern of the double-slit experiment. Reprint from C. Philippidis et. el., Nuovo Cimento B 52, 15 (1979) with permission of Springer.

spatially varying frequency distribution corresponding to the interference pattern of the waves. [Bohm & Hiley 1993, Philippidis 1979].

The quantum potential U depends only on the form of the wave function ψ, but not on its magnitude, since the amplitude of ψ occurs both in the denominator and in the numerator of the quantum potential. This is an expression of the actions of matter waves over very large distances. Thus, the quantum potential is something qualitatively new compared to the known classical interaction potentials, which drop rapidly with distance. In matter–wave states of several particles entangled by interference, the quantum potential depends on all particle motions. Therefore, a main objection to Bohm's theory is its property of non-locality with "instantaneous actions at a distance," which Einstein condemned as "eerie." In this point, however, it is equivalent to Schrödinger's and Heisenberg's quantum theory. An advanced theory of matter waves and the quantum potential caused by them must be based on experiments on the range and speed of propagation of matter waves, which are discussed in the following Section 4.3.

David Bohm was a Marxist and at times a proponent of a dialectical-materialistic world view. He saw his theory only as a first preliminary step beyond the Schrödinger theory of matter waves. In 1957, a critical analysis of the mechanistic world view [Bohm 1957] appeared. In it, he classified his approach of 1952 as a rough description of a new structural level of matter below particles, which produces the quantum phenomena. He developed the suggestion of the "qualitative infinity" of nature, which occurs in infinitely many structural levels. He developed the idea that the "vacuum" is full of extremely fast activity of matter and that physics to date only covers the relatively more stable movements. In his later years, however, Bohm increasingly distanced himself from his materialistic-dialectical world view and developed holistic ideas that replaced the existence of real material particles with "holograms," which opened the door to esoteric ideas.

4.3 Quantum entanglement: common matter waves

The core of the idealistic interpretation of quantum theory is the declaration that quantum–mechanical objects do not have any reality independent of the measurement or the measuring apparatus (see for example critical review in [Selleri 1990]). The object can then only be described as a phenomenon in connection with the measuring apparatus. A phenomenon is described while the observer is "directly interwoven into it by the act of measuring". The wave function ultimately becomes a quantity that describes the knowledge of an observer, that is, consciousness and not objective reality. Albert Einstein criticized this idea:

> There is such a thing as the "real state" of a physical system, which exists objectively independently of any observation or measurement and can be described in principle with the means of expression of physics. (...). Now there is no doubt that the ψ function is a kind of description of a "real state". But the question is whether this description characterizes the real state completely or incompletely.   [Einstein 1955, p. 14]

According to Einstein, a theory is complete when "*every element of physical reality has a counterpart in the physical theory.*" [ibid]

Einstein recognized quantum theory only as a preliminary theory. However, he mainly criticized its "incompleteness" because it did not allow a complete spatiotemporal description of the motion of a quantum object, which ultimately denied the qualitatively new laws in the microcosm. In order to prove the "incompleteness of quantum theory," Einstein and his colleagues tried to uncover internal contradictions in quantum mechanics.

For this purpose, they developed a thought experiment known as the Einstein–Podolski–Rosen (EPR) paradox. The original form of the paradox consisted of a thought experiment in which two particles from a source are emitted into a state with a common matter wave, and measurements are made on both particles. The common matter–wave state has a well-defined total momentum P and center-of-mass coordinate X.

Now a momentum measurement is to be carried out on particle 1 and a position measurement on particle 2. The measurement of the momentum p_1 at particle 1 also determines the momentum of particle 2 as $p_2 = P - p_1$ via the total momentum. The position measurement at particle 2 also determines the position of particle 1 as $x_1 = X - x_2$ via the center-of-mass coordinate. Thus, according to Einstein, the locations and momenta of both particles are known and fixed, and thus a more complete characterization would be achieved than allowed by quantum theory and its uncertainty principle.

Bohr contradicted Einstein: a measurement on particle 1 has an immediate effect on particle 2 via their common matter wave, and vice versa, according to Bohr, so that the "complete characterization" of the two particles as independently existing objects presented by Einstein is not possible. So what Einstein doubted was the existence of the entangled states discussed earlier, in which two or more particles form a common matter wave.

The thought experiment developed by Einstein and his colleagues was realized in a similar form as an experiment in the early 1980s (Figure 40). In the experiment, photon pairs are emitted from a common source in opposite directions. The total polarization of both photons is determined by the state of the source from which

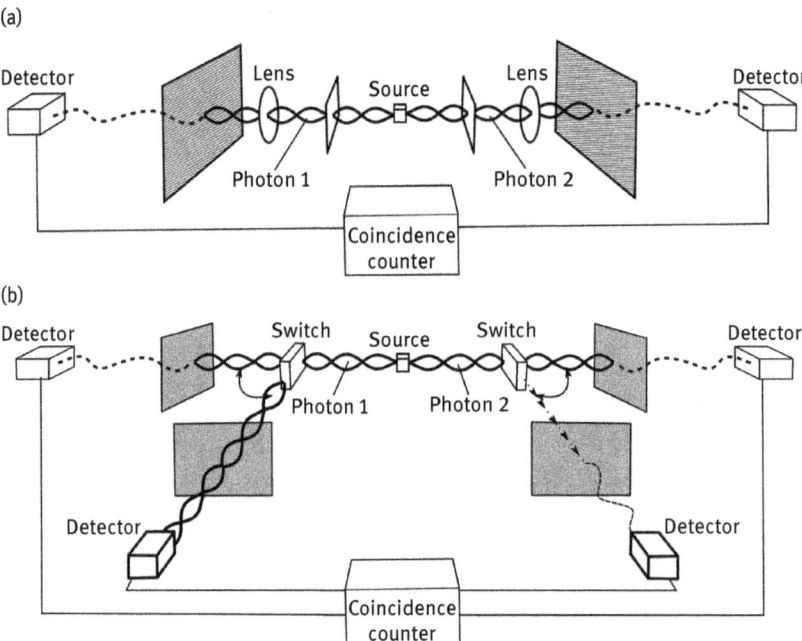

Figure 40: Investigation of entangled matter waves by measurement of two-photon coincidences. (a) Measurement of the coincidence of the polarization of two photons emerging from a source in which they were generated in a common process and thus have a common matter wave [Aspect 1981]. (b) Extended experimental setup in which additionally random switches are used to switch back and forth between different detectors in order to make the formation of a correlation between the two detector arms more difficult [Aspect 1982].

they were emitted and cannot change, due to the law of conservation of angular momentum. The result of the experiments clearly showed that the two photons are not independent [Aspect 1981]. Although the measurement of polarization on one photon at a time follows a statistical law according to which the probability of a photon either passing the polarization filter or being absorbed depends on how the polarization filter is aligned with the light source, the other photon does not behave independently. The probability that the second photon passes or is absorbed by the second polarization filter depends on the measured polarization of the first photon.

In order to exclude the possibility that the two-photon coincidence measurements are influenced by an unknown correlation between the detectors, additional switches were placed in the beam path in 1982 to switch between different detectors (Figure 40b). Here, too, there are two-photon coincidences that confirm the result of the first experiments that a common matter–wave state of the two particles exists [Aspect 1982].

Modern experiments on long-range correlations of photons in entangled matter–wave states show further remarkable properties. Figure 41 shows an experiment on the interference capability of entangled matter waves at a double slit. Although only one branch of the beam splitter BS contains the double slit and thus propagates only one

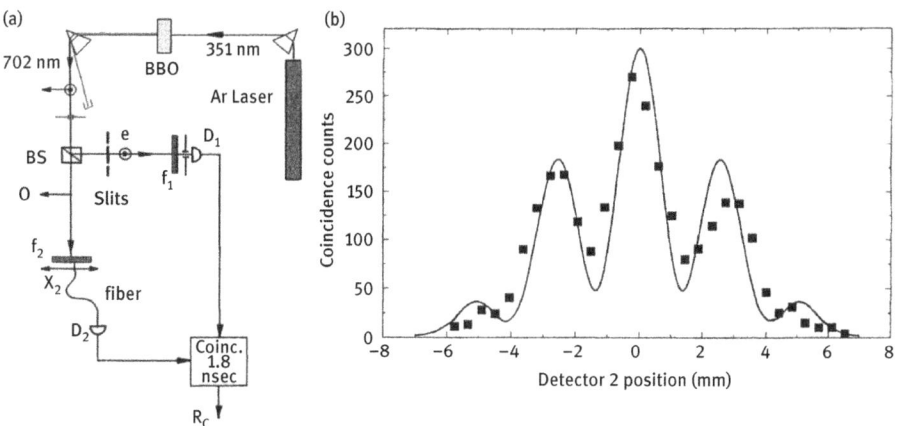

Figure 41: Investigation of the interference capability of common matter–waves of two-photon states. (a) Experimental setup with laser, generation of entangled two-photon states in a nonlinear crystal (BBO), and a beam splitter (BS) and double slit (slits). The interference pattern is recorded by two-photon coincidence measurements for mutually perpendicular polarization of the photons depending on the position of detector 2; result is shown in (b). Although only one of the two photons is exposed to a double slit, an interference pattern occurs in the frequency that two photons impinge simultaneously at the end of both arms of the beam splitter (BS). Reprinted with permission of D. V. Strekalov et. al., Phys. Rev. Lett. 74 (1995) 3600; copyright (1995) APS.

particle of the photon pair through the slit arrangement, an interference pattern can also be seen in the entangled partner. Interference at the double slit modifies the common matter wave of the photon pair. What is remarkable about the experiment is that no interference pattern is observed if only those photons are detected that pass through the double slit and hit detector D_1. In this experiment, the coherence of the wave of the single photons is not sufficient to interfere with itself with the chosen slit geometry. In contrast, the coherence of the entangled matter–wave of the photon pair is large enough to produce an interference between the photon passing through the double slit and its partner without the double slit.

The entangled two-photon states can even be used to transmit an optical image of an object over a considerable distance [Shih 2004]. This effect is also called "quantum teleportation" in popular literature or science fiction novels. However, this term is misleading because there is no transfer of objects or matter. The previous experiments are in complete agreement with quantum theory and disprove Einstein's concept of local polarizations associated with the individual photons that occur independently of each other. With his rejection of entangled matter waves as "eerie actions at a distance," he remained rooted in classical physics with its local, separable building blocks and interactions. Such an idea is not part of the dialectical-materialistic epistemology, which only assumes that material processes exist objectively and in reality in the microcosm, as well.

An open question at present is whether the modification of photon 1 by polarization filter and detector actually transfers a material effect by a "collapse" of the common matter wave to photon 2 (Einstein's spooky action at a distance). Alternatively, the higher correlation of the particle properties in the entangled matter–wave state could already be present during generation. The second point of view is supported by the fact that an entangled matter–wave state would couple part of the zero-point fluctuations, so that quantum fluctuations in entangled states of particles exhibit a higher correlation than in independent states. In fact, entangled two-particle states have a temporal distance of incidence at the two detectors $\Delta(t_1 - t_2)$ that is more sharply defined by orders of magnitude than the temporal fluctuations of the individual photons selected Δt_1 or Δt_2 [Shih 2003]. Since such two-particle coincidences are also observed over distances of more than 15 kilometers [Yin 2013], the coherence length of the entangled matter waves is enormous. The alternative theory of a "collapse" would mean that matter waves change at speeds greater than the speed light. In quantum electrodynamics (QED), such super-light-speed processes in the quantum æther are predicted explicitly for the unstable "off-resonant" excitations [Dittrich 2000].

Abrupt changes of matter-wave states through interaction with other matter systems such as detectors are real material processes and have nothing to do with the idealistic fata morgana of a "collapse through the act of observation."

4.4 Stability of the electron shell of atoms

All "mini planetary systems" atomic models with electrons orbiting the atomic nucleus developed at the beginning of the twentieth century had a fundamental theoretical contradiction: It is the fact that the laws of classical electrodynamics contradicted the stability of the electron shell of the atoms. Every type of electron orbit around an atomic nucleus represents an accelerated movement of electrical charges which, according to the laws of classical electrodynamics, should lead to a continuous emission of electromagnetic waves. This would cause the electrons to continuously lose energy on their orbits and their orbits to spiral in towards the nucleus until they finally fall into the nucleus (Figure 42a). So how could the atoms be stable?

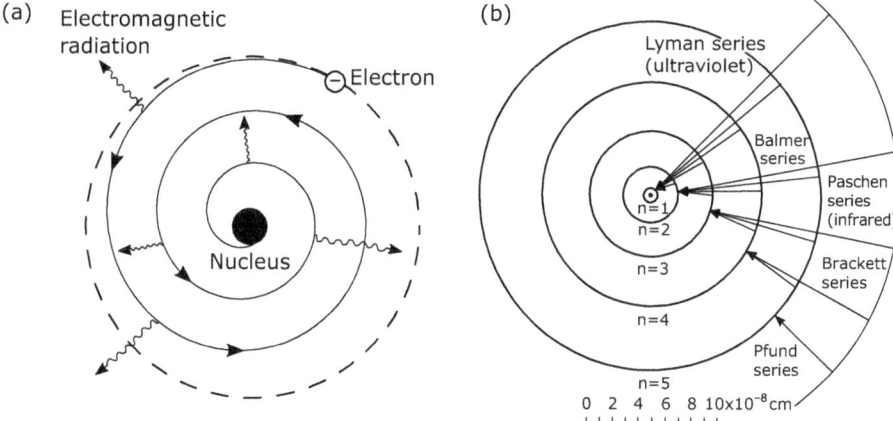

Figure 42: Stability of the electron shell of atoms. (a) According to the laws of classical electrodynamics, an electron moving in a circle emits light. Therefore atomic models based on the classical laws of motion of mechanics and electrodynamics cannot explain the stability of the electron shell. (b) Bohr's atomic model with circular trajectories of electrons assumed ad-hoc with quantized energy. This results in an emission spectrum of light with discrete spectral lines through transitions between electron states.

The Bohr atomic model of 1913 "solved" the problem by assuming that electrons may only move in certain discrete orbits (Figure 42b). It explained the spectral lines by quantizing the energy. Only those electron states were to be allowed for whose energy difference is $\Delta E = E_2 - E_1 = h\nu$, where E_1 and E_2 were the energies of two stable electron orbits. One could now deduce from the line spectra found in the energy of the permitted and stable orbits of the electrons themselves. Thus the quantum of action h also entered directly into the theory of the electron shell of the atom. The discreteness of the interaction of light with matter by means of the "atom of action," Planck's constant h, was extended to the inner structure of atoms with quantized energy.

However, this theory is completely incompatible with the laws of classical physics. Why don't electrons constantly radiate energy? Even the quantization of the light field by photons alone does not provide an explanation since it would be possible in principle that electrons constantly emit photons of a frequency as small as necessary, and thus energy hv, and thus gradually crash into the atomic nucleus despite the quantum nature of the electromagnetic field.

An early key to the solution of this problem, from today's perspective, is Walter Nernst's idea in 1916. He attempted to derive the structure and stability of the electron shell from the existence of the electromagnetic zero-point field. According to him, the stationary states of the electrons are nothing more than those states in which the electron movement is in a resonance of the constant emission and reabsorption of quanta of the zero-point field. Modern computer simulations of the movement of electrons in the field of the atomic nucleus, taking into account the statistically fluctuating electromagnetic zero-point field, show that indeed stationary electron states are formed which largely correspond to the electron-density distribution resulting from standing matter waves in the hydrogen atom [Cole 2003].

The structure of the electron shell of the atom as a self-organization process

In fact, the electrons in the atom constantly exchange unstable photons with the zero-point field. This exchange is part of the attracting electromagnetic force between the different electrical charges of the atomic nucleus and the electrons [Feynman 1989, p. 117].

So the stable electron states are then states of resonant absorption and emission of unstable photons by electrons. Electrons change their state of motion briefly by emission of unstable zero-point photons until they absorb the emitted unstable photons again. The fluctuating zero-point field thus receives a stationary spatial modulation, which is expressed in matter waves. The quantization of the light field and the existence of stable stationary electron states in the atomic shell, which change from one to another abruptly by emission and absorption, are two sides of the self-organization process of moving charges and of the zero-point field. In atoms, the zero-point field is modified by the electron motion in the electrostatic potential of the nucleus. The change from one stable state to another is determined by the action quantization.

The decisive modification of the law of classical electrodynamics ("accelerated charges radiate") by QED, therefore, consists in the quantum of action. For processes below the threshold of the quantum of action, each emission of radiation is compensated by a corresponding absorption, so that this type of interaction with the unstable photons of the zero-point field does not lead to a lasting change in the energy of the electron. The self-organization of zero-point field and electron movement to a resonant process guarantees the stability of the atom. If, however, the threshold of the quantum of action is exceeded, the electron state changes by emission or absorption of stable photons. However, this theoretical description is mathematically much more complicated than that of the Schrödinger theory of matter waves (see Figure 43), which reflects the result of the self-organization process by means of novel statistical laws of stationary electron states.

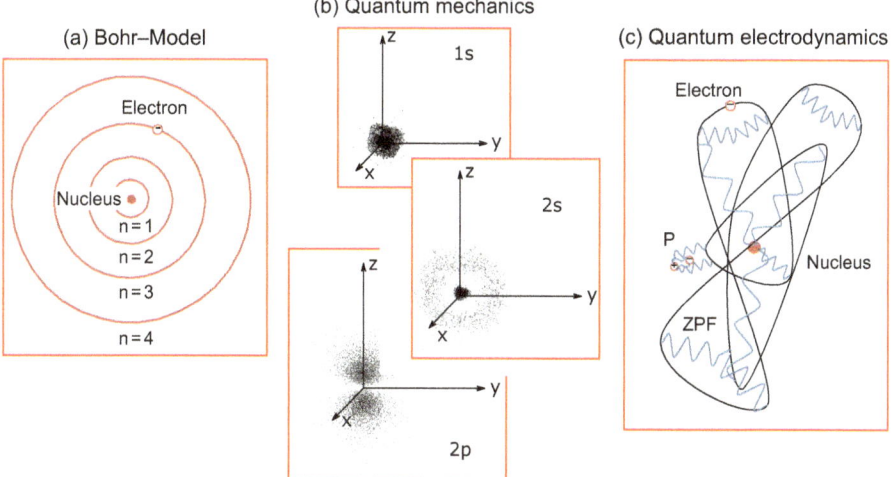

Figure 43: Stages in the modeling of the electron shell of a hydrogen atom. (a) Bohr model. (b) Schrödinger model with the representation of the different stationary modes of oscillation of the matter wave of an electron in the hydrogen atom as a point cloud, which is a measure of the probability of the electron's location. Shown are the ground state 1s and higher excitation states 2s and 2p. (c) Quantum electrodynamics of electron motion in the field of zero-point excitations (NPF), taking into account polarization effects of the quantum æther by unstable electron–positron pairs (P).

This materialistic explanation is seldom to be found in physics textbooks – quite contrary to the thesis advanced by the German fascist Philipp Lenard in 1905 that "the atom is as empty as the universe" (for example, the empty atom is propagandized in [Völker 2010, Kacher 2013]).

Schrödinger's theory of the structure of the electron shell

The Schrödinger theory deals with the structure of the electron shell from the point of view of the matter–wave states of the electrons that arise. Compared to the treatment by Nernst or the QED developed later, it is an effective description of the electron shell of atoms (see Figures 43b and c). It yields the following results for the state of the electron in the simplest atom, the hydrogen atom:

- Stable discrete energy states of electrons in the electromagnetic field of the atomic nucleus that are accompanied by stationary matter–wave states.
- In addition to the ground state as the state of lowest energy, there are discrete excitation states of electrons, which are numbered with a quantum number n. They correspond to more complex forms of motion, which can also include states with orbital angular momentum, which are expressed in nodes and maxima of matter waves (see Figure 44).
- The quantization of the orbital angular momentum of the electrons is caused by interference of rotating matter waves that pass into themselves and are characterized by the quantum number of the orbital angular momentum l (see figure 44). In addition to the quantization of the magnitude, a directional quantization of the angular momentum also occurs, so that a fixed axis of angular momentum axis is distinguished despite statistically fluctuating electron movement.
- In Max Born's statistical interpretation, the matter wave described in Schrödinger's wave equation represents a probability, determined by the square of its amplitude, of finding the electron at a certain location in a certain state.

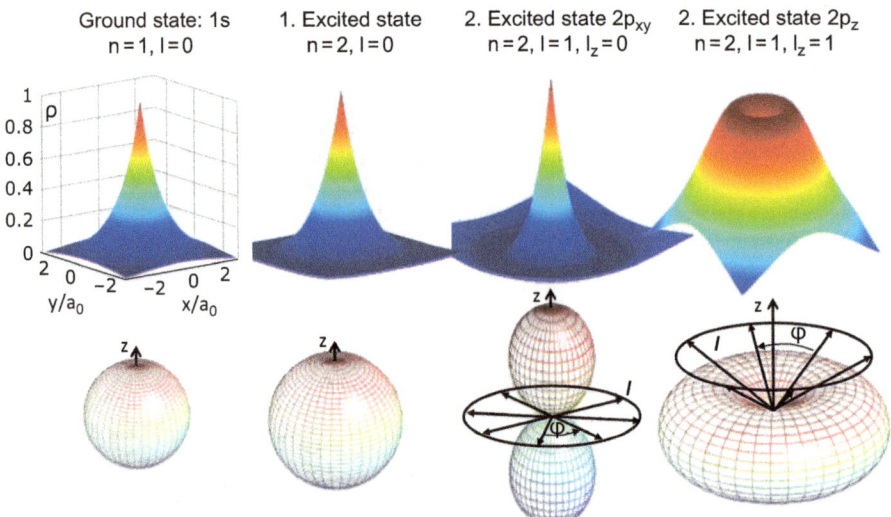

Figure 44: Electron density distribution ρ in the H atom. Top: Cross-section of the electron density through a (x,y) plane, normalized to the Bohr radius (a_0). Below: Surface of constant electron density. The ground state 1s ($n = 1$) and the first excitation state 2s ($n = 2$) of the electron have no orbital angular momentum, the standing matter-wave is spherically symmetrical, and the electron density has a maximum at the nucleus location. The second excitation state 2p ($n = 2$) has one quantum of angular momentum ($l = 1$). The angular momentum axis l precesses around a preferred direction z and the angular momentum can have three different values $l_z = -1$, $l_z = 0$ and $l_z = +1$ relative to this direction.

The connection between the stationary state of the matter wave and the electron density can be made directly visible in experiments today (Figure 45) [Stodolna 2013]. The figure shows the electron density distribution of the ground state ($n = 0$) and two different excitation states ($n = 1, 2$). In order to allow for the influence of the electric field of the microscope, the superposition of the states $n = 29, 28$ and 27 induced by this has to be taken into account.

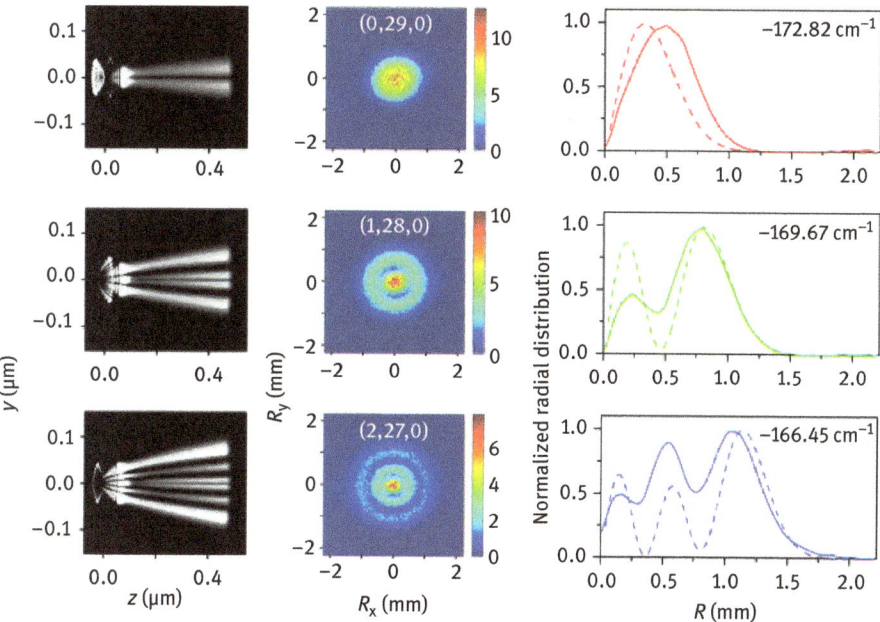

Figure 45: Experimental visualization of the electron density distribution of hydrogen atoms in different excitation states. The electrons are emitted from the electron shell by photoionization and observed in a microscope of high magnification. Experimentally observed electron-density distribution (middle column) and calculated electron paths (left column). The right column compares the measured and the theoretically calculated radial electron density (solid/dashed lines). Reprinted with permission of A. S. Stodolna et. al., Phys. Rev. Lett. 110, 213001 (2013). Copyright (2013) by APS.

The organization of the fluctuating electron movement in the shell of atoms into orbitals as an expression of the dialectical unity of particle movement and matter wave has been comprehensively confirmed experimentally today.

However, even applied to the simplest atom, the hydrogen atom, Schrödinger's theory requires various extensions in order to describe the energy of the electron movement exactly. These include the consideration of the self-rotation of the electron, its "spin," which produces a magnetic spin moment. The magnetic coupling between the orbital motion of the electron and its spin changes the energy of the stationary states, causing a fine structure splitting of the lines. It is further modified by the mass increase of the electrons at high velocities (see Chapter 5). Both the electron spin and the mass increase of the electron can only be satisfactorily explained within the framework of the Dirac theory. In addition, there are further splits and shifts of the spectral lines due to the change of the field of the zero-point fluctuations in the vicinity of the atomic nucleus (called the Lamb shift), which was demonstrated in the 1950s, as well as the hyperfine splitting due to the interaction between the magnetic moments of the electron and the nucleus.

The electron shells of atoms that possess more than one electron also form stationary states of matter waves. However, complicated electrical and magnetic multiparticle interactions between the charges and the orbital and spin moments of all electrons and the atomic nucleus must be taken into account. In addition, matter–wave states are formed for each electron which, due to their destructive interference, cause an additional repulsive contribution, known as the "Fermi pressure," in the interaction between the electrons.

The transformation of the quantity of the number of electrons into the quality of the structure, forms of motion, and properties of the electron shell is the basis for the structure of the periodic table of the elements: Atoms with a similar orbital structure of the electrons in the outermost shell have similar chemical properties.

4.5 Dirac sea, quantum fields, and infinities

Significant advances in the theoretical understanding of quantum effects as an expression of the existence of a new structural level of matter developed with the "relativistic" extension of quantum mechanics by Paul Dirac, as well as in the description of interacting fields and particles within the framework of quantum field theories. The Dirac theory takes into account modification of the relationship between wavelength and frequency (dispersion) of matter waves at speeds close to the speed of light. This led to the theoretical prediction of a "sea of negative energies." With this "Dirac sea," a new concept of the æther developed since the early 1930s. In an article "Is there an ether?" on the development of QED , Dirac concluded in 1951: "... *with the new theory of electrodynamics we are rather forced to have an ether*" [Dirac 1951, p. 906].

The Dirac theory of the electron

When the speed of particles with rest mass approaches the speed of light, an increase in mass occurs. In 1928, Paul Dirac succeeded in establishing an extended wave equation for matter waves, taking into account the resulting change in the energy–momentum or frequency–wavelength relation of the electron. This was only possible by establishing a connection between the structure of the matter wave and the "spin" of the electron. In 1925, Uhlenbeck and Goudsmit had already theoretically interpreted splits in the spectral lines of atoms in a magnetic field as a consequence of the electron spin.

And already in 1915, Einstein and de Haas had measured an "abnormal value" of the magnetic moment of the electron. It contradicted the theoretical explanation of the magnetism of the electron by the circulating current of a rotating spherical electron charge. The Dirac theory for the first time calculated approximately correctly the abnormal magnetic moment of the electron by considering the reaction of the "sea" to its spin. This required a further overcoming of simplified mechanical notions of the electron as a tiny "rotating billiard ball": If the electron spin were created by the pivoting of a rigid sphere with the classical electron radius of 3×10^{-15} m, the surface of the sphere would have to rotate faster than the speed of light . In fact, the electron spin is inextricably linked to the properties of the revolving matter–wave field [Ohanian 1986].

In quantum theory, allowing for the mass increase of particles at speeds close to the speed of light inevitably leads to the existence of states of "negative energy." These are separated from the states of positive energy by an energy gap (see Figure 46). The size of the energy gap E_g is precisely the energy equivalent of twice the rest-mass m_0 of the electron $E_g = 2\,m_0 c^2$. The existence of negative energy states, however, posed a fundamental problem: A particle of positive energy could always gain energy by plunging to lower and lower (negative) energy states by emitting light, and

> . . . so all particles with positive energy would have to end by falling into this energy abyss. Dirac now had the brilliant idea of assuming that all negative energy states are already filled with electrons, in accordance with the Pauli principle, whereby each energy state can be occupied by at most two electrons with antiparallel spin. The infinitely high negative charge of this

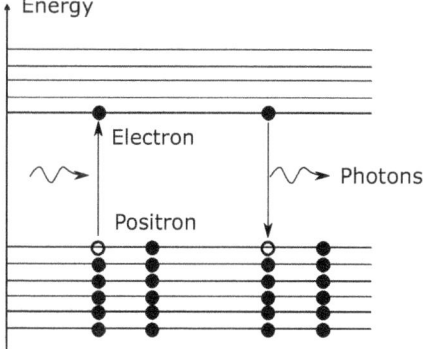

Figure 46: Representation of the Dirac Sea and its excitations. In the ground state, all states of negative energy are occupied by electrons. They are separated from those of positive energy by an energy gap of the double rest mass of the electron. When an electron–positron pair is generated, an electron changes from negative to positive energy by absorption of two photons. When a pair is destroyed, two photons are produced.

"Dirac sea" could be thought of as compensated by the positive charges of the protons, which also satisfy the Dirac equation and have to fill a corresponding Dirac sea. The vacuum would then be interpreted as these two occupied Dirac seas. [Haken 1987, p. 252]

It was precisely through the combination of the results of special relativity theory and quantum theory that the "vacuum" was again filled with matter after an idealistic interpretation of special relativity theory had largely banished the concept of æther from the work of physicists. An important prediction of the Dirac theory consisted, among other things, of the existence of the antiparticles of the electron. In 1931, Dirac suggested interpreting a "hole" in the sea of negative energy filled with electrons as antiparticles of the electron. In fact, shortly afterwards, in 1932, Anderson discovered the antielectron, also called positron.

Pair generation and polarization of the Dirac sea

The Dirac theory also predicted the generation of pairs of electrons and positrons by photons with sufficient energy, so that an electron is lifted over the energy gap and a hole remains. In the case of pair annihilation, the reverse process takes place. In fact, pair generation does not represent the generation of matter from nothing, but the transition from one state of matter to another, triggered by the absorption of photons by a sea electron. The only reason "negative energy" is attributed to the sea states is because the zero point of the energy scale was arbitrarily set to the upper edge of the energy gap before the development of the Dirac theory, and thus the state of lowest energy is an electron with velocity $v = 0$. In a Fermi fluid, this would equate to setting the energy zero-point of the electrons to the Fermi energy, and regarding only all excitations above E_F as really existing, with positive energy. In fact, the Dirac theory of the electron represents the effective description of the quantum æther as a Fermi fluid with energy gap $E_g = 2m_0c^2$ and electron–positron quasiparticles.

Electrons with positive energy can polarize the Dirac Sea. Under the ideological influence of positivism, this effect is also called "vacuum polarization." It manifests itself in the occurrence of fluctuating unstable electron–positron pairs, which surround the electrons and other electrically charged particles, in addition to the electromagnetic zero-point field, as a polarization cloud. The excitation energy of unstable electron–positron pairs is smaller than the energy gap. The energy gap is the threshold for pair generation of stable electron–positron pairs. Therefore, spatial movements of the electron are exposed not only to the influence of the zero-point field but also to the influence of this unstable polarization cloud, which cause what is called a "Zitterbewegung" (trembling motion). In the Dirac theory, the "trembling motion" becomes part of the matter-wave of an electron. Due to the higher energies of polarization involved, compared to the zero-point field, the trembling motion of the electron mainly appears at high speeds and small wavelengths of matter waves [Bjorken 1987, p. 52]. The effect of the polarization of the Dirac Sea on the movement of electrons is relevant and clearly proven for inner, rapidly moving electrons in the field of heavy atomic nuclei.

The development of quantum electrodynamics

In the late 1920s, the Dirac theory was generalized to the interacting system of electromagnetic fields and particles. QED describes the electromagnetic field as quantized excitations of a spatial continuum of harmonic oscillators. The amplitude of the electric and magnetic field strength at a particular location is determined by the number of excitation quanta, the photons. QED distinguishes between unstable photons and freely propagating stable photons by means of the critical threshold of the quantum of action h; see Figure 47. It reflects the fact that the electromagnetic field is in a state of continual "zero-point activity" even if no stable photons are present as quantized excitations.

At low field strengths/photon numbers, this is also noticeable in the statistical properties of the photon fields: There are fluctuations in the particle number ΔN and the phase position $\Delta\varphi$ of the photons, which cannot fall below a minimum (see Figure 34 and related text). Therefore in QED, photon states are also described by means of matter-wave states: Stable propagating photons influence the field of electromagnetic zero-point fluctuations as excitations below the threshold of stable photons, which affect photon propagation.

A classical electromagnetic wave with well-defined amplitude and phase is emerges in the limiting case of the superposition of very many photons.

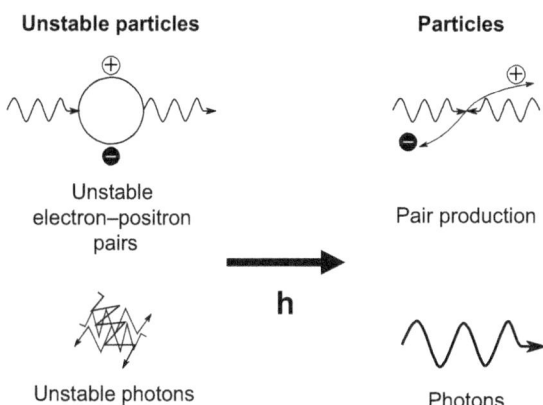

Figure 47: Simplified diagram of the unstable and stable electromagnetic excitation stages of the quantum æther. Unstable excitations occur below the threshold of the quantum of action h. Their energy–momentum relationship differs from that of the stable particles and thus also their mass. On the other hand, stable particle excitations develop when the effect of the excitation exceeds the threshold h. These have exactly quantized energy, rest mass, intrinsic angular momentum, and angular momentum.

Depending on whether the wave states of the photons are superimposed coherently or incoherently, different light states, such as thermal light or laser light (see Figure 34), arise which are correctly described inQED.

Since photons are generated or annihilated by accelerated electrical charges, a consistent quantum theory of the electromagnetic field had to be a theory of the electron from the outset. QED is, therefore, a statistical theory of the electromagnetic field in interaction with the field-generating particles. This theory includes both the electromagnetic zero-point field of the unstable photons and the polarization of the unstable electron–positron pairs. During its journey, a propagating stable photon transforms briefly again and again into an unstable electron–positron pair.

As early as 1930, Robert Oppenheimer pointed out that QED not only has an infinitely high energy of the zero-point field but also mathematical divergences (infinities) in the calculation of the interaction between electrically charged particles and the light field. This would also cause the self-energy, and thus the rest mass, of electrically charged particles to grow without limit. Since the infinities occur at short lengths, this showed a failure of the theory for small distances.

Thereupon, various proposals were made to eliminate the divergences of the theory, for example, an artificial cut-off of the interaction at small lengths. But at what length should this happen without being arbitrary?

So in the 1930s and 1940s, theoretical work on quantum field theory stagnated until new impulses came in 1947, stimulated by the experiments of W. E. Lamb and R. C. Retherford [Lamb and Retherford 1947]. They detected a tiny shift of 10^{-6} of the energy of a spectral line in the hydrogen atom, which was too large to be interpreted as the effect of the polarization of Dirac Sea alone. This energy shift is caused by the change of the zero-point field near the atomic nucleus. Thus, even before the discovery of the Casimir effect in 1958, a direct experimental clue to the reality of the zero-point field had been provided.

The experimental discoveries of the Lamb shift and the Casimir effect led H. Bethe, S. Tomonaga and J. Schwinger, among others, to work on the problem of quantum field theory in the 1950s. Their main focus was on eliminating infinities from the previous theory. It turned out that it was possible to introduce infinities into the theory in a certain way which just made up for the existing infinities, so that a correct finite charge and energy of the electron resulted from the theory, which corresponded very well with the experiment. But was it just a mathematical trick? Dirac himself criticized this process, which is called "renormalization":

> These equations lead to infinities when one tries to solve them: these infinities ought not to be there. They remove them artificially. (. . .) Most physicists say that these working rules are, therefore, correct. I feel that is not an adequate reason. Just because the results happen to be in agreement with observation does not prove that one's theory is correct. After all, the Bohr theory was correct in simple cases. It gave very good answers, but still the Bohr theory had the

wrong concepts. Correspondingly, the renormalized kind of quantum theory with which physicists are working nowadays is not justifiable by agreement with experiments under certain
conditions.   [Dirac 1987, p. 195/196]

Renormalization as an expression of a structure formation in the quantum æther

If, however, a mathematical procedure such as renormalization produced correct
results in the description of reality, there must be at least an element of truth in it.
This insight has been widely accepted, since renormalization also brings fruitful results in other areas of physics (e.g., in the theory of phase transitions). A pragmatic
explanation of renormalization as a suitable addition and subtraction of infinities is
based on the argument that in real experiments only energy differences can be measured and never absolute energy.

However, the physical significance of renormalization lies deeper: It is a direct
reflection of the dynamic partial shielding of the electrical charge of particles due
to the electromagnetic zero-point excitations and fluctuating polarization of the
quantum æther. The partial shielding of the electric charge is in accordance with
the experimental observation that the electromagnetic structure of the electron depends on the distance. When studying the inner structure of the electron with photons of ever higher energy, more and more new contributions of the interaction
with the polarization cloud of unstable excitations become visible [Levine 1997,
Berger 2015]. This is the basis of the theoretical structural model that the "naked"
electrical charge of the electron is partially shielded outwards by the polarization
cloud of unstable excitations that surround the electron as a "cloud" of unstable
electron–positron pairs and other unstable particles. Renormalization theory regards the shielding as a function of the distance from the electron. If one imagines
taking a closer look at the immediate surroundings of the electron with a microscope, that is, to look at the unstable particle cloud around the electron under continuous scale transformations, one would see a self-similar image, as with a fractal.
This leads to the effect that the negative electron charge that an observer would see
under such scale transformations increases until the naked charge is reached
[Wilson 1979].

However, in the limit of QED as an effective theory, particles are described as
infinitely small points. So within the framework of the renormalization techniques
in QED, the scale transformations have to be continued mathematically into the infinitely small and the naked charge of the electron modeled as a point would, therefore, have a value of minus infinity.

The electron actually turns out to be a finitely extended complex structure in
the quantum æther. The question of a cut-off length for the diverging self-energy of
the zero-point field becomes identical with the question at which length scale something qualitatively new occurs in the structure of the electron (Chapter 6).

The magnetic moment of the electron and its angular momentum structure

The fascinating agreement between experiment and QED in determining the magnetic moment of the electron clearly shows how limited the idealist notion of the electron as a point-like object without structure is. Rather, its spin and the associated magnetic structure are created by the motion of a complex, dynamic charge distribution, which also contains parts of the polarization due to unstable excitations in the electron's surroundings.

The magnetic moment and the spin of the electron are therefore not clearly separated from the motion or excitation of the quantum æther. These dynamic excitations and their contribution to the angular momentum structure are rather an inseparable part of the electron. The electron must, therefore, be seen as a self-organizing structure of the quantum æther, similar to that of a topological defect in quantum fluids.

It was a success of the work of Bethe, Feynman, Schwinger, and Tomonaga in 1947 and 1948 to illustrate the dynamic structure of the cloud of unstable excitations of the quantum æther using a diagram technique. This was also a strategy for approximate solutions of the mathematical expressions (Figure 48).

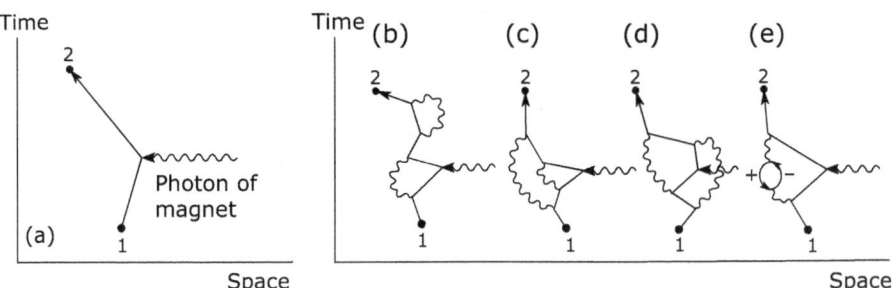

Figure 48: Some contributions of the unstable excitations to the magnetic moment of the electron. (a) Direct interaction of the electron with a magnetic field by a photon. (b–d) Contributions to the interaction taking into account unstable photons. (e) The polarization of the quantum æther by an unstable electron–positron pair. According to [Feynman 1985, p.115–117].

By means of the diagram technique, the contribution of ever higher orders of unstable excitations to the magnetic moment can be analyzed and compared with ever more precise experiments. The deviation from the classical value of a rotating charge distribution is expressed by a factor $g/2$, the gyromagnetic factor g. For a punctiform electron, $g/2 = 1$. Current experiments yield a value of $g/2 = 1.001\ 159\ 652\ 180\ 73$ (±28 last 2 digits) [Hanneke 2008].

No deviation from the QED has yet been found within the scope of this extreme accuracy. In theory, contributions must be considered that contain up to eight nodes (vertex points) in a graph, which includes tens of thousands of different interaction diagrams. In the calculation, the strength of the electromagnetic coupling constant α is included in all vertex points. The experimental value is [Gabrielse 2006]

$$\alpha^{-1} = 137,035\,999\,070\,(\pm 98 \text{ last 2 digits}),$$

and thus agrees to an accuracy of 1:1 billion with the theory. In this high-precision comparison between theory and experiment of the magnitude of the magnetic moment, it must already be taken into account that the fluctuating shielding cloud around the electron also contains unstable excitations of the nuclear forces ([Hanneke 2008] see also Chapter 6). Even taking these extensions into account, QED remains one of the most accurate physical theories. It reflects the dynamic structure of the electron down to a length scale of 10^{-18} m very accurately. To this day, high-energy experiments that explore the electron down to a length scale of 2×10^{-20} m do not observe any other substructure in the sense of being composed of stable "sub-particles."

With renormalization, one cannot get rid of infinities, because the interaction in every vertex is described as "point interaction." However, with the associated concept of a fractal dynamic electromagnetic structure, these infinities cancel one another out. Therefore, the demand for "renormalizability" has strongly influenced the theoretical development of quantum field theories of further interactions (weak and strong nuclear force). Steven Weinberg writes as follows:

> Quantum electrodynamics had almost been killed off by the problem of infinities but had been saved by the idea of canceling the infinities in a redefinition or renormalization of the electron mass and charge. But in order for the problem of infinities to be solved in this way, it is necessary that the infinities occur in calculations in only certain very limited ways, which is the case only for a limited class of especially simple quantum field theories. Such theories are called renormalizable. [Weinberg 1992, p. 120]

However, Meissner points out that quantum field theories of electromagnetic and nuclear forces with renormalization are only a preliminary, approximately valid step toward a microscopic theory that describes the different aspects of the excitations and structure formation of the quantum æther comprehensively and may have to be replaced in the future.

> The infinities are the guideline, are the "fundamental contradiction," which particle physics, at least as far as the majority of physicists are concerned, has been swinging along. Science is always guided by its contradictions. In post-war theoretical physics, infinities have assumed the role of a leader, but perhaps he is a pied piper. In Feynman's own words, the renormalization he has promoted so much is only a method of cleverly wrapping up the infinities and sweeping them under the carpet. Whether this is enough for an all-encompassing theory is more than questionable. [Meissner 1992, p. 308/309]

Today, not only electromagnetic but also nuclear forces are successfully described using what are called relativistic quantum field theories (RQFT), similar to QED. They already take into account the "relativistic" change of the energy–momentum relation (dispersion) of the particles (increase in mass) at high energies. However, the comparison with the excitations of quantum fluids shows that RQFT are not micro-scopic theories of the condensed system. Rather, they are "emergent" effective media theories for relatively small excitation energies, compared to the condensation energy of the quantum fluid [Volovik 2003]. They are obtained by averaging over extremely high-energy intrinsic motion forms of the quantum fluid. Although an RQFT with the formation of unstable particle clouds and stable particle pairs already takes into ac-count the effects of higher energies than the nonrelativistic quantum theory, it repre-sents a low-energy approximation in relation to the energy and length scales of the inner degrees of freedom of the quantum æther. Their infinities can only be elimi-nated when a microscopic model of the quantum æther has been developed.

Feynman's positivism and pragmatism

Although R. Feynman made important contributions to the development of the RQFT, his reactionary ideological stance is opposed to the understanding of the real processes taking place in the interacting system of electron and electromagnetic field. With his "Feynman Lectures on Physics" he has exerted a strong influence on the education of entire generations of physicists worldwide, thus contributing to the consolidation of the enormous ideological influence of positivism and pragma-tism on the natural sciences. He writes:

> The most shocking characteristic of the theory of QED is the crazy framework of amplitudes, which you might think indicates problems of some sort! However, physicists have been fid-dling around with amplitudes for more than fifty years now, and have gotten very used to it. Furthermore, all the new particles and new phenomena that we are able to observe fit per-fectly with everything that can be deduced from such a framework of amplitudes, in which the probability of an event is the square of a final arrow whose length is determined by combining arrows in funny ways (with interferences, and so on). So this framework of amplitudes has no experimental doubt about it: you can have all the philosophical worries you want as to what the amplitudes mean (if, indeed, they mean anything at all), but because physics is an experi-mental science and the framework agrees with experiment, it's good enough for us so far.
>
> [Feynman 1985, p. 129]

This quotation clearly reveals the pragmatic world view of physicist and Nobel Laureate Feynman. The theory does not agree so well with the experiment because it (at least approximately) reflects reality correctly, but because it can be used to calculate numerical values that can be found experimentally. The calculability of a number is its criterion of truth.

Based on the discovery of the Maya that five Venus cycles correspond to ap-proximately eight years of 365 days each, Feynman explains as follows:

> Taking the example of the Maya one step further, we could ask the priest why five cycles of Venus nearly equal 2,920 days, or eight years. There would be all kinds of theories about why, such as, "20 is an important number in our counting system, and if you divide 2,920 by 20,

you get 146, which is one more than a number that can be represented by the sum of two squares in two different ways," and so forth. But that theory would have nothing to do with Venus, really. In modern times, we have found that theories of this kind are not useful. So again, we are not going to deal with why Nature behaves in the peculiar way that She does; there are no good theories to explain that. [Feynman 1985, p. 17]

This pragmatism is a variant of positivism: This ideology is not interested in the material causes and mechanisms and is an essential cause of the ongoing crisis of physics in the search for a unified description of the microcosm. The scientific method and world-view that seeks to understand more deeply the qualitatively new level of matter of the quantum æther in dialectical unity of theory and practice, induction and deduction, is combated.

Feynman's treatment of antimatter also corresponds to this. For example, there is a contribution to the Compton scattering of photons by electrons in the indirect process of the generation of an unstable electron–positron pair by the photon (Figure 49). The resulting unstable positron is then annihilated by the existing stable electron, so that the unstable electron created during pair formation transforms into a stable electron by absorbing further energy and continues on its path, while the original stable electron is annihilated. For mathematical reasons, in the formalism of quantum field theory, a positron can be represented as an electron running backwards in time. Feynman writes as follows:

I'd like to show you what this backwards-moving electron looks like to us, as we move forwards in time. [Feynman 1985, p. 103]

Time

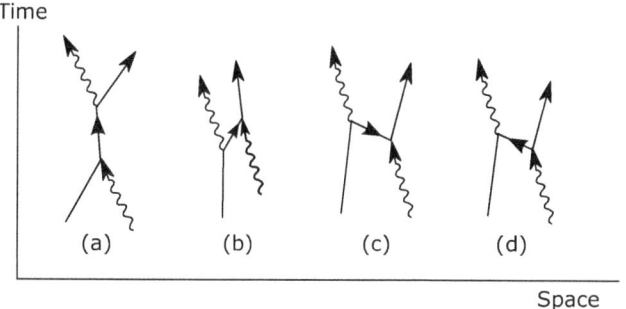

Space

Figure 49: Representation of various contributions to Compton scattering of a photon (wavy arrow) at an electron (straight arrow) according to Feynman. (a) Absorption of a photon and then emission. (b) Spontaneous emission of a photon and then absorption. The contribution of unstable electron–positron pairs according to Feynman with electrons running backwards through time (c) and realistically (d), with antiparticles running forwards through time. (a–c) [Feynman 1985 p. 97–99].

For him, the electron running backwards in time is real because it is simple in the formalism. The positron running forward in time is only its appearance for us. This point of view not only negates the obvious irreversibility of time, it also has further consequences: The interpretation of the antiparticles as "particles running backwards in time" stands in contrast to the materialistic hole model of the Dirac Sea, which regards the antiparticle as a real "hole state" in a sea otherwise occupied by electrons. The production of pairs is recast as the formation of matter from nothing, instead of as excitation and structure formation of an underlying material background. Feynman's theory of antiparticles as particles running backwards in time erects a dam against a materialistic view of the new level of matter below the structural level of the particles.

Even though the model of the Dirac Sea is only a low-energy approximation to the quantum æther within the framework of an effective media theory, it is far superior to Feynman's *empty space* in which crazy particles move forward and backwards in time.

A further main effect of positivist ideology is to interpret the unstable excitations of the quantum æther as "virtual particles" in an "empty space." Virtual particles are only apparently present, imaginary constructs and not matter. Thus, the mechanistic clarification of one of the main questions of quantum physics is avoided: How can the world of relatively stable particles and steady motions emerge from the sea of enormously energy-rich fluctuating motions? How can the extreme discrepancy between the energy density of the zero-point field in quantum theory and gravitational theory be explained? (see, for example, Hawking 1997, p. 93).

4.6 Matter waves as organizational states of the zero-point field

The existence of matter waves during the motion of microscopic particles is confirmed on all sides by a wealth of experimental observations. They range from interference experiments, which allow a direct determination of their wavelength, to the spatial structure of stationary matter-wave states of the electron shell in atoms and their chemical bonds, to the more subtle effects on spin and magnetic moment of particles. In the different forms of quantum theory, from the Schrödinger theory of matter waves to RQFT, physical properties of free and interacting particles in unity with their matter-wave states are recorded in a statistical form. All these theories have in common is that they make no statement about the deeper nature of matter waves.

Since the 1970s, criticism of the ideologically motivated banishment of the material causes of matter waves has led to the formation of an approach to theoretical physics called "stochastic electrodynamics," which investigates the effects of the real existence of an electromagnetic zero-point field and the difference to the electromagnetic field of stable photons. It draws conclusions from this on the material causes of matter waves [de la Pena 1996].

The invariance of the zero-point field under uniform motions

How does a free particle with an electrical charge move in the electromagnetic zero-point field? Walter Nernst's considerations in 1916 led to the problem that moving particles in the zero-point field should experience a friction force that depends on the velocity. The Doppler shift of the frequency of the unstable photons arriving from the front would lead to a frequency increase so that the fluctuating forces of the zero-point photons acting from all sides would no longer cancel each other out on average. Thus an "æther wind" would occur and the zero-point field would not be Lorentz-invariant, that is, invariant under uniform motions. However, in a theoretical work Timothy Boyer [Boyer 1969] showed that a zero-point field with a mode density increasing with the third power of the frequency is Lorentz-invariant, that is, its energy flux density is isotropic, independent of the state of motion of the observer. A uniformly moving particle would therefore not experience any velocity-dependent frictional force.

Isn't the absence of a friction force for a particle exposed to a fluctuating field contradictory to observations in other fields of natural science? In the Brownian molecular motion of particles on a liquid surface, or the diffusion of atoms in a solid by thermal random motion of the surrounding atoms, no inertial motion is possible. Any particle that starts with an initial velocity v would lose its kinetic energy due to the frictional forces and thermalize after a short time. Its expected average velocity value would then be $<v> = 0$, and only statistical fluctuations of the velocity of the order of magnitude of the thermal energy of the bath would occur. In such a system, there would be no possibility for inertial movements, the movement in the "vacuum" would be connected with a friction.

Microscopically, the disappearance of friction during movement in a sea of fluctuating excitations is synonymous with the occurrence of a detailed energy balance [de la Pena 1996]. The interaction with the fluctuating zero-point field must be such that each energy–momentum change of a particle due to absorption of an unstable zero-point photon is compensated after a short time by a corresponding emission. Such processes are the central subject of the summation of interaction diagrams in quantum field theories. For sufficiently long times and for sufficiently long distances, particle energy or particle momentum are preserved as long as a critical threshold of action is not exceeded. This threshold for the transition between reversible fluctuations, which are subject to a detailed energy balance, and the abrupt lasting change of the quantum state is determined by the quantum of action.

The quantum of action is thus an expression of an organizational law. It determines, expressed in the Heisenberg uncertainty relationships, on which time or energy scales fluctuations bring about a lasting change in quantum states. Conversely, it is also a measure of the relative stability of motion states. A change of motion states by external forces can only take place in certain portions of energy, time, location, momentum, or angular momentum. The fluctuations in the motion of stable particles, which occur spontaneously due to the interaction with the zero-point fluctuations, remain below the threshold set by the quantum of action. The fluctuations of the energy and the momentum can assume very large values for short times or short distances. The greater the energy–momentum fluctuation, the smaller the time span or spatial interval in which they cancel each other out.

Let us look at the movement of a topological defect in the background of the frictionless superfluid for comparison. As explained in Chapter 3, it possesses unstable quasiparticle excitations as low-energy collective motions of the superfluid condensate. If a topological defect, for example, a quantized vortex tube, moves as a stable particle in the field of these excitations, there should be no friction. However, this is not correct, as the Soviet physicist Kopnin showed in 1991: A friction force occurs due to the transfer of quasiparticle excitations in the normal-fluid vortex nucleus into the condensate [Kopnin 1991]. This friction, also called the Kopnin force, occurs because the density of the unstable excitations in fermionic superfluids is not Lorentz-invariant and because the gap in the excitation spectrum is destroyed in the vortex nucleus. Both are fundamental differences to the excitation spectrum of the quantum æther (see Section 6.6).

The origin of matter waves

De Broglie already suggested in his theory of matter waves that each particle is accompanied by both a high-frequency and a low-frequency wave. For speeds far below the speed of light, the high-frequency Compton wave is largely independent of whether the particle is at rest or moving relative to a laboratory system. Its frequency is the Compton frequency determined by the self-energy of the particle and thus by its rest mass $h\nu_c = m_0 c^2$. For electrons it is in the range of $\nu \approx 10^{20}$ Hz. This corresponds to the Compton wavelength of $\lambda \approx 3 \times 10^{-13}$ m. At these frequencies and wavelengths, the Dirac trembling motion due to unstable particle–antiparticle pairs also begins to make itself noticeable.

The low-frequency de Broglie matter-wave, on the other hand, is linked to the velocity of the particle and at velocities below the speed of light is significantly lower than the Compton frequency. The dependence of the frequency and wavelength of the matter-wave on the velocity of the particle v is given by

$$h\nu = \frac{1}{2} m_0 v^2 \text{ and } \lambda = \frac{h}{m_0 v}$$

(see mathematical appendix).

In a remarkable proposal by Kracklauer [Kracklauer 1992] and de la Pena and Cetto [de la Pena 1996], the matter wave arises as a modulation wave of the zero-point field due to its interaction with intrinsic motion forms of the electron. Under the assumption that this is subject to a resonance-like amplification at the Compton frequency v_c, there is a coordination of the zero-point excitations at frequencies $v = v_c$ and wave numbers $k = k_c$ with the intrinsic motion of the particle.

If the motion of the particle relative to a laboratory system is analyzed, it is found that the zero-point modes in resonance with v_c in the forward direction are shifted to larger frequencies according to the Doppler effect. The reverse effect occurs in the reverse direction of particle motion, that is, a Doppler shift to smaller frequencies. Seen from the comoving system of the particle, there is still a wave propagating from the front and the back at the resonance, which overlaps to a standing wave with the wavelength of the Compton wavelength and is largely independent of the velocity of the particle.

Seen from the laboratory system, however, something new emerges: The Doppler shift in the resonance leads to a modulation of the zero-point modes. A low-frequency oscillatory beat of the zero-point field in space and time is created in the vicinity of the particle. Due to the particle motion, the zero-point excitations form a wave-like modulation of their fluctuations, whose frequency and wavelength depend on the velocity of the particle according to the de Broglie relations (Figure 50).

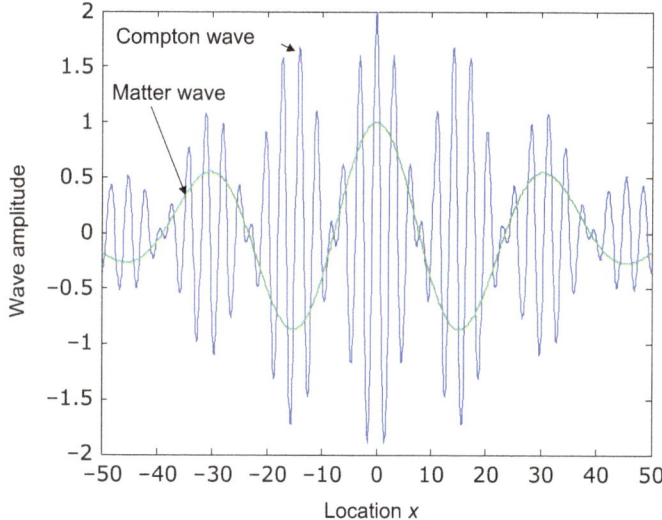

Figure 50: Illustration of the formation of matter waves by the interaction of an electron with the fluctuating excitations of the zero-point field. At the Compton energy a resonant amplification of the interaction between the zero-point field and the electron evolves. This results in a standing high-frequency wave of the zero-point field with the Compton wave number k_c (blue curve). If the electron moves, the resonance in the forward and backward direction is shifted differently, according to the Doppler effect, so that a modulation wave with the de Broglie frequency and the de Broglie wavelength is generated, in addition to the rapidly oscillating wave (green curve). A velocity $v/c = 0.1$ was selected for the calculation. The axis x is given in units of the Compton wavelength λ_c.

In accordance with the random character of the zero-point excitations and the relationship between wavelength and particle velocity governed by the de Broglie relations, the modulation wave formed has both a random and a deterministic character. In this model, the matter wave is created by a self-organization process of the zero-point excitations in the resonant interaction with the intrinsic forms of motion of particles. The modulation wave induced by it in the field of unstable particle excitations reacts on its state of motion. This results in a stability of the motion on sufficient length and time scales, which is determined by the quantum of action.

Mathematically, the difference between a dissipative random Brownian motion and the motion of particles in the random field of quanta is not large [Roepstorff 1994].

Optical transitions: the jump to a stable photon

The transition from unstable to stable photons requires accelerated electrical charges and, as far as is known today, does not take place spontaneously from the zero-point field in free space. Figure 51 shows the birth of a photon as a quantized wave packet through the transition of an excited electron to its ground state. The dipole oscillation of the electrical charge of the electron induces a temporal superposition of the two matter-wave states involved, the initial and the final state.

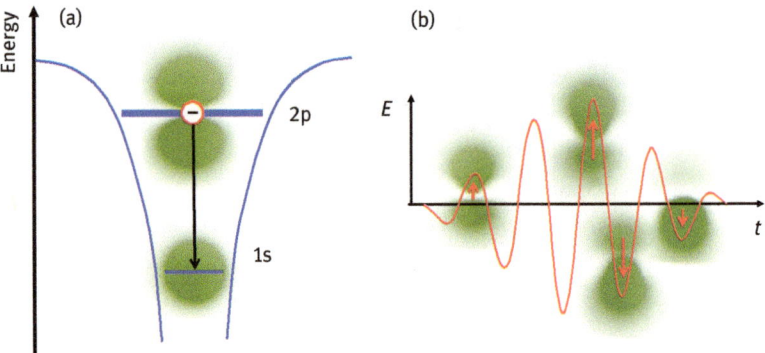

Figure 51: Birth of a photon by optical dipole transition of an electron from an excited state to the ground state (a). The formation of the electromagnetic wave (only the electric field strength E is shown here) is identical with the superposition of two stationary matter-waves of the electron (here 2p and 1s) to form a temporally oscillating electrical dipole in the electron shell of the atom (b).

In this process, a temporally or spatially oscillating matter-wave becomes identical with an electromagnetic field. In QED, this identity is expressed by means of a gauge symmetry: A phase jump in the development of matter waves is identical to a photon. But how does the quantization of action, the jump, arise? Dipole fluctuations, occur as part of the stochastic electron movement in a stationary matter-wave state, are obviously not sufficient. The jump requires at least one complete phase cycle of the oscillation, both in the dipole moment of the electrical charge density and in the matter-wave field moving with the particle. An exact understanding of the mechanism of action quantization does not yet exist. In Chapter 6, within the framework of the modeling of the quantum æther by a quantum fluid, a possible microscopic mechanism of the quantization of action is discussed.

The quantum of action as an expression of an organizational law

To summarize: the development of the law of inertial motion of particles, which in the absence of external forces maintain their state of motion in space and time, from chaotically fluctuating movements in the zero-point excitations of the quantum æther is determined by the quantum of action as the threshold for the transition from unstable fluctuating to stable forms of motion.

The exact process of the effect of the zero-point field on particles and the effect of particles on the zero-point field is understood quite accurately in QED today.

However, a clear picture of the processes is avoided for ideological reasons. In fact, the interaction between chaotically fluctuating fields and particles leads to stationary states of motion: for example, a state with a defined velocity of the particle is only possible through a delocalized, that is, spatially extended modulation of the zero-point field induced by the motion of the particle. The conservation of energy and momentum of motion is an expression of the reversible exchange of energy and momentum below the level of the quantization of action. Conversely, a closely localized particle cannot form a long-range stable modulation wave, since it interacts with many modulation modes simultaneously.

The states of matter waves are also dependent on the macroscopic environment. The Casimir effect shows that the field of zero-point excitations is significantly influenced by the presence of macroscopic bodies. The modulation of the unstable excitations induced by the particle motion must be compatible with the boundary conditions enforced by the macroscopic environment. If these are not compatible, the matter wave and thus the organizational state of the unstable excitations, are modified. This naturally has a retroactive effect on the particle, which must adopt a new state of motion which is compatible with the environment. If, for example, a slit or a double slit forces the matter wave to propagate only through the openings, this influences the matter-wave field, and thus also the zero-point field, on a macroscopic scale.

The transition between different matter-wave states must therefore be accompanied by a transformation between different organizational states of the zero-point field.

The synthesis of particle and wave concepts results in an "atomism of action" with Planck's constant h. It no longer divides matter only into the smallest discrete building blocks in the sense of finished particles but into "smallest stable processes" that control the structure formation of matter in the form of the smallest action. The quantum of action is also a measure of the influence of the unstable, fluctuating excitations of the zero-point field on the structural level of the stable particles. The Göttingen physicist Friedrich Hund [Hund 1975] gave a fitting definition of quantum theory, namely as *"The theory of the role that h plays in nature."* This also makes it clear that quantum theory is by no means a "fundamental theory" which, as a prototype of a "world formula," should be evaluated differently than electrodynamics, mechanics, or thermodynamics. Rather, it is a statistical theory of cooperative behavior at the interface between two different structural levels of matter.

5 The influence of deeper structural levels of matter on motion and gravity

Relative and absolute movement form a dialectic unity of opposites. The absolute movement of a body consists of its inner motions as well as the totality of its movements relative to other bodies and forms of matter. The progression of human knowledge from single relative movements of certain bodies and certain forms of matter to the overall context of all movements is an infinite process that will never end.

The various spatial movements to which the Earth is subject as part of different matter systems in the cosmos can be determined today by measuring the Doppler shift of the cosmic microwave background radiation. This forms the largest reference system in the cosmos known today to describe the spatial motion of systems.

Our planet Earth is subject to various local movements – the Earth's own rotation within 24 hours around the axis of rotation. The associated speed at the equator is about 0.5 km/s. In addition, there is the movement of the Earth around the Sun, with a speed of about 30 km/s, which varies along its elliptical orbit. However, in addition, the Earth participates as a component of the solar system in the latter's motion and our entire solar system revolves around the center of our galaxy, the Milky Way, at a speed of about 230 km/s. The Milky Way in turn moves as part of an even larger cosmic system, the local group of galaxies, at about 620 km/s, relative to the reference system formed by microwave background radiation.

Friedrich Engels writes in *Dialectics of Nature*: *"All nature accessible to us forms a system, an coherent whole of bodies, and here we understand by bodies all material existences from the star to the atom, even to the æther particle, if its existence be admitted. The fact that these bodies are connected already implies that they interact with each other, and their mutual influence is movement. It is already evident here that matter is unthinkable without motion."* [Engels, 1885, p. 355]

To deny the objective reality of this overall connection of the motion of all matter systems to each other is the ideological core of the various physical models which emphasis the "relativity of motion" or even the "relativity of gravity." In fact, the fundamental questions as to how deeper structural levels of matter affect the motion of particles and macroscopic forms of matter have not been satisfactorily answered to date. Shouldn't the movement of the Earth at high speeds lead to a change in the speed of light? And how is the quantum æther influenced by the macroscopic movements and gravitational fields of masses? The impossibility of answering these questions within mechanical æther models was an essential initiator of the crisis of physics at the beginning of the twentieth century. The crisis of mechanical æther models led to the establishment of theories of special and general relativity. Albert Einstein elevated the "relativity of motion," the (alleged) impossibility of determining the state of motion of the æther experimentally, into principles above nature:

https://doi.org/10.1515/9783110644203-005

The "invariance of the laws of physics under uniform motion" (special relativity principle) and "under accelerated motion" (general relativity principle): The materialistic claim of the gradual recognition of the overall context of all movement from the study of relative movement gave way to relativism.

Einstein's special and general theory of relativity, its revolutionary materialistic essence and its idealistic form in the four-dimensional Minkowski space appear today in a new light on the basis of new experiments and quantum-fluid æther theories.

5.1 Movement at high speeds

If atomic building blocks and thus also more complex forms of matter consisting of atoms are development products of a deeper structural level of matter, a movement of these particles, at least at high velocities, must have an influence on their structure and physical properties.

The increase in mass of the electron and the existence of a limiting velocity

There is clear experimental evidence for the increase in electron mass at high velocities. The investigation of the fine structure of the spectral lines of atoms and their interpretation in the context of quantum theory already showed the necessity to take into account an increase of mass of the electron at fast movement in the electron shell [Sommerfeld 1916]. Direct investigations were carried out by means of the deflection of fast electrons in magnetic fields. The Swiss physicists Guye, Ratnowsky, and Lavanchy [Guye et al 1921] were able to show, through experiments in the early 1920s, that the relationship between electron momentum and electron velocity is modified from the classical Newtonian relation $p = m_0 v$ (rest mass m_0) to a relation $p = m(v) v$ with

$$m(v) = \frac{m_0}{\sqrt{1 - \frac{v^2}{c^2}}}$$

While Hendrik Lorentz assumed an actual velocity dependence of mass based on the theory of electron motion in an æther, Albert Einstein interpreted this relationship based on the special theory of relativity as a *"modified relationship between velocity and momentum"* [Hecht 2009, p. 336].

In fact, a velocity dependence of mass and the existence of an upper limit for velocity is identical with an alteration of the energy–momentum relation of classical mechanics $E = p^2/(2m_0)$ or of the corresponding energy–velocity relation $E = \frac{1}{2}m_0 v^2$ to

$$E = \sqrt{(m_0 c^2)^2 + (pc)^2} = \frac{m_0 c^2}{\sqrt{1 - \frac{v^2}{c^2}}}$$

The investigation of the relationship between electron velocity and energy by parti-cle accelerators [Bertozzi 1994] clearly confirmed the increase of mass with energy (Figure 52c). The existence of a velocity limit for electrons and other particles was also clearly proven in particle accelerators.

The changed energy–momentum relation corresponds to the dispersion of free electrons in the Dirac model with an energy gap for the excitation of an electron–positron pair of $E_g = 2m_0 c^2$. It illustrates the nature of the electron as an excitation in the quantum æther. The speed of light is a material constant of the quantum æther as the maximum propagation speed of stable excitations.

Although the velocity-dependent mass of subatomic particles has been proven clearly [Farago 1957, Geller 1972] and is applied every day in particle accelerators or high-voltage electron microscopes, this effect is swept under the carpet under the influence of an idealistic world view by geometrizing space-time. Physicist Carl Adler explains this in the prestigious *American Journal of Physics*:

> In fact, most of the recent criticism of relativistic mass is presented in the context of the four-vector formulation of special relativity. ... The use of relativistic mass can mislead students into believing that the structure of moving objects is actually affected by their motion, as was actually the case in the Lorentz theory, when, in fact, the observed effects are due to an alteration in space-time.   [Adler 1987]

This is how the myth developed that velocity-dependent mass is a pseudo-effect of special relativity [Hecht 2009]. The principle of special relativity on which it is based demands just this, that a state of motion relative to the æther cannot be de-termined, and therefore the change in the properties of bodies, such as their mass, does not exist objectively and in reality during motion, but is a result of their "rela-tionship to a moving reference system" [Ryckman 2005, p. 41]. The positivist method of using formulas without creating awareness of their physical meaning has led to the paradoxical concept of "relativistic mass increase." In fact, however, the increase in the mass of moving bodies is not a pseudo-effect of their relative motion to any reference system, but the result of their dispersion, modified at high speeds with respect to the quantum æther.

Changing atomic clocks by motion

The deceleration of electron oscillations in moving atoms was predicted by H. Lorentz in 1899 as part of his æther theory. The first experimental indications of a deceleration of atomic processes were provided by the experiments of H. E. Ives

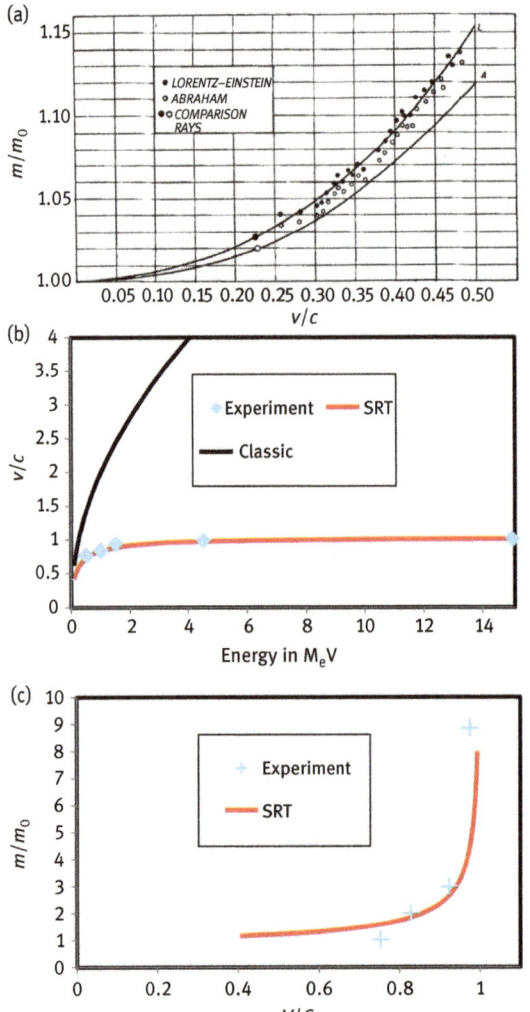

Figure 52: (a) Determination of the velocity dependence of the electron mass and comparison with the theories of Lorentz and Abraham [Guye 1921]. (b) Velocity of electrons as a function of their energy determined by a calorimeter. (c) Resulting velocity dependence of the electron mass. The deviation from the theory results from inaccuracies in the determination of the electron mass. Data in (b) and (c) from [Bertozzi 1964].

and G. R. Stilwell in 1938 [Ives 1937–1952]. They studied deviations of the Doppler shift of the light emitted by ion beams (canal rays).

The Doppler effect is caused by the finite speed with which light propagates. When the light source or an observer moves relative to the light field, either a shortening (blue shift) or a lengthening (red shift) of the wavelength of the light occurs, depending on the direction of the movement. This effect, observed by Christian Doppler in sound waves as early as 1842, is also part of the classical wave theory of light.

What was new in the experiments of Ives and Stilwell was that the frequency of the emitted light also changes transversely to the direction of motion of atoms due to the deceleration of internal oscillations of the electron shell. This also results in a modification of the positive and negative frequency shifts in the longitudinal forward and backward directions in comparison to the classical theory.

In the course of time, experiments were carried out that proved the change in frequencies of different subatomic processes during local motion.

This includes the drastic extension of the lifetime of unstable particles in rapid motion. Experiments on negative and positive muons at 99.94 % of the speed of light show an increase of their lifetime to about 64 µs compared to 2.19 µs at rest [Bailey 1977]. In 1971, Hafele and Keating first sent two atomic clocks with airplanes in opposite directions around the Earth. They measured a time shift of −40 ns for the eastward journey and 275 ns for the westward journey [Hafele and Keating 1971], compared to an atomic clock stationed on the rotating Earth. Today, clock decelerations due to local movement and the changed gravitational field are regularly taken into account in GPS navigation systems [Su 2001]. Recent experiments with high-precision atomic clocks are able to detect tiny clock decelerations down to a relatively slow movement of 10 m/s [Chou 2010]; see Figure 53. All these

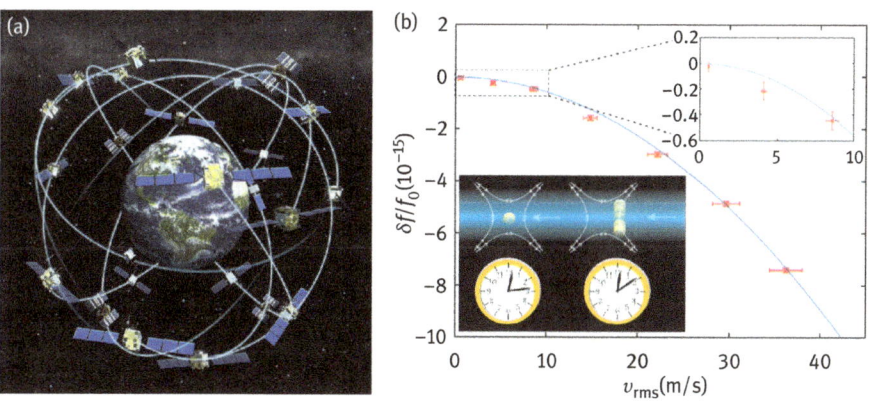

Figure 53: Changes in the running of atomic clocks due to local movement and gravity. (a) When synchronizing the atomic clocks of the GPS satellite navigation system, the following effects must be taken into account: The Earth's reduced gravitational field at an altitude of 20.2 km allows the clocks to advance by 45.7 µs per day. The slowing down of the clocks by movement of the satellites at a speed of approximately 3.9 km/s causes the clocks to follow by 7.1 µs per day.
(b) Measurement of clock deceleration at low speeds by comparing two atomic clocks in the laboratory. The atomic movements here are controlled by electric fields. Pictures (a) NOAA.
(b) From C. W. Chou et. al., Science 329 (2010) 1630, reprinted with permission of AAAS.

decelerations of atomic clocks by local movement are not expressions of the "relativity of time" but have their cause in the change of material processes during movement.

The search for æther drift effects on the speed of light

From the idea of a mechanical æther at rest within space as the medium for light, there inevitably follows that the speed of light, measured by an observer moving relative to the æther, must depend on the motion.

This was investigated in a famous experiment conducted by Michelson and Morley in Ohio in 1887 [Michelson and Morley 1883]. Here, the relative change of the two-way speed of light (outward and return path) for two perpendicular interferometer arms was measured (Figure 54). Any change in the speed of light along one of the two arms of the Michelson–Morley interferometer would result with great precision in a change in the interference pattern of the light waves of the two interferometer arms. The experiment resulted in a very small displacement of the interference fringes of the two light beams, which, however, did not correspond to the Earth's motion of 30 km/s around the Sun [Michelson and Morley 1883; see also Vigier 1997].

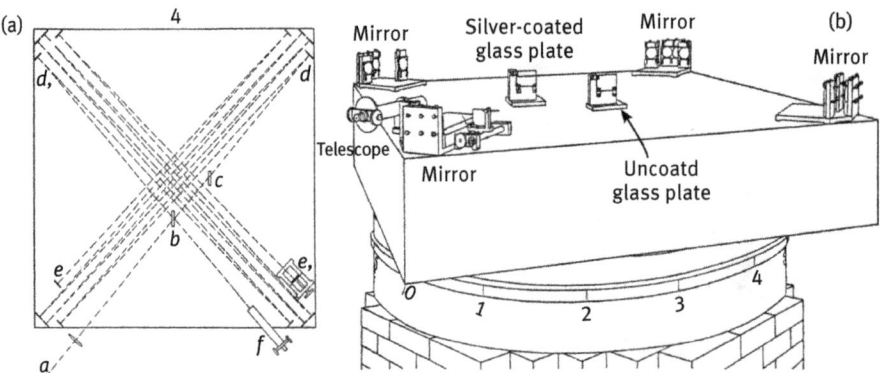

Figure 54: Michelson–Morley experiment to determine a change in the speed of light due to Earth movement. (a) Schematic diagram of the beam path of the interferometer. (b) Figure of the apparatus used [Michelson and Morley 1887].

Since then, this and similar experiments have been repeated many times with increased accuracy, at different seasons (to avoid a chance coincidence of the motions of the Earth and æther), at different times of the day, and at different altitudes above sea level. To date, however, no change in the speed of light due to the Earth's own motion has been detected, with a sensitivity down to 10 m/s [Baynes 2012]. Besides the Michelson–Morley-like two-way experiments, in which an anisotropy of the speed of light is searched for in two interferometer arms perpendicular to each other, recent experiments search for differences in the forward and backward directions of a light beam by means of resonators. No changes in the speed of light have been detected down to an accuracy of 10^{-17} [Eisele 2009].

Such experiments, however, do not clearly prove that the speed of light does not change relative to the movement of the Earth. H. Lorentz already explained the negative outcome of the Michelson–Morley experiment with the slowing down of atomic oscillators and the "Lorentz contraction." He assumed that moving bodies shorten in the direction of motion by a factor that depends on their speed. By shortening the arms of the Michelson–Morley interferometer in the direction of motion relative to the æther, the displacement of the interference pattern is compensated for by the lower speed of light when moving toward the light, if also the change in frequency is taken into account.

This hypothesis, which at first seems rather strange, is the immediate precursor of the theory of relativity, which assumes a contraction of space during rapid motion [Barone in: Selleri 1998]. The Lorentz contraction theory of the physicists G. F. FitzGerald and H. Lorentz, which was already put forward at the end of the nineteenth century, is however, in contrast to the theory of relativity, a theory of real length changes of bodies that move absolutely relative to the æther. For the Lorentz contraction, it was assumed that the basic atomic building blocks of all bodies compress their electron shells in the direction of motion.

This theory can explain the results of the Michelson–Morley experiment. The hypothesis of an associated motion of the æther due to the Earth's motion can also explain the outcome of the Michelson–Morley experiment, but is in contradiction to the observation of the aberration of starlight.

Deflection of starlight (aberration)

The observation of the aberration of light shows very clearly that the speed of light changes when the observer moves transversely to the light field. Aberration describes the apparent change of position of all the stars when the observer moves, due to the finite speed of light (Figure 55). In the particle concept of light, this effect can be explained simply by the vector addition of the velocity of the particle and the observer. In the image of an electromagnetic wave, aberration occurs if the directional character of the radiation flux is taken into account. The consideration of the law of constancy of the speed of light in Einstein's theory of relativity results in a small correction in the deflection. According to this theory, the magnitude of the velocity vector $c + v$ should not be greater than the speed of light.

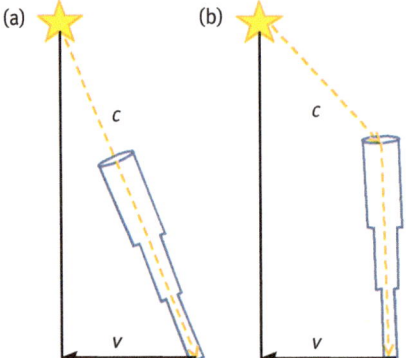

Figure 55: Aberration of light. (a) Observation of the apparent change in the position of stars in the sky due to the movement of the observer on Earth. This changes the speed of light from c to c + v. (b) The hypothesis of entrainment of the æther would lead to double refraction of the light, which is not observed.

If the speed of the light source were relevant, the stars in binary star systems, for example, would appear with a temporally oscillating deflection, which is not observed. *"The theory of relativity states that no absolute velocity exists. Only the relative velocity between objects has a physical meaning."* [Marmet 1996]. However, Marmet makes it clear in his critical remarks that Einstein's principle of relativity may only be applied to the relative motion between the observer and the resting system of the light field, and not to the relative velocity of the observer and the radiation source.

The Sagnac effect in the global positioning system (GPS)

The question of relative to which electromagnetic system at rest the Earth actually moves is answered by the "Sagnac effect" in the synchronization of atomic clocks in the GPS. In addition to the effects of the local movement and gravity on the running of the atomic clocks, the Earth's own rotation must also be taken into account.

Their synchronization according to GPS guidelines is based on the assumption of an electromagnetic system at rest with the Earth's core as its center, in which the Earth revolves in 24 hours. In order to synchronize the clocks on the satellite and the Earth station, in addition to the various contributions to the timing of the clocks, what is called the "Sagnac correction" (Figure 56) must be taken into account. The physical interpretation of this is the subject of much controversy.

It consists in the change of the speed of light for signals propagating with or against the rotation of the Earth [T. Flandern in Selleri 1998 and Gift 2014]. The change depends on the tangential speed of the Earth's surface and thus on the latitude of the Earth station: at the Equator, the correction is 0.5 km/sec. However, this materialistic explanation calls into question the relativistic principle of a constant

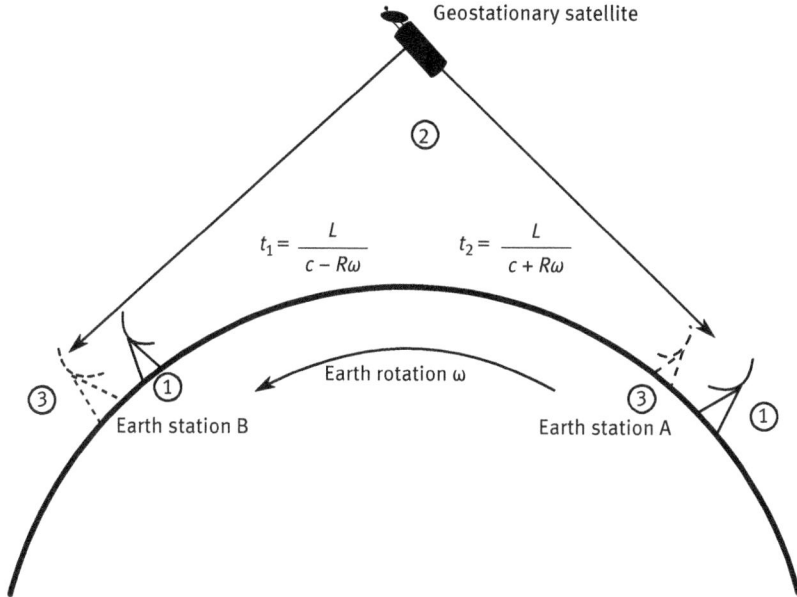

Figure 56: Sagnac effect: Effect of the Earth's own rotation with angular velocity ω on the propagation of electromagnetic signals. A geostationary satellite detects a change in the speed of light $c \pm R\omega$, where R is the radius of the Earth and ω die is the angular frequency of the Earth's rotation, relative to the two Earth stations moving in opposite directions relative to it. The difference in the arrival time of the signals is $\Delta t = t_1 - t_2 \approx 2LR\omega/c^2$, at a distance L between the two Earth stations and the satellite.

speed of light [Engelhardt 2013]. Because rotational movements are not uniform but accelerated movements, the Sagnac effect is alternatively also described in general relativity theory [Ashby 2010]. In the limiting case of very large radii, however, the circular motion approaches a uniform motion. Since the Sagnac effect depends solely on velocity and not on acceleration, it should also be included in the special theory of relativity as a limiting case.

But it is not because the theory assumes a contant speed of light with respect to any moving observer. Therefore, its material cause is obscured by the relativists with the statement: *"The Sagnac effect is described by the fundamental scalar invariant coupling space and time"* [Ashby 2010, p. 1]. In other words, it can be calculated with the mathematical construct of Minkowski space as *"loss of absolute simultaneity"* due to curved space-times.

The synchronization of the atomic clocks in the GPS by light signals, however, confirms the change in the "one-way" speed of light due to the movement of the Earth relative to geostationary satellites. Since both the Earth and satellites perform the Earth's movements around the Sun and the galactic center together, they are not suitable for measuring the change in the one-way speed of light caused by such movements.

5.2 The special theory of relativity

The question as to what mechanisms are responsible for the increase in mass, slowing of clocks, and length contraction at high velocities has not yet been answered. The investigation of the change of the dispersion of quasiparticles during their local motion in quantum fluids, however, shows a way. Even more confusing was the initial situation at the end of the nineteenth and beginning of the twentieth century, when attempts were made to explain the new observations with ad-hoc assumptions within the framework of mechanical æther models. In this confusion, the special theory of relativity brought a simple order. It did not attempt to model the æther at all and reduced all velocity-dependent effects to the relativization of space and time for motions between bodies. The materialistic approach of H. Lorentz, of setting up empirical laws in interaction with the concrete microscopic modeling of the æther, was however aborted due to the increasing impact of the idealistic world view, and/or its crisis due to its being entrapped in mechanical analogies.

The paradoxical contradictions of a mechanical æther

The æther conception of the nineteenth century was that of a fine medium, which was supposed to surround or penetrate all atoms and bodies. It was assumed to possess elastic or viscous properties, and to be a mechanical medium for electromagnetic waves, similar to solids or liquids as mechanical media for sound waves. For this, it would have to have the following properties:
- Since the elasticity of a medium determines the speed of its elastic wave, the æther would have to be thousands of times more rigid than steel to produce light waves as elastic waves propagating at the speed of light.
- However, since light is a pure transverse wave, the high rigidity should only affect the transverse elasticity, while the volume elasticity is vanishingly small.
- In order to be compatible with the laws of classical mechanics, it should have no noticeable influence on the inertial movement of bodies, that is, no frictional force. Otherwise, the planets would have long since fallen into the sun or the electrons into the atomic nucleus.
- In addition, it would have to be the medium of the gravitational field, and show itself as an inertial force for accelerated motions (but not constant motions).
- Furthermore, it would represent an absolute reference system for the movement of bodies and for electrodynamics. A motion relative to it would have to lead to an altered speed of light and thus to changes in the laws of electrodynamics and gravity.

Theoretical construction of special æther models based on mechanical analogies was a dead end. æther was supposed to be either elastic or viscous, depending on which effect was considered, frictionless but also inertial, sometimes entrained, sometimes not. Attempts to interpret mass electromagnetically led to contradictions – a metaphysical concept, shaped by prejudices from previously known physics was at a dead end.

The special theory of relativity as an idealist dialectical negation of the æther

The special theory of relativity combined a large number of experimental observations of the influence of motion on the processes of electrodynamics and mechanics into two principles:
- The speed of light is a natural constant for all systems, independent of their state of motion.
- The principle of relativity: No physical measurement can determine the absolute motion of uniformly moving systems (inertial systems). Only relative movements between moving bodies are physically detectable.

Albert Einstein was able to formulate a uniform theory from this, which successfully described many experiments and revealed the statement that mass is an expression of the total energy content of a body. That this cannot be true is critically discussed, see e.g. Grit Kalies [Kalies 2019]. However, both principles, materialistically interpreted, by no means contradict the existence of an æther, as Albert Einstein repeatedly admitted [A. Einstein 1922]. As a dialectical synthesis of electrodynamics and mechanics at high speeds, they avoid that arbitrariness of models of a mechanical æther. The first principle is the expression of light being perceived as a kind of natural oscillation of a material system of some kind and speed of light represents a materials constant. The second principle says that physical laws of motions, including velocity-dependent effects, only include relative motions of bodies and absolute motions, for example motions relative to the æther, cancel out for many properties.

The special theory of relativity gives the same mathematical connections for the effects of clock deceleration, the Lorentz contraction and the modification of the laws of mechanics by mass increase at high speeds, as the Lorentz theory of movement in the æther. In addition, it makes new approximate statements about the equivalence of energy and mass, modifies the energy scale of motion by taking into account the energy of formation (self-energy) of particles, and predicts individual corrections in kinematics (e.g., the Doppler effect) that go beyond the Lorentz model. The two principles can be mathematically combined in a new symmetry principle: the invariance of the laws of physics under Lorentz transformations. Materialistically interpreted, Lorentz symmetry is a reflection of properties of the quantum æther, in particular, the dispersion of its excitation spectrum. The idealistic interpretation of Lorentz symmetry leads to the four-dimensional construct of Minkowski space.

Due to the symmetry of the Lorentz invariance, the influence of an absolute motion relative to an assumed rest system of the Aether can be eliminated in special relativity theory precisely because a transition from an inertial system S to the æther system A, and from A to another inertial system S', by means of a Lorentz transformation for each case, can be replaced by a single Lorentz transformation (between S and S', without reference to A). The velocities of the two systems S and S' relative to A are thus eliminated and therefore also the absolute motion of the two systems S and S' relative to the æther A as the material cause of the Lorentz contraction and the slowing-down of clocks.

Thus, from a positivistic point of view, absolute speeds of both systems relative to the æther seem to be superfluous. This, however, is valid only as long as one is not interested in the physical mechanisms.

When limited to relative velocities, effects of clock deceleration can be attributed either to the system S or to the system S', since they are apparent effects of the observer at the other velocity. This leads to the twins paradox: Each of the two twins, which moves relative to the other at high speed, could rightly say that the other has stayed younger by "slowing down the clock." Only a mechanistic study of the causes of the change in the speed of clocks at high speeds resolves this paradox.

Conflicting interpretations of the special theory of relativity

The effect of the positivist world view in physics led to the fact that the special theory of relativity was not regarded as a first step forward in order to describe the different phenomena uniformly and to temporarily dispense with an explanation of the material causes of the "relativist effects." The prevailing doctrine became that relativity theory had made a medium of electromagnetic interaction (and other quantum fields) superfluous. It was replaced by the "relativity of space-time."

Einstein was also influenced by positivism, which is expressed in his work "Grundzüge der Relativitätstheorie" [Einstein 1922, p. 5]: *"All science, be it natural science or psychology, seeks in a certain way to order our experiences and to bring them into a logical system"*. Accordingly, he wavered in the question of the existence of an æther. While in 1911 he still stated: *"The theory outlined below is not compatible with the æther hypothesis"* [Einstein 1911, p. 2], he changed his attitude and influenced, among other things, by an intensive correspondence with Hendrik Lorentz. In the course of the First World War, Einstein also became increasingly critical of idealism. In a letter to Lorentz in 1916, Einstein wrote as follows: *"The general theory of relativity is closer to the æther hypothesis than the special theory of relativity. However, this new æther would not violate the relativity principle, because its state would not be that of a solid body in a state of motion independent of matter, but its state of motion would be a function of position, determined by material processes."* [according to Kostro 1998, p. 137] The rigid æther concept of the mechanistic world view of the nineteenth century, with its mechanical properties, began to develop into a dynamic material structure in constant motion and change.

Einstein went through further metamorphoses in the question of the materiality of space. In the 1930s, he again held an idealistic position: *"We may continue to use the word æther, but only to express the physical properties of space"* [Einstein and Infeld 1938, p. 204]. In contrast, in the 1940s and 1950s he developed clearer materialist ideas: *"Space-time has no existence of its own as something independent, but only as a structural property of fields (. . .). an empty space, i.e. a space without a field, does not exist."* [Einstein 1960, p. 137]

In the Soviet Union during its socialist period, there was a broad debate about the interpretation of Einstein's theory, ranging from defense (L. Landau) to crudely materialist rejection e.g. by A. Maximow "Against reactionary Einsteinianism in modern physics" (see for example [Thiessen 2009], who rightly criticisms the lack of dialectics in this attitude). In a 1942 study, W. Dickhut, a theoretician of the workers movement in Germany, developed a critique of the geometrization of matter:

> You can't see the forest of reality for all the mathematical spaces. The mathematical formulas become the primary, the objective reality the secondary. (. . .). Thus, Minkowski's space-time world is an unreal super-constructive structure, an empty formula framework of mathematical symbols. Space is only real if material, filled with substance, with matter. Space itself, pure abstract space, is unreal.
> [W. Dickhut 1987, p. 289/290]

Among the important critics of the geometrization of matter are Herbert E. Ives [in: Hazelett 1979], the discoverer of the slowing of atomic processes, and the Italian physicist Franco Selleri [Selleri 1996, 2004b]. Their materialistic interpretation of the law of Lorentz invariance consists of the following theory:

- The speed of light is an expression of a natural oscillation of the æther and is constant relative to it.
- For an observer moving relative to the æther, the speed of light is changed, but he is not able to measure this change based on two-way experiments (Michelson-Morely interferometer), unlike one-way experiments (Sagnac effect). This is due to the following two effects:
  - Through the motion of bodies relative to the æther, a Lorentz contraction, that is, a shortening of atomic distances in the direction of motion is initiated, probably caused by a change of electromagnetic forces.
  - The frequency of atomic and subatomic processes is also slowed by a motion relative to the æther.

Limits of the Lorentz invariance

The dialectical method must detect in the constant aspects of the phenomena the deeper laws of nature. While the metaphysical method aims to explain the effects at high speeds by means of special models, the dialectician Einstein took up important new phenomena and generalized them in an empirical law, but on an idealist basis of the relativity of space and time. The law of Lorentz invariance expresses essential new properties of the quantum æther, and it reflects its excitation spectrum, in particular the dispersion of the low-energy excitations of the subatomic particles as excitations of the quantum æther. Consequently, the effect of the motion on the spatial structure (Lorentz contraction) and on the frequencies (clock slowing) has a universal velocity dependence. The immediate expression of this is the connection between magnetism and electricity in classical electrodynamics. Magnetic fields are created by "Lorentz contraction" of the electrical field in the direction of motion or by temporal changes of the electric field.

So in the further investigation of the structure and excitations of the quantum æther, it is necessary to understand the law of Lorentz invariance materialistically as an important expression of its low-energy properties, and based on this, to find the expected limits of this law, such as in phase transitions of the quantum æther at very high energies. There the Lorentz invariance may no longer hold [Volovik 2003, p. 3].

5.3 Accelerated motion, inertia, and gravity

Accelerated movements of bodies play an important role in our everyday life. Everyone has experience with the inertial or centrifugal forces involved. But few people are conscious of the fact that inertial forces are a counterpart of gravity, and both are direct expressions of the effect of deeper structural levels of matter. The dialectic of gravitational attraction and centrifugal forces/inertia determines the formation of bound systems of masses such as the Earth–moon system, the solar system, and up to the structure of galaxies. A multitude of new discoveries in recent decades have led to a deeper understanding of the effect of gravity and inertia on all other structural levels of matter and their motions, be it the propagation of light, the modification of matter waves, or the formation of structures in the macrocosm. They are of great importance for the understanding of the development of the forms of matter in the cosmos.

Role of mass in the laws of classical mechanics

The laws of motion of particles with rest mass differ from those of mass-less particles. Today, the law of inertial motion of masses has been comprehensively confirmed. They resist a change in their state of motion in the form of inertial force. The principle of inertia is anchored in Newton's first law of mechanics: It says that a uniformly moving body retains its uniform motion, in particular a body at rest remains at rest, if no force is exerted on it. Newton's second law treats phenomenologically the fact that an inertial force arises when the state of motion of a mass is changed. With

$$F = ma,$$

the inertial force F is proportional to acceleration a and to inertial mass m. Newton's third law deals with the dialectic of forces.

"*Action is equal to Reaction*" applied to the accelerated motion of a mass under the action of a force means that the latter is always equal to the inertial force.

In addition to inertia, the mass of bodies appears in a second way in physical effects as the source of gravity. Newton's law of gravity between two masses M and m states that the force of attraction is proportional to the two masses and the inverse square of their distance apart r. G is the constant of the gravitational field.

$$F = G\frac{Mm}{r^2}$$

Due to the exclusion of "action at a distance without physical cause," Newton already concluded from the law of gravity force the existence of a gravitational field, which emanates from the masses and via which they interact. This field exists objectively and in reality, as Newton's bucket experiment (Figure 57) proves. The rest system of the gravitational field generated by the distribution of masses in the universe forms an absolute reference system for accelerated motion. Inertial forces and centrifugal forces occur during accelerations relative to this reference system.

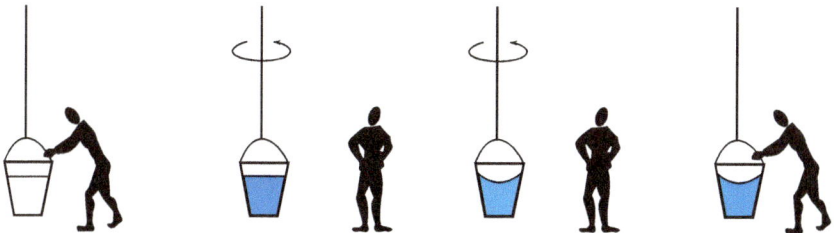

Figure 57: Newton's "bucket experiment" on the gravitational field as an absolute system of reference for accelerated movements. A bucket is suspended from a string that is by turning it. Then water is poured in and the bucket is released, which then rotates. In the beginning, the water does not rotate with the bucket, as it first has to be accelerated by frictional forces on the bucket wall. The water surface remains even. Only when the water rotates relative to the celestial sphere, and not relative to the bucket wall, centrifugal forces occur and the surface is curved. This is also the case when the bucket is suddenly held and the water continues to turn due to inertia.

Attraction and repulsion of planetary motion

The interaction of gravitational attraction and inertia is the basis for the formation of solar systems with stable planetary orbits (Figure 58). With circular orbits, gravitational attraction and centrifugal force would cancel each other out exactly at any point in time. In the actual elliptical orbits of the planets, their orbital speed also

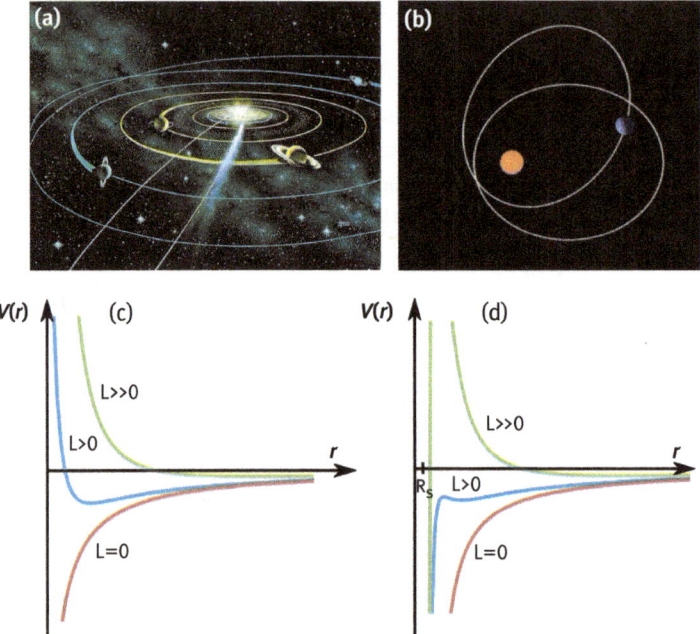

Figure 58: Stability of the solar system. (a) Artistic illustration of the solar system as seen from the outside. The elliptical planetary orbits are the result of the unity of gravitational attraction and centrifugal repulsion by inertial forces. (b) The perihelion precession of the vertex of the elliptical orbits is caused by the gravitational attraction between the planets. (c) The interaction of gravitational attraction and centrifugal repulsion produces an effective potential V(r) as a function of the distance from the Sun. With sufficient angular momentum L, stable orbits are formed in the potential minimum. (d) Modification of the effective gravitational potential at small distances r by corrections of the general theory of relativity. They lead to an additional contribution of the perihelion precession of Mercury and a gravitational collapse for radii smaller than the Schwarzschild radius $r \leq R_s$.

changes. Thus, the sum of gravitational attraction, centrifugal force acting perpendicular to the trajectory and tangential inertial forces cancel one another out.

This unity of attraction and repulsion can be expressed for $M \gg m$ in an effective gravitational potential:

$$V_{eff}(r) = -\frac{MmG}{r} + \frac{L^2}{2m\,r^2}$$

(Figure 58c). The strength of the repelling portion is determined by the total angular momentum L of the moving masses. If no other forces act, the orbital movements of the planets should be elliptical orbits whose positions in space do not change. The existence of other planets in the solar system generates additional gravitational forces among themselves, which lead to a perihelion rotation of the elliptical orbits. The vertex of the ellipse rotates every year by a small angular amount.

For Mercury, the rotation of the vertex is 575 seconds of arc per century and for the Earth, 1170 seconds of arc per century. As early as 1859, Urbain Le Verrier recognized, by precisely measuring the orbit of Mercury, that the perihelion rotation of Mercury deviated from the value given by Newtonian theory, 530 seconds of arc per century, by about 40 seconds of arc per century.

In 1915 Albert Einstein used general relativity to explain this deviation very precisely by modifying Newton's gravitational law at small distances from the sun, which was of considerable importance for the acceptance of his theory. The exact calculation of 42.98 seconds of arc gives a very exact agreement with modern precision observations by radar, which give a value of 43.13 ± 0.14 as the deviation from Newton's theory [Anderson 1992]. And there is good agreement between the potential curve of general relativity theory and the observations for the perihelion rotation of other planets or in binary star systems. The modification of Newton's law of gravity for strong fields and small distances is an expression of the fact that the energy of the gravitational field itself gravitates. This produces a nonlinearity in the inverse-square law of the gravitational field at small distances and a collapse of the theory at the length scale of the Schwarzschild radius R_s.

The deflection of light in the gravitational field

According to Newton's theory of gravity, light should be deflected in the gravitational field if photons have a mass. Even the most modern high-precision measurements show that photons have no rest mass. However, their frequency is associated with a kinetic energy and thus a mass according to $E = mc^2$. Einstein applied this connection in 1911 to the deflection of light in a gravitational field. Only the consideration of the modification of the gravitational law by the general relativity theory resulted in the double, and thus correct, value for the light deflection. The experiments that Eddington and Dyson carried out in 1919 at a total solar eclipse [Eddington 1919] to measure the deflection of light of stars in the gravitational field of the Sun confirmed the predictions of general relativity.

Further influences of the gravitational field on the propagation of light that have been experimentally confirmed today are the gravitational red-shift of photons that propagate against the attracting potential of a gravitational field. They were first interpreted as an expression of the "apparent weight of photons" [Pound 1960], which, however, would curtail the general effect of the gravitational field on all other structural levels of matter. In fact, the gravitational field not only changes the energy of particles and thus the frequency of their matter waves but also the speed of light.

Such a change of the value of the speed of light as a propagation delay of photons was first demonstrated [Shapiro 1968] by reflection measurements of Venus. This was confirmed in 2003 by investigations of the signal transmission from the Cassini space probe [Bertotti 2003]. All these effects are in agreement with predictions of the theory of general relativity, which describes the effect of the gravitational field by means of the "metric tensor" $g_{ij}(x)$ as a location-dependent "change of the refractive index" for the light propagation (Figure 59). Thus, the isotropic speed of light c in the presence of a gravitational field transforms into an anisotropic tensor parameter.

$$c \rightarrow \sqrt{g_{ij}(r)}$$

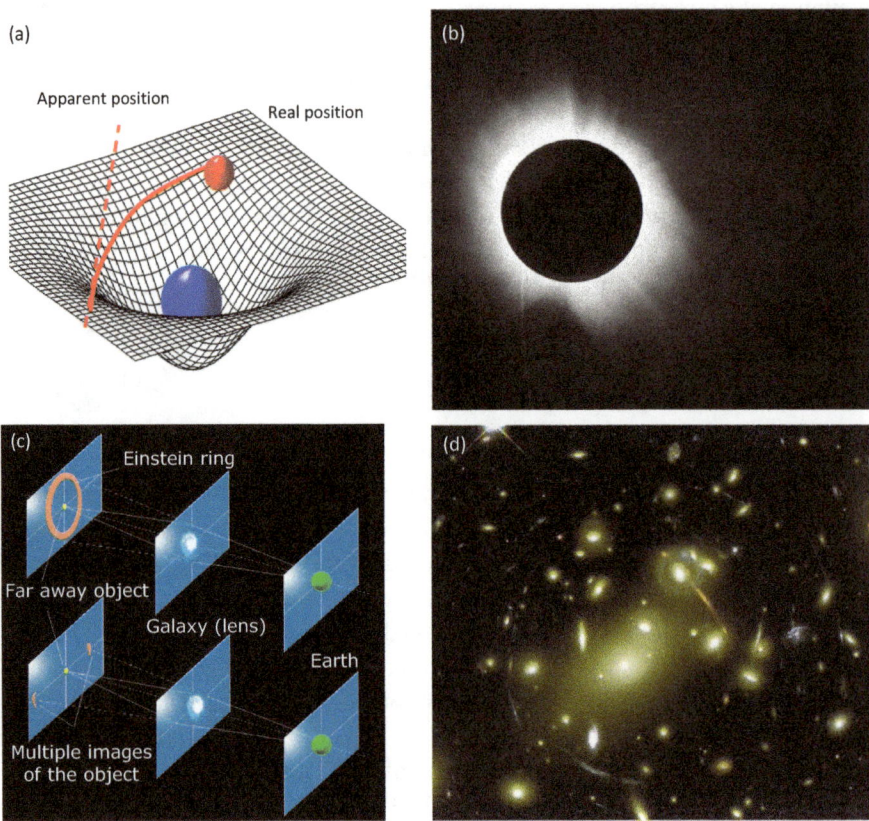

Figure 59: Deflection of light in the gravitational field. (a) The Sun's gravitational field exerts a deflecting force on photons, changing the apparent position of a star behind it. The change of the refractive index for the propagation of light is geometrized in general relativity by means of a "space curvature." (b) Photography of the total solar eclipse used by Eddington in 1919 to measure light deflection. (c) Formation of Einstein rings or partial rings by the deflection of light at massive objects. (d) Gravitational lens effect due to light deflection at the galaxy cluster Abell 2218, which is about 2 billion light-years away from Earth [NASA 1999].

Since the speed of light also depends on the material parameters of the dielectric polarizability ε_o and the magnetic permeability μ_o of the quantum æther via

$$c = 1/\sqrt{\varepsilon_0 \mu_0}$$

there must be a deep connection between the gravitational field and changes in the electromagnetic forms of motion of the quantum æther.

The identity of gravitational and inertial mass

The identity of gravitational mass and inertial mass is not self-evident. The gravitational mass determines the strength of the gravitational field generated or the magnitude of the weight of a mass in the gravitational field of other masses. The inertial mass determines its resistance to accelerations. Why should both qualities be identical? The identification of gravitational and inertial mass is the basis of Newtonian mechanics but could not be explained by it. The weak equivalence principle of classical mechanics based on the equivalence of inertial and gravitational mass states that in sufficiently small regions of space (approximately constant gravitational field) the mechanical laws of motion of uniformly accelerated bodies and those located in a gravitational field are identical. For example, the occupants of an elevator falling freely in the gravity field of the Earth cannot distinguish their situation from that of floating in zero-gravity space as long as they study only the motion of bodies inside the elevator. The gravitational field has the remarkable characteristic that it accelerates all bodies equally regardless of their mass and state of motion and is therefore independent of the accelerated bodies.

The identity of gravitational and inertial masses was repeatedly verified with ever greater accuracy. Suitable for this purpose are pendulum experiments in the gravitational field, in which both the gravitational force of the Earth and the inertial force of the acceleration of the pendulum mass act. Experiments by Newton already showed an agreement between inertial and heavy mass of better than 10^{-3}, and improved experiments by the Hungarian physicist Eötvös between 1890 and 1909 showed an agreement of better than 10^{-9} [Eötvös 1890]. Through experiments with laser reflectors set up on the moon during Apollo missions (Lunar Laser Ranging), Irwin Shapiro was able to prove the validity of the equivalence principle with an accuracy of 10^{-12} in 1976 [Shapiro 1976].

5.4 The general theory of relativity

In fact, there is a deep connection between inertia and gravity. Albert Einstein placed this connection at the center of his theory of general relativity. He postulated the strong equivalence principle, also known as the principle of general relativity. It states that in sufficiently small regions of space all laws of physics (including laws of electromagnetism, etc.) should be independent of the state of motion of the bodies. The mathematical formulation of this principle is the basis of a geometric gravitational theory, in which the effect of gravitational and inertial forces is expressed by a metric tensor, which comprehensively influences all material processes, from the locomotion of masses to the propagation of light down to atomic and subatomic processes.

In this sense, the theory of general relativity that Einstein and the mathematician Grossmann developed between 1912 and 1916 is an ingenious dialectical synthesis and answer to the shortcomings and problems of Newton's classical mechanics and gravitational theory. Newton's theory of gravitation was a theory of action at a distance of the attraction of masses. Although it formally contained a gravitational field, this had no material substance, due to the instantaneous action at a distance. The theory of general relativity took into account the propagation-time effects of the gravitational field (at the speed of light), as well as the effect that the energy present in a region of space (including the gravitational energy) is also a source of the gravitational field. Thus it not only described more precisely the structure and dynamics of the gravitational field in the vicinity of large masses and for fast movements but also predicted a number of new effects.

The comprehensive and all-inclusive effects of gravity and motion on all forms and movements of matter at higher structural levels are captured by the metric tensor as an expression of changes at a lower structural level of matter, without proposing a microscopic model of this structural level. This also affects the propagation of light; the speed of light is anisotropic in the gravitational field, atomic processes slow down, and the spatial structure of matter is contracted. Taking into account the energy of the gravitational field as a source of gravity leads to the prediction of qualitatively new states of the gravitational field at extremely large masses, up to the complete suppression of light propagation within the Schwarzschild radius of extremely massive objects, called "black holes". There, a limit of the validity of the theory of general relativity is also reached.

Absolutization of the principle of relativity and geometrization of physics

Through its geometrization of the material effects, however, the general theory of relativity promoted the spread of the idealistic world view in the natural sciences. One of Einstein's decisive driving forces was his desire for a generalization of the special theory of relativity guided by "symmetry principles." The principle of special relativity of the equivalence of uniformly moving systems was "unsatisfactory" from his point of view since it distinguished uniformly moving systems from accelerated movements. Einstein's goal was to generalize the principle of relativity in such a way that all bodies or reference systems, no matter in what kind of motion they are, should be physically equivalent [Einstein 1915]. Therefore, it should in principle not be possible (at least locally, in small regions) to distinguish between the physical effect of a gravitational field and accelerated motion. All effects that a gravitational field produces on matter, including subtle effects such as the slowing down of atomic processes, must also occur by accelerations. This led Einstein to an idealistic relativization of the material effects of motion and gravity into effects of the reference to a four-dimensional curved coordinate system.

The metric tensor changes from an effective description of material processes by means of a field to a "curvature of the space-time continuum."

Space and time are general forms of existence of matter. They exist in reality as properties of the spatial extension and processes of matter. The description by means of a coordinate system is an abstraction from the concrete forms of existence. The recognition of the objectivity of space and time by the materialistic world view is not to be equated with the recognition of the objectivity of "empty space without matter" and an "empty time without processes." The latter are pure abstractions of human thought, which progresses from concrete objects and processes to abstract space and time. Space and time are bound to matter, to structures as well as to processes. It is, therefore, a fundamental methodological error to identify abstract space and time with physical properties, as the idealistic interpretation of the general theory of relativity does.

Steven Hawking would like to derive something apparently materialistic from the geometrization of matter when he writes: *"Einstein's general theory of relativity transformed space and time from a passive background in which events take place to active participants in the dynamics of the universe."* [Hawking 2001, p. 21] However, it was not space and time but the conception of the æther that was transformed from a passive substrate to a highly dynamic, active system of matter, whose state of motion is determined as a function of position by structures and processes of matter at a higher level.

The modification of the objectivity of space and time to something subjective, dependent on the state of an observer, is a fundamental direction of physical idealism. There can be no objection to the use of curved coordinate systems to describe the dynamics of material continua, which are a very powerful mathematical tool in the description of fields. However, to declare these curved coordinate systems to be the reality and to replace the gravitational field by relations of coordinates depending on the state of motion of the observer, is idealism.

Problems with the conservation of energy of the gravitational field

According to the Einstein equations, the metric tensor depends not only on the mass content but also on the energy–momentum content, that is, the movement of the system. But in the question as to which parts of the different motion forms of a matter system contribute to gravity, the general theory of relativity shows inadequacies; this is reflected in the debate about the structure of the energy–momentum tensor of this theory, which continues to this day. [Landau & Lifshitz 1939, Baryshev 2006, Mannheim 2006]. In Einstein's theory, the latter is the source of the gravitational field and contains not only the density of matter but also its kinetic energy, momentum, and thus the pressure. In 1939, Landau and Lifshitz showed that a theory that fulfilled the principle of general relativity could not be consistent with the demand for conservation of energy–momentum. This means that a gravitational field, and thus its energy–momentum content, cannot be transformed away arbitrarily by the choice of a corresponding accelerated coordinate system. The energy–momentum tensor of a system is not independent of its state of motion. To deny this in order to absolutize a "symmetry principle" dogmatically is an ideological expression of idealism.

The unresolved problem of the mutual relationship between gravitation and the energy–momentum content of matter has considerable theoretical consequences. In theory, for example, singularities (infinities) occur in very large gravitational fields. This would not be a problem if these singularities were simply accepted as the validity limit of the theory. Paradoxical contradictions, however, develop when two "effective æther theories," quantum theory and general relativity theory, are combined mathematically:

If one assumes that the entire energy density of the zero-point field gravitates according to Einstein's equations, the gravitation and thus the "space-time curvature" in the geometric theory would be unimaginably large – in stark contradiction to observation. Moreover, the fluctuations of the zero-point fields would be associated with extreme fluctuations of the "space-time continuum," some scientists speak of a "foam-like space-time." Section 5.5 deals with the connection between zero-point fields and gravity or inertia. It is precisely the length and time scales of the spatial and temporal structure of the fluctuating zero-point field at which the "effective" theories lose their validity and must be replaced by a microscopic æther theory.

Einstein also saw problems with the question of which energy contributions gravitate. The presumptuous attempt to apply his general theory of relativity to the "entire universe" initially resulted in a finite universe, no larger than the Milky Way. In order for it to be stable and not collapse immediately under the attraction of the masses, he had to introduce a hitherto unknown repulsion force. This was what is called the "cosmological constant," which was supposed to represent the energy of the "vacuum." After the advent of the Big Bang theory, he called this ad-hoc assumption "his greatest folly."

The Big Bang theory was presented by Friedmann in 1922 as a further solution to the equations of general relativity, abandoning the repulsion. For the calculations to be mathematically feasible, very idealized assumptions had to be made such as "homogeneity" and "isotropic" of the universe. Already at that time, there were many observations which disproved such simplifying assumptions (compare e.g. [Lutz 1991]).

Dialectics of attraction and repulsion in galaxies and superclusters

A proper law of gravity must determine the relative stability of galaxies and forms of matter on the next higher structural levels of matter as a unity of opposites. However, the law of gravity of general relativity contradicts the observation of the gravitational bonding of masses in large structures in space. While it correctly describes the structure and dynamics of our solar system except for smaller anomalies in the outer solar system [Turyshev & Anderson 2004], larger and larger errors occur on higher structural levels of matter.

For example, the study of the rotational speed of the stars in the spiral arms of galaxies (see Figure 60) shows that the mass of the luminous matter of the stars and of the interstellar gas of a typical galaxy is far from being able to explain the gravitational bond of the galaxy. While both Newtonian and general relativity expect the rotary speed outside the galactic disk to decrease with an inverse square law, the opposite is true: the rotational curves are very flat and sometimes even decreasing with distance. The question of the dialectic of attraction and repulsion of matter is raised on the next higher level of structure formation.

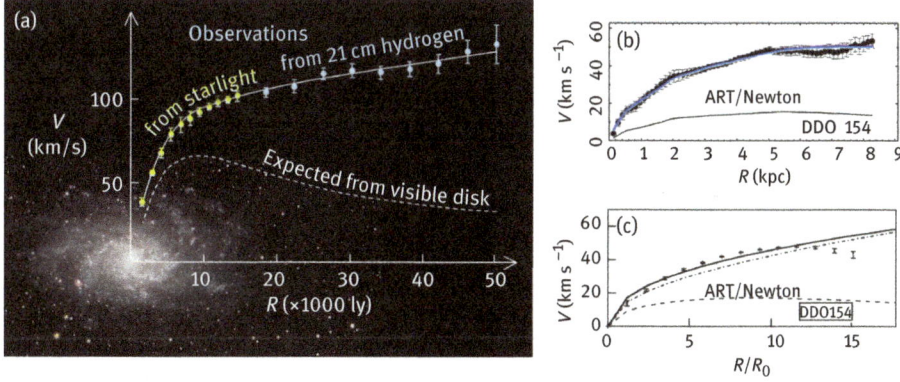

Figure 60: Contradictions of the general theory of relativity (GTR) to the stability of galaxies by gravitational binding. (a) Measured rotary speed of a typical galaxy (M33) as a function of radius compared to the theoretical expectation based on the law of gravity of the GTR, which under these conditions is identical to Newton's law, and the mass of visible matter as the source of gravity. (b + c) Successful description of the rotation curve of a typical galaxy (DDO154) by modified gravitational laws (b) Modified Newton Dynamic (MOND): blue points and (c) conformal gravitation: solid line (dot-dash line: linear approximation). R_0 is the extension of the visible optical disk of the galaxy. Images: (a) S. de Luca (b) B. Famaey (c) Reprinted from P. D. Mannheim, Prog. Part. Nuc. Phys., 56 (2006) 340 with the permission of Elsevier.

In order to solve this contradiction between theory and observation, "dark matter" must be introduced if the gravitational law is to be retained, that is matter which compensates the missing gravitational force of the luminous and thus visible mass fractions of the galaxies. The alternative is a modification of the law of gravity.

In principle, there can be no objection to the search for signs of new forms of matter. For example, already known forms of matter such as neutrinos could contribute to gravitational attraction by their nonvanishing rest mass or new, previously unknown forms of matter could actually occur. In order to explain the observed rotation curves, however, this "dark matter" would have to form a halo, whose main mass is arranged outside the visible luminous mass distribution of the galaxies. This poses serious problems for the stability of such rotating mass distributions and requires for each galaxy a "fine-tuning" of the parameters characterizing such a "dark mass distribution" [Mannheim 2006].

The unbiased researcher becomes more and more suspicious when more and more "missing dark matter" arises at ever larger matter systems. On the scale of galaxy clusters, the gravitational effect of the "dark matter" to be introduced would already be ten times that of luminous matter. For a finite cosmos, as calculated by the general theory of relativity in the form of the Friedmann solutions of an "expanding space-time bubble", the discrepancies at cosmic length scales become extreme: it must be assumed that nearly 30% of the universe consists of "dark matter" and nearly 70% of additionally assumed "dark energy" in order to obtain a correspondence of the cosmic expansion assumed from the red-shifts with Friedmann's solutions. Less than a 1% share remains for the actually observed "bright" matter. The dominant idealistic world-view of the cosmos could hardly isolate itself more blatantly from the scientific method of investigation through observation.

The theory of dark matter was already shaken further in the summer of 2007 when the observation of the colliding composite galaxy cluster Abell 520 revealed that the galaxies were positively being repelled from the region of "dark matter" [Mahdavi 2007]. Conversely, in regions with the highest concentration of galaxies, no significant accumulation of "dark matter" was detected by the gravitational lensing effect.

This manner of introducing "dark matter" also ignores the real observed matter in the microcosm, such as the zero-point field. Instead of being interested in the connection of gravitation with the quantum æther at all, "new matter" is introduced ad-hoc in order to save laws whose validity has not been demonstrated at all on the length scales of the macrocosm. This is extreme scientific arbitrariness.

Changes of the law of gravity at larger structural levels of matter

Dialectical materialism understands laws of nature such as gravity as the expression of regular forms of motion, which can change with the self-organization of matter on different structural levels. There is no reason to assume an exception for the law of gravity – that it is valid at all levels of matter.

A theory known as Modified Newtonian Dynamics (MOND) offers a systematic description of the deviations of real gravitation from the Newtonian (and thus, in this case, the theory of general relativity). It assumes that below a universal critical strength, the inverse-square law of the gravitational force of a point mass is modified from $1/r^2$ to $1/r$, and thus decreases less at large distances. This phenomenological theory can correctly describe the gravity of large objects such as galaxies and clusters of galaxies, without assuming "dark matter", with a single new natural constant (see example in Figure 60b). On the other hand, it cannot indicate any material mechanism for the "amplification of gravity" at small accelerations. Rather, such a modification would have to result from an improved microscopic theory of gravity.

An alternative to Einstein's general theory of relativity was already developed in 1918 by Herrmann Weyl with the "Conformal Theory of Gravitation" [Weyl 1918], in an effort to unite gravity with electromagnetism. It solves the problem of energy–momentum conservation, reproduces many observed effects in the gravitational field, and is able to correctly describe the rotation curves of galaxies in Figure 60. With a modified inverse-square law of gravitation at great distances, it is also able to explain the Hubble law of the cosmic red-shift of photons without an expansion of space [Mannheim 2006] (see Chapter 7). However, there are indications that this theory leads to deviations from Newton's law at the length scale of laboratory experiments, and is thus in contradiction to experiment [Yoon 2013].

This shows that modern physics is far from understanding the modification of the law of gravity at the different structural levels of matter. Like other laws of nature, the law of gravity will evolve from the smallest length scales of the microcosm to length scales of galaxy clusters and beyond as an "emergent law." Wanting to apply a gravitational model directly to the whole universe, in order to calculate a Big Bang, is more than overestimating one's own capabilities. It serves only to develop a desired world view that does not originate from natural science but from the ideological needs of society.

5.5 On the origin of inertia and gravity in the quantum æther

In this section, we want to discuss which theoretical conclusions can be drawn from the known excitations of the quantum æther about the unity of gravitation and inertia and their microscopic origin.

Inertial forces due to the effect of zero-point fields

The experimentally proven real existence of zero-point fields as unstable excitations of the quantum æther obviously raises the question of what effects they have in accelerated movements and in a gravity field. The identification of the gravitational field via the metric tensor with an anisotropic change in the propagation of light reveals a deep connection between inertia, gravity, and electromagnetic phenomena.

In the theoretical works of Haisch, Rueda, and Dobyns [Haisch 1999, Haisch 2000, Haisch 2000b] it is shown that under an accelerated motion, the Lorentz invariance of the spectral density of the zero-point excitations no longer applies. A higher mode density of the zero-point excitations occurs in the direction of motion than in the opposite direction.

This results in an energy and momentum flow of the zero-point excitations to the accelerated particle, which produces a resistance force against the change in the state of motion. This inertial force has the characteristic that, for velocities far below the speed of light, it does not depend on the velocity but only on the acceleration.

So far, there is no experimental evidence for this origin of inertia. It could only be demonstrated if the excitation spectrum of the zero-point field were suitably modified to study a change in inertial force. However, this theoretical direction is promising because the study of the interaction of masses with the zero-point field would lead to a uniform mechanistic understanding:

- With uniform motions (see Chapter 4), matter waves arise from the interactions of particles with the zero-point field as stable organizational states of the fluctuating excitations. For increasing mass and decreasing de Broglie wavelength, they produce an inertial motion in the borderline case of the transition from quantum mechanics to classical laws of motion and explain Newton's first law.
- With accelerated motions, the "empty space" of Newtonian mechanics is transformed into an entity full of forces, into an absolute space, which with its enormous activity of unstable excitations opposes a reaction force, in the form of inertial or centrifugal forces, to any kind of change in motion. The change of motion arises as a developed contradiction between accelerating and inhibiting forces. This must, of course, express itself at the microscopic level in a transformation of the matter-wave state. With accelerated movements, the de Broglie wavelength of the matter wave is modified by means of continuous or abrupt phase shifts of the matter wave.

Is there a connection between gravity and the zero-point field?

A direct relationship between matter waves and gravitation is demonstrated by interference experiments with neutrons passing through different gravitational potentials [Rauch 2000]. They show that matter waves in the gravitational field undergo a phase shift that can be detected by a change in the interference pattern of the neutrons. A phase shift of the matter waves of the neutrons is an indication that the zero-point field itself undergoes a modification in the gravitational potential, and that its excitation spectrum may change gradually.

In recent years, a very old theory of gravitation by Le Sage (1724–1803) has been taken up again [Edwards 2002], which in 1748 already attempted a microscopic explanation of Newton's law of gravitation. It was based on a mechanical æther model, which assumed that all space was filled by a gas of rapidly moving particles. Two macroscopic bodies attract each other because they cause a mutual "shadowing effect" of the particle stream. The pressure of the "non-shaded" particle stream would create an effective force of attraction (Figure 61b).

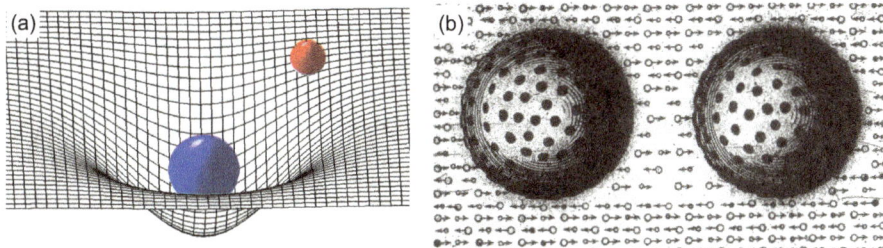

Figure 61: Comparison of the geometric theory of gravitational attraction according to the general theory of relativity (a) and the theory of gravitation by anisotropic flows of particles according to Le Sage (b).

According to this theory, gravity is not a real force of attraction, but the reflection of the anisotropic pressure of a particle stream on the bodies. Such a theory reproduces Newton's law of gravity with a decrease of gravity by the square of the distance of the bodies. By considering the finite velocity of the particle stream, "runtime effects" arise, which can correctly explain the progression of the perihelion of Mercury. The metric tensor of general relativity in this model would be nothing more than a geometric expression of the existing anisotropies of the particle stream filling all space. [Edwards 2002, p. 137 ff]. But a mechanical theory of gravitation by particle flows leads to paradoxical contradictions: bodies would have to heat up from the impacting particles; for planet Earth this would lead to a warming of several billion degrees [Poincare 1908].

So the question of the possible nature of the particle flow filling all space is debated hotly. Some physicists believe in a new type of particle, called "gravitons." An obvious idea would be to study the connection of these particles to the unstable particle excitations of the various quantum fields, whose existence has been proven. According to the views of Haisch [2000], a mass would cause an energy and momentum flow density of the zero-point excitations, just like an acceleration. An effective momentum flow of the zero-point excitations directed at the source of the gravitational field results just from the geometric effect of the occupation of a certain part of the volume of the quantum æther by massive bodies. They would reduce the density of the zero-point excitations locally.

In a fully developed theory, such an anisotropy in the flow of the zero-point excitations should be able to explain the effects of the gravitational red-shift of light, the curvature of light rays in gravitational fields, or the clock deceleration of atomic clocks via their influence on the matter-wave fields.

The details of processes may be more complex than treated in the model of Haisch and colleagues. Its revolutionary core, however, is to demystify the property of inertia and weight of masses from a property of "bodies per se" to a result of interaction with the movements of a deeper structural level in which mass distribution and zero-point movements of the quantum æther are interdependent. There may be another effect of accelerated motions in the zero-point field, the Davies–Unruh effect. Davies and Unruh theoretically predicted that an accelerated motion in the zero-point field leads to black-body radiation and thus to an increase in temperature [Davies 1975, Unruh 1976]. The effect is tiny at accelerations of 10 m/s^2, such as in the Earth's gravitational field, (in the range of 10^{-20} K), but it is currently being searched for experimentally.

Quantum fluids as model systems for inertia and gravity

The investigation of the change of the excitation spectrum of quasi-particles (collective excitations) in quantum fluids opens up further deep insights into the effect of motion and gravity on an "internal inhabitant" [Volovik 2003]. In this way, the material causes of the effects described geometrically in the special and general theories of relativity become clear. This also illuminates their role as "effective theories" of the quantum æther.

Let us imagine a universe all of whose matter is formed by a quantum-fluid medium, the quantum æther, and its diverse excitations. For simplicity's sake, in Figure 62 we consider only unstable and stable photons, which in simplified form are both subject to the same linear dispersion. Let an "internal inhabitant" be composed of matter of stable excitations (e.g. topological defects). He sees photons propagating at the speed of light. The energy–momentum transfer of the zero-point field to the internal inhabitant will average out on sufficiently large time and length scales (Figure 62a).

With uniform motion with a velocity v, the dispersion of the excitations changes as seen from the moving system of the observer. This makes the speed of light of the photons arriving with and against the direction of motion different. However, the mode density of the zero-point field in and against the direction of motion does not change for a uniform motion. For all effects that depend on the mode density of the zero-point field, a uniform motion relative to the superfluid background would therefore not be detectable for the internal inhabitant, and the law of motion would be subject to the "special relativity principle." For velocities approaching the speed of light (speed of sound of the quantum fluid), modifications would occur which, however, only reflect the dispersion of the excitations.

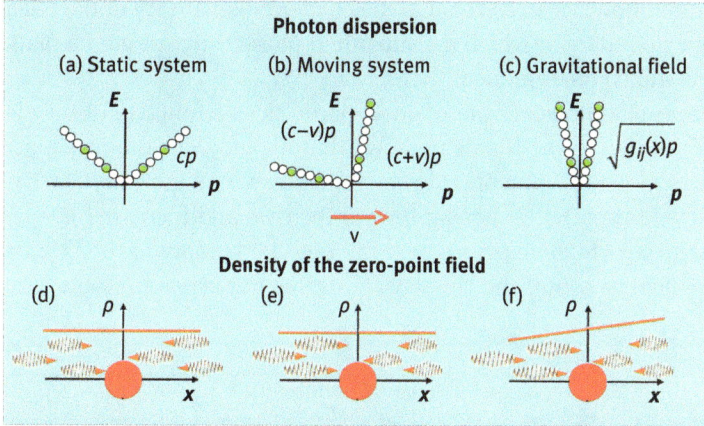

Figure 62: Modification of the excitation spectrum of a quantum fluid by motion and gravitation using the example of photons. (a) Linear dispersion (energy–momentum dependence) in the reference system at rest of the quantum fluid. c is the propagation speed of the excitations (speed of light). (b) Change of the propagation velocity by uniform motion with velocity v relative to the system at rest of the superfluid condensate. The resulting velocity of the excitations decreases or increases in and against the direction of motion respectively, but not their density (d, e). (c) Gravity manifests itself as a local anisotropic change in the speed of light, for example by superfluid currents. Below: Density of the unstable quasi-particle excitations ρ for an "internal inhabitant" of the superfluid cosmos at rest (d), uniformly moving (e) and accelerated motion.

A gravitational field changes the speed of light. The slope of the dispersion now becomes a function of the metric tensor, which describes the anisotropic propagation of light through the gravitational field. With the location-dependent speed of light (change of the slope of the dispersion), the density of the zero-point field also becomes location-dependent, and its momentum flow density anisotropic. This leads to an energy–momentum transfer to an object. It also applies to accelerated movements of the object with the resulting inertial forces. The identity of inertial and gravitational forces in this concept is the result of the uniform effect of a flow of unstable excitations on stable excitations and forms of matter.

Event horizons and black holes in quantum fluids

In quantum fluids, changes in the dispersion of quasiparticles occur due to movements of an "internal inhabitant" relative to the homogeneous superfluid condensate. In the vicinity of topological defects (particles with mass), gradients of condensate density also occur in connection with superfluid flows. One example is a vortex filament in superfluid helium. The flow velocity increases toward the center of the vortex filament until the critical limit of the Landau velocity is reached, at which the kinetic energy of the superfluid flow becomes equal to the condensation energy of the quantum fluid. This is the boundary at which the superfluid phase collapses and a phase boundary to a normal-fluid core is formed.

The study of the excitation spectrum of quasi-particles as a function of the distance from the core of the vortex filament shows that a change in the dispersion occurs for particles in directions of motion toward and away from the core. The particle energy is reduced for particles moving toward the core and increased for particles moving outwards. At high velocities near the phase boundary to the normal-fluid core, even the energy gap of the condensate begins to decrease (Figure 63c).

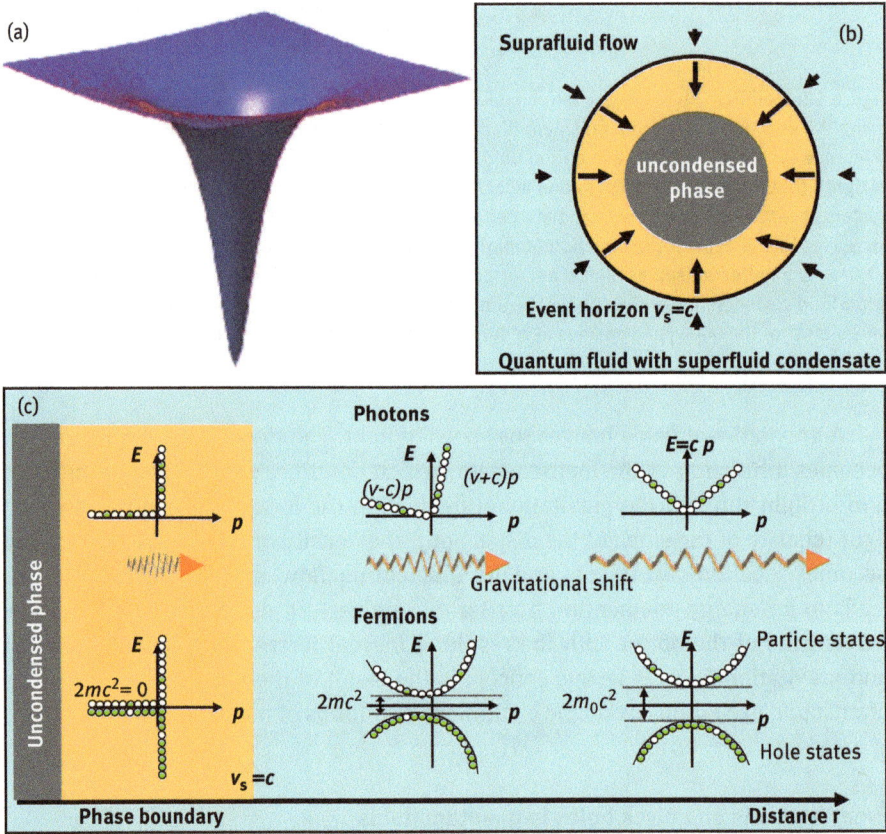

Figure 63: A black hole as a "singularity of the space-time continuum" (a) or phase boundary in the quantum æther (b). When the Landau velocity $v_s = c$ of a superfluid flow with velocity v_s is reached, an event horizon develops, since quasi-particle excitations cannot escape from the interior. (c) Change of the excitation spectrum of photons and fermions (energy E via momentum p) as a function of the distance to the event horizon. The excitation spectrum of the particles changes as a function of their direction of motion when approaching the "black hole." The Dirac energy gap for particles with mass decreases and disappears beyond the event horizon.

When the Landau velocity is reached at the phase boundary, the energy gap disappears and the slope of the dispersion for escaping quasi-particles becomes infinite. Near the phase boundary to the core of vortex filaments, the condensed fraction is constantly converted to a normal phase and vice versa. This results in a radial component of the superfluid flow, compensated by the reflux of quasi-particle excitations. This is a general property of topological defects in condensates.

At the phase boundary, something qualitatively new, an "event horizon", arises for the "internal inhabitant" of the quantum fluid. If the velocity of the superfluid condensate exceeds the maximum achievable velocity of its excitations at the Landau velocity, the quasi-particle excitations are inevitably drawn across the phase boundary. The "internal inhabitant" could not obtain any information by observing the excitations from the region of the normal-fluid core, and would perceive all regions beyond the "event horizon" as a "black hole" [Volovik 2003]. In this model, the gravitational potential in the quantum fluid is directly connected to the kinetic energy of the superfluid condensate. The change in the dispersion of the unstable and stable quasi-particles also influences all temporal processes of "quasi-particle matter" (clock deceleration and gravitational red-shift).

Conclusions

Applied to the quantum æther of our real cosmos, the following conclusions result:
- The change and anisotropy of the speed of light in the gravitational field is an expression of the change of the dispersion of all excitations of the quantum æther.
- The metric tensor g_{ij} is therefore, an expression of the change of the dispersion, probably not only of the photons, but also of all other excitations of the quantum æther in the gravitational field.
- In electromagnetic excitations, the gravitational field becomes identical to an anisotropy of the speed of light and thus to a location-dependent, anisotropic "refractive index" $n_{ij}(r) \sim \varepsilon_0(r)\mu_0(r)$.
- In very strong gravitational fields, the value of the speed of light can become zero, which denotes an event horizon and a phase boundary to another aggregate state of the quantum æther.
- The change of the dispersion for particles with rest mass in a sufficiently strong gravitational field is accompanied by a decrease of the Dirac energy gap of the particles $E = 2m_0c^2$, which disappears within the phase boundary.
- In Einstein's geometric gravitational theory, the phase boundary represents the validity boundary, a "singularity" of the theory. It indicates a new state of matter.

In astrophysics, the experimental evidence for the existence of extremely massive, non-luminous bodies in the centers of galaxies is increasing. Any over interpretation of geometric gravitational theory beyond its validity limits, either in very strong gravitational fields, or for very small length scales in the microcosm, or on cosmic length scales, is unmaterialistic. The eerie conclusions derived from this, as to the existence of *"wormholes in space-time"* or an *"expanding multi-dimensional cosmos"* are ideologically motivated, constructed concepts of the world. They have nothing to do with the real cosmos.

The new concept of forces that emerges in modern physics by overcoming the dictate of geometrization of physics is the dialectical negation of the concept of force by an incipient understanding of material processes in the quantum æther. Friedrich Engels already criticized the limited value of the concept of force:

"Not because we have completely recognized the law, but precisely because this is not the case, because we are not yet clear about the rather complicated conditions of this phenomenon, precisely for this reason, we sometimes resort here to the word 'force'. So, we are not expressing our science, but our lack of science of the nature of the law and its mode of action. In this sense, as a short expression of a causal nexus that has not yet been established, as a makeshift of language, it may be passable in everyday use. Anything more than that is an abuse." [Engels 1885, p. 365]

6 Self-organization in the microcosm: the zoo of the "elementary particles"

Almost all forms of matter known today consist of discrete particles and continuous fields. Bodies consist of atoms, atoms of electrons and nuclear building blocks, and so forth. So far, new subcomponents and forms of motion have been discovered in every newly discovered deeper structural level of matter, which was initially regarded as "elementary." Can a "lower end" of the structure of matter in the microcosm actually be expected? Or is "*The electron is as inexhaustible as the atom, nature is infinite, but it infinitely exists,*" as Lenin formulated it in his main philosophical work *Materialism and Empirio-Criticism* [Lenin 1908, p. 262].

In fact, in addition to the electron and the nuclear components proton and neutron, a "zoo" of more than 200 other particles is known today, whose masses (expressed in the energy units E/c^2) cover a huge range from eV/c^2 to 200 GeV/c^2. This diversity of subatomic particles and their mutual transformation is, according to current knowledge, the result of the self-organization of a relatively small number of "elementary" basic building blocks: 6 leptons and 6 quarks, as well as continuous fields with their excitations such as photons, gluons, and weakons, which mediate their mutual transformation. However, experiments over the last 50 years in high-energy physics have shown that no end has been reached here, either. To interpret the experiments, not only the zero-point field with unstable particle excitations already discussed, but also other deeper structural levels of matter, such as superfluid Higgs and gluon condensates, must be assumed as part of the quantum æther. From a materialistic point of view, the unity of discrete matter (excitations, particles) and continuum (fields, condensates) is fundamental for an understanding of matter, its evolution and transformation.

The prevailing positivist world-view takes the existing quantum field theories of electromagnetic, weak and gluonic forces of nature and seeks their unity through mathematical constructions by means of symmetry. However, all fields in these effective theories have a "vanishing vacuum expectancy value," so that the existence of superfluid condensates is artificially grafted onto the theory. The unity and differentiation of the "elementary particles" and natural forces in the forms of motion of the quantum æther is replaced by the symmetrization of the particles: Theories on which thousands of physicists worked over decades, such as "supersymmetry" or "grand unification," postulated for each quark or lepton observed a "symmetrical partner." These are all supposed to have occurred equally in the Big Bang. However, the symmetrization approaches to the particles have been falsified since 2012 by the absence of the predicted "supersymmetric" particles in the high-energy experiments at CERN. With the crisis in high-energy physics, the crisis in big-bang theory has therefore also deepened further.

In contrast, the dialectical-materialist development theory of "elementary particles" through the collective self-organization of superfluid forms of motion in the quantum æther has made significant progress. The differentiation of electromagnetic,

https://doi.org/10.1515/9783110644203-006

gluonic, and electroweak motion forms in the quantum æther is interrelated with the formation of quantized particles as highly structured topological defects with local phase transitions. Without recognition and further investigation of the real existence of the quantum æther and its internal motion patterns, the resulting crisis in particle physics cannot be resolved.

6.1 The discovery of the "zoo of elementary particles"

The existence of different chemical elements with qualitatively different properties was successfully attributed at the end of the nineteenth and beginning of the twentieth century to different numbers of atomic components in the atomic nuclei and the electron shell. Physics then increasingly turned to the investigation of the structure of the atomic nucleus and of cosmic-ray particles. It turned out that there is a whole zoo of "elementary particles," but which in the course of further research almost all turned out not to be "elementary."

The structure of the atomic nucleus

As early as 1919, Marschen and Rutherford proved that the positively charged proton was a component of the nucleus by bombarding atoms with alpha rays. But the assumption that the atomic nuclei consist solely of protons resulted in a completely wrong atomic weight. On the other hand, the positively charged protons accounted for the entire electrical charge of the atomic nucleus. Therefore, Rutherford postulated the existence of additional, electrically neutral components of the nucleus, the neutrons. In 1932, they were detected by Chadwick by bombardment of beryllium with helium nuclei, for which he received the Nobel Prize in Physics in 1935.

The discovery of radioactivity by Becquerel, Mr. and Mrs. Curie, and Rutherford in the years 1896 to 1898 inspired further investigations to better understand the structure of atomic nuclei and nuclear transformations. In 1932 Cockcroft and Walton bombarded lithium with protons with a kinetic energy of 800 keV. Lithium disintegrated into two helium atoms. The first artificial nuclear transmutation had succeeded.

Otto Hahn, Fritz Straßmann, and Lise Meitner triggered the first artificial nuclear fission releasing energy, by neutron bombardment of uranium, in 1938. This research was continued by Hahn and Heisenberg after Lise Meitner emigrated due to German fascism. They did not shy away from attempting to develop atomic bombs for the Nazi regime, which fortunately failed. Uranium fission provided the basis for the later development of the atomic bomb in the Manhattan Project in the USA, and the nuclear generation of electrical energy in nuclear reactors.

The understanding of the structure of the atomic nucleus, which emerged by the middle of the twentieth century, revolutionized the concept of atoms:

- The atomic nucleus accounts for more than 99.9% of the atomic weight, but only 10^{-12} of the atomic volume.
- The electron shell, including the matter waves, fills almost the entire space of the atom and determines its chemical properties, but contributes almost nothing to the atomic mass.
- While atoms have a diameter of about 10^{-10} m, the components of the nucleus, the proton and neutron, have a size of only about 10^{-15} m. For the electron, a classical radius of about 2.8×10^{-15} m can be estimated. However, the electron itself is a complex object.

A contradiction in the explanation of the stability of atomic nuclei was the question of how the strong electrical repulsive forces of the protons of same electrical charge can be overcome. In 1935, Yukawa postulated a strong attractive nuclear force for this. From its short range, limited to dimensions within the atomic nucleus, Yukawa deduced a "carrier particle" of this force with a mass of about 250 electron masses, the pion. Such a particle was actually discovered in cosmic radiation in 1946.

While neutrons bound in stable atomic nuclei are stable, free neutrons decay into a proton p, electron e and antineutrino $\bar{\nu}_e$ after an average lifetime of about 10 minutes, releasing kinetic energy:

$$n \rightarrow p + e + \bar{\nu}_e$$

In the absence of a better explanation of such transmutations of components of the nucleus, the weak nuclear force has been postulated as another force that only acts at short range within components of the nucleus.

The study of cosmic radiation

The Earth's atmosphere is constantly exposed to bombardment by high-energy particles from outer space. In addition to photons, cosmic radiation consists of atomic nuclei that have lost their electron shell. This primary radiation generates electrons, muons, and neutrinos in the upper layers of the atmosphere. Positrons, pions, neutrons, protons, and other particles also occur. Compared to the radiation of radioactive elements (1 gram of radon radiates 10^{12} particles per second), their intensity is much lower with about 20 particles per second and square centimetre. Cosmic radiation was discovered as early as 1911 by V. Hess from Austria, who used an electrometer in a hot-air balloon to demonstrate the increase in the number of ionized particles with increasing altitude.

In 1932 Anderson discovered the positron in cosmic rays, the antiparticle of the electron, which had been predicted by Dirac in 1928. This was followed by the discovery of the "generation and destruction" of electron–positron pairs by transmutation of photons of sufficient energy.

The pion was discovered in 1946 by Powell in cosmic rays. There are three types of pions with different electric charges: The positively charged π^+, the negatively charged π^-, and the neutral π^0. All three are not stable but transform into other particles such as muons and photons under the influence of electromagnetic and weak interaction. Today it is known that pions are composite particles and each consist of a quark–antiquark pair. The description of the strong nuclear force as the exchange of pions is only a rough approximation. Today the strong nuclear force is successfully interpreted as gluon exchange.

Muons, the heavy (210 electron masses) "twins" of electrons, were discovered between 1930 and 1936 in Cambridge. After an average lifetime of about two microseconds, they decay into electrons and other particles under the effect of the weak nuclear force.

Study of particle transmutations in particle accelerators

Starting with the construction of particle accelerators in the 1940s, research into "elementary particles" increasingly shifted in the 1950s to the investigation of transmutation products from collisions of high-energy particles which are analyzed by particle detectors. A modern detector is visible in Figure 64.

Thus, a real "explosion" of the discovery of new particles, featuring by different masses, charges, spins, and lifetimes, began. These included the kaon (950 electron masses), the lambda (2250 electron masses), the xi, and sigma particles, all of which occur in different variants. This paved the way for the insight that there are qualitatively new properties of the subatomic particles that characterize their different states. These were designated by new quantum numbers with imaginative names such as Charm and Strangeness.

Figure 64: Photo of the interior of the Atlas detector at CERN. It is used to detect the conversion products of particles after high-energy collisions in the accelerator and to derive conclusions about the structure and interactions of subatomic components. The Higgs particle, among others, was found with this detector.

The discovery in the 1950s that there is an antiparticle to each particle was significant. Each antiparticle has the same mass and spin as its particle, but an opposite electrical charge. For example, the antiproton was observed in 1955 by E. Segré in a proton–antiproton annihilation reaction in which the two particles transform into other particles in a positive cascade of particles.

The picture of the Dirac Sea of electrons had to be extended to the nuclear building blocks and insights were gained into the laws of universal transmutation of all particles into one another. The concept of the nuclear building blocks, proton and neutron, as elementary particles was abandoned when Enrico Fermi discovered resonances in 1952. He investigated the interaction of electrically charged pions with protons by making the two particles react with one another at different kinetic energies. He discovered that in a certain energy range a strong maximum occurs in their interaction cross-section. It characterizes the interaction of particles by means of an "effective cross-sectional area" of the two particles during the collision. At first, his explanation that the pions drove the protons into an excited state (delta resonance), which returns to the ground state after an extremely short lifetime of 10^{-23} seconds, was rejected, because the proton was regarded as elementary. However, the delta excitation showed that the proton is a compound system (Figure 65).

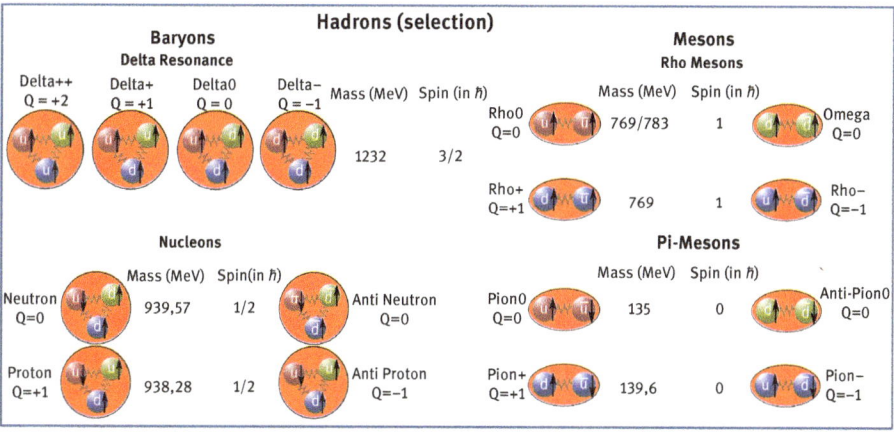

Figure 65: Structure of simple hadrons from d- and u-quarks: Baryons consist of 3 quarks and mesons of 2 quarks. Left: The ground states of the triple compounds of quarks are protons, neutrons, and their antiparticles. The delta resonance forms the first excited state. Right: The ground state of quark-antiquark pairs of d and u quarks is formed by the pi mesons, and their first excited state is the rho mesons. Q is the electric charge, red, green, and blue are their color charge, u and d denote different quark types, and ↑ their spins.

Quarks as the substructure of nuclear building blocks

The discovery of further resonances and unstable particles in the nuclear transformations forced a rethink. The first step was taken in 1960/61 by Murray Gell-Mann and Yuval Ne'eman. They proposed a **classification scheme** for the particles known at that time, which was the "eight-fold path", a name taken from Buddhism. The arrangement of the particles according to their electric charge and a newly introduced quantum number, "strangeness," led to a systematization.

With the help of the scheme, a new particle was proposed in 1962 to complete the scheme, the Ω. A few weeks later it was indeed found at CERN with the mass predicted. The classification scheme indicated an underlying substructure of matter. It paved the way for the quark model, which traced a large number of particles of the "particle zoo" back to composite systems of a few quarks. First, 3 different quarks were introduced, the u-, d- and the s-quark. Each quark was supposed to have a certain number of thirds of the electron's charge.

Further evidence of nucleon substructures was provided by the scattering experiments of high-energy electrons on protons carried out in the late 1960s at the large accelerator facilities SLAC (USA) and CERN (Switzerland). With a higher energy of the electron, the resolution of ever smaller structures becomes possible due to the smaller de Broglie wavelength. These experiments clearly showed that the proton has a substructure consisting of 3 quarks. Later experiments showed that the inner structure of nuclear building blocks is highly dynamic: In addition to the 3 quarks present on average, there are fluctuating unstable quark excitations. The quarks are held together by exchange particles of the strong nuclear force, the gluons. In a nucleon (proton or neutron) there are 3 stable quarks and a fluctuating swarm of unstable quark–antiquark pairs and gluons.

Another important experiment for the clarification of the substructure of "elementary particles" was the electron–positron annihilation reaction at high energies (Figure 66): Different conversion reactions occur depending on the energy. If the resulting high-energy photon is destroyed by a quark–antiquark generation, two particle jets are formed from its decay, producing two beams of particles in opposite directions. These experiments prove that the building block of the electron shell of atoms, the electron, can transform itself under suitable circumstances into building blocks of the atomic nuclei, the quarks.

This revealed new properties of the quarks, which were later confirmed by other experiments such as electron–proton scattering: A single free quark or antiquark cannot exist under normal conditions. It is observed in these reactions that two, sometimes even three quarks are always formed, which form particle jets in different directions as they decay into other particles, and in this way can be indirectly detected experimentally.

In 1974, with the discovery of what is called the J/Psi particle, another fourth quark, the c quark, had to be assumed. As a further quantum number, besides

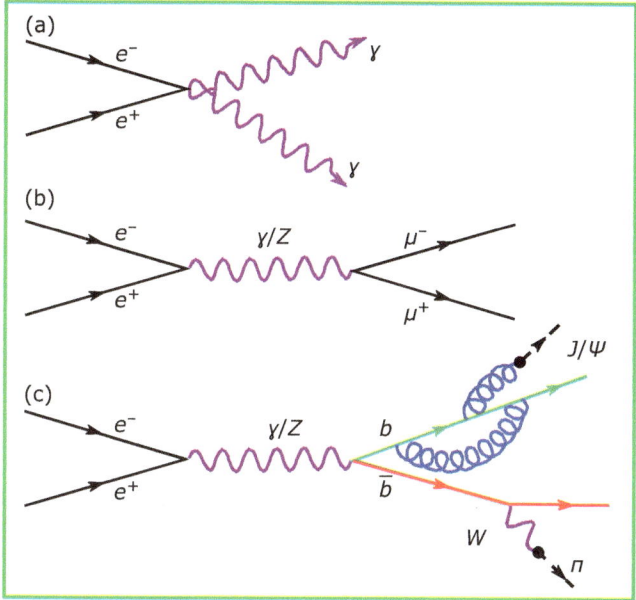

Figure 66: Universal conversion of leptons in electron–positron radiation. (a) Conversion into a pair of high-energy photons. (b) At higher energy, a new process occurs, the transformation into a muon–antimuon pair. It takes place via an unstable photon γ or a Z boson. (c) At energies in the range of 80 GeV, additional transformations into hadrons occur, the building blocks of nucleons, for example into a quark–antiquark pair b and $\bar{b}$.

"strangeness," "charm" was introduced. The formal introduction of new quantum numbers for novel particles helped with the logical ordering of the possible ground states and excitation forms of the bound states of quarks, but did not replace the necessary materialistic investigation of the cause of these new qualities as an expression of differentiation and structure formation in the microcosm.

The number of newly discovered particles virtually exploded in the years after 1974, so that today a "particle zoo" of far more than 200 different subatomic particles is known.

6.2 The Standard Model and its limits

The Standard Model, the development of which began in 1974, did not gain general acceptance in physics until the early 1980s. It attributes all presently known particles (atoms, atomic nuclei, stable and unstable subatomic particles, and resonances) to twenty-four fermions and 4 bosons (Figure 67). The twenty-four fermions consist of twelve particles and twelve antiparticles. The twelve fermions are further

Three generations of matter (fermions)

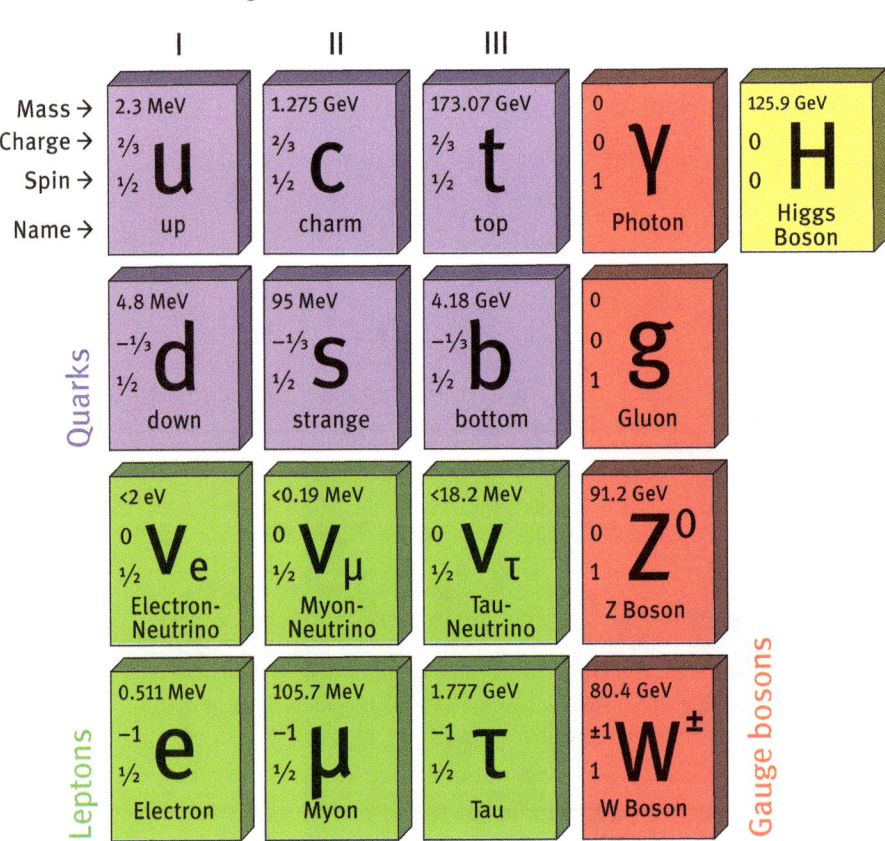

Figure 67: Standard Model of the elementary particles. The "zoo of particles" in the microcosm can be reduced to these twenty-five particles that are considered "elementary." This contains information about their electric charge, spin, and rest-mass in energy/c^2.

subdivided into a group that is not subject to strong nuclear force (i.e. do not exchange gluons), the leptons (Table 1). The other group of six particles are subject to the strong nuclear force, exchange gluons and form the group of quarks. All fermions have in common is that they have a spin of a certain number of halves of the angular momentum quantum. In this model, the four basic forces of nature are explained by bosons as exchange particles with integer spins. These are the photon (electromagnetic force), gluon (strong nuclear force), weakon (weak nuclear force), and a graviton (gravitational force), which has only been postulated until now, but which has not yet been demonstrated experimentally, and probably does not exist (see Chapter 5.5).

Table 1: Types, strengths, and ranges of the 4 known types of forces. Gravitons are still pure speculation.

Force type	Charge	Exchange particles	Relative strength	Range	Rest mass
Strong	Color charge	Gluons	1	10^{-15} m	$0–1.5$ GeV/c^{2*}
Electromagnetic	Electrical charge	Photons	10^{-2}	4×10^{25} m**	$<10^{-18}$ eV/c^{2***}
Weak	Weak charge	Weakons (W+, W−, Z)	10^{-13}	10^{-17} m	80 GeV/c^2
Gravitation	Mass	Gravitons?	10^{-39}	partially shielded?	not known

*The gluon rest mass is often simply assumed to be 0 in the Standard Model. Refined theoretical analyses of experimental data assume a mass of about 1.5 GeV/c^2 [Consoli & Field 1997].
**From Hubble law, see Section 9.4.
***Upper limit from magnetic field measurements [Amsler 2008]. From the Hubble law it follows 5×10^{-33} eV/c^2.

Its great success consists of the accurate qualitative and also quantitative description of the various excitations and transformation processes of the simple and composite particles. It has successfully predicted the existence of new particles with the right properties: the W and Z bosons, the gluon, the top quark, and the charm quark. The predicted properties were found with good accuracy. Its fundamental weakness is its focus on particles and excitations and this its ignorance of the underlying material continuum, from which particles evolve and transform.

The elementary building blocks of matter: leptons and quarks

In the family of 6 leptons, the two heavy unstable "sisters," the muon and the tauon, resemble the electron. All 3 have an integer negative electric charge (in units of the "elementary charge").

The 3 neutrinos are electrically neutral. They have a very low rest-mass (presence of rest-mass is not yet certain for the electron neutrino), and they interact only very weakly with other components of matter. For example, they are formed in large numbers in the nuclear transmutations in the Sun and every second about 50 billion neutrinos fly through a human being. There are clear indications that neutrinos periodically change their character during their propagation. During such a neutrino oscillation, for example, an electron neutrino v_e can transform into a muon neutrino v_μ and back again. Since neutrinos propagate at almost the speed of light, the conversion period is several kilometres long. In 2015, the Nobel Prize for Physics was awarded to T. Kajita from Japan and A. McDonald from Canada for their proof of neutrino oscillation in the years 1998–2002.

All 6 components of the family of quarks have electric charges with a value in thirds of the electron charge and differ with regard to other internal forms of motion and their rest-masses. The three quarks "charm," "bottom," and "top" were only detected experimentally after their theoretical prediction in 1974 (c), 1977 (b), and 1997 (t).

Three basic forces of nature and their exchange particles

The interactions between these basic building blocks include not only attracting and repelling "forces" but also mutual transformations mediated by four types of exchange particles. The strength of the "force" or of its transformation is determined by different types of charges.

The strongest "force" is the strong nuclear force transmitted by the exchange of gluons. Gluons were first detected in an experiment in Hamburg in 1979 in what is called a "three-jet event." The experimental investigation of conversion reactions between hadrons is in agreement with a quantum field theory of color charge (quantum chromodynamics) in which eight different types of gluons occur which differ in what is called their "color charge." Color charges characterize the strength of the strong nuclear force between quarks and gluons. In contrast to the electrical charge, which is characterized by a bipolarity, there are six different "color charges": red, anti-red, blue, anti-blue, green, and anti-green. This color coding is of course completely artificial and has nothing to do with the physical nature of the charge. However, it reflects the contrary nature of quark and antiquark, as well as the laws governing the action and screening of strong nuclear force approximately correctly.

The strongest force is the nuclear force mediated by gluons. The eight gluon types mediate the mutual transmutation of different quark "color charges". For example, the exchange of a red-antigreen gluon transforms a quark with the color charge "green" into a quark with the color charge "red," which creates an attractive binding force. The range of this force is very small and limited to the size of the nuclear building blocks. The formation of correct concepts, which reflect the objective reality of the processes and forms of movement on the structural level below 10^{-15} m, seems to be still quite rudimentary.

The second strongest interaction is the electromagnetic interaction mediated by photons. To the best of our knowledge at present, there is only one type of photon. They do not carry any charge and therefore do not interact directly with each other (except in media with a nonlinear refractive index). Therefore, the electromagnetic interaction transmitted by photons is long range. The Standard Model assumes an infinite range. Realistically, a finite range must be assumed, which can be deduced from the observation of a cosmic red-shift of the photon frequency.

The third strongest natural force is the weak nuclear force, which is imparted by the weakons. There are three types of weakons, the positively and negatively electrically charged W^+ and W^- bosons, and the electrically neutral Z particle. They mediate certain transmutation reactions of the leptons and nuclear building blocks with each other, which lead to the radioactive decay of nuclear building blocks. Their strength is expressed by a "weak charge." One of the achievements of the Standard Model is that a step has been made toward the uniform description of photons and weakons as two different sides of a unique natural force.

The Standard Model as a theory of particles and field-like interactions

The Standard Model describes all three natural forces (excluding gravity) successfully by quantum field theories of electromagnetic, weak and strong nuclear forces, following the model of quantum electrodynamics. They reflect qualitatively different aspects of the excitations of the quantum æther as an interacting system of particles and fields with exchange quanta. All 3 quantum field theories have in common is the occurrence of infinities in the interaction of particles and exchange quanta, which are eliminated by "renormalization." This reflects approximately the finite spatial extension of all particles with a complex structure. Thus the value of all 3 types of charge (electric, color charge, and weak charge) depends on the distance from the particle.

This means that the infinities introduced into the theory by describing the particles as infinitely small punctiform particles are eliminated by taking into account the fluctuating clouds of unstable particles (see Chapter 4.2). The "shielding clouds" of unstable excitations around a charge are self-similar at least on certain length scales, like the fluctuations at a phase transition.

The fluctuating field of the zero-point excitations causes all 3 types of interactions to be influenced by random fluctuations. The motion of nuclear building blocks and quarks also produces matter waves that change during transmutations. Each exchange of a boson becomes identical with a sudden change of the matter-wave states of the particles. In fact, the zero-point field of the quantum æther, when all forces of nature are taken into account, not only displays unstable photons but also fluctuating unstable gluons and weakons. However, their effects only become noticeable at small length scales and high energies. Zero-point excitations of the gluons and weakons have a considerable influence on particle and nucleus transmutations and are the cause of the characteristic statistical laws of radioactive decay.

The rigid distinction between particles, forces and fields in the Standard Model is an expression of the pronounced simplifications in the concepts. Depending on the situation, the particles may also play the role of interaction fields, creating forces, such as in photon–photon scattering, which is mediated by the exchange of unstable electron–positron pairs. Conversely, fields such as the nuclear force or the matter-wave fields are composed of particles that form a collective dynamic state. The great success of the Standard Model lies above all in the correct description of the probabilities of conversion and interaction as a statistical theory. However, it does not explain the mechanisms of the formation of fields and particles by self-organization processes in the quantum æther.

6.3 Structure formation by partial shielding of charges

The formation of "elementary" basic building blocks of matter is subject to the law of dynamic partial shielding of charges. For example, the "bare" electrical charge of electrons or protons is shielded by the dielectric polarization of the surrounding regions of the quantum æther. At long distances, the negative effective electron charge is smaller than at short distances. Under the aspect of electromagnetic interaction, the quantum æther therefore represents an electrically polarizable medium (Figure 68a). The electron thus consists of the bare core with a large negative electric charge, plus its polarization cloud. Due to the shielding, the bare core and its properties remain largely hidden to us, and they can only be indirectly revealed by high-energy scattering experiments.

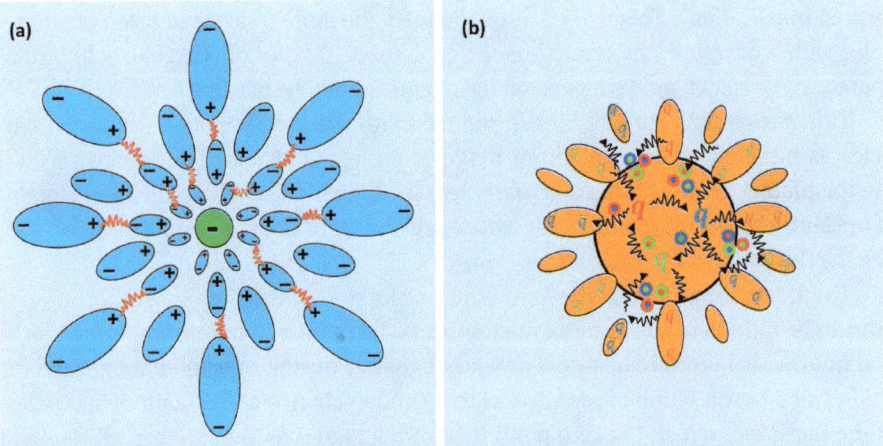

Figure 68: Shielding of electrical and color charge. (a) The bare electrical charge inside particles is shielded from the outside due to fluctuating polarization clouds of unstable electron–positron pairs and other unstable particles. The charge visible from the outside is, therefore, smaller than the bare charge. (b) Antishielding of the color charge inside a quark. Since the gluons also carry color charges, in addition to the clouds of unstable quark–antiquark pairs, self-interaction of the gluons occurs. Unstable gluons increase the color charge of the quarks.

The remaining external electrical charge of the electron is the basis for the structure formation at the next larger length scale: the organization of electrons and building blocks of the nucleus into atoms. Through the formation of atoms, the remaining remnants of the electrical charge are shielded to a large extent, so that the atom as a whole is electrically neutral. But here, too, there is room for further self-organization processes through partial shielding: Both small fluctuating dipole distributions of the electron shell as well as rearrangements of electron distributions between several atomic nuclei are the basis for the formation of molecules and solids by means of chemical bonds. This is the starting point for the structure formation of our macroscopic world.

Partial shielding of color charge

Structure formation also occurs with other types of charge in conjunction with partial shielding. The nucleons and other particles subject to the strong nuclear force are composed of quarks in such a way that their color charge is shielded and thus neutralized. Quarks thus organize themselves into "colorless" systems. These are, on the one hand, mesons consisting of quark–antiquark pairs (such as the pion), where the color charge of the quark is compensated by the antiquark's anticolor charge. Alternatively, the color charge is shielded by triple compounds of quarks, the baryons, which include the basic building blocks of the atomic nucleus, the protons and neutrons. The combination of a red, green, and blue quark (=white) creates a bound stable state of quarks in the nucleons (Figure 68b).

In this case, too, the shielding of the charge is not perfect, since the spatial arrangement and fluctuations in the color-charge distribution of the nucleons give rise to an effective residual force, the strong nuclear force. It binds the nucleons in an atomic nucleus. The further shielding of this residual force is the driving impetus for the formation of the enormous variety of atomic nuclei with different masses and electrical charges by agglomeration and organization of nucleons.

As is the case with the electromagnetic interaction, a distance-dependent coupling parameter of the strong nuclear force reflects the shielding effects of the bare color charge by clouds of unstable gluons and quarks. Thus, the effective color charge of the baryons also depends on the distance to their core. A major difference, however, is that the exchange particles of the strong interaction, the gluons themselves, carry a color charge.

This has a far-reaching consequence: unlike photons, gluons can interact directly with other gluons. While the photon–photon interaction only occurs indirectly via the electrical polarization of the quantum æther, gluons can attract each other directly and clump together to form gluon balls. This self-coupling of the gluons leads to an anti-shielding of the color charge of quarks: The binding of quarks by the exchange of gluons becomes weaker with decreasing distance and stronger with increasing distance. Consequently, quarks with partially shielded color charges in nucleons or mesons behave like gas bubbles in a liquid: The particles are held together by the external pressure of the gluon sea, as shown in Figure 69.

In a discussion about the "structure of the vacuum", the theoretical particle physicists B. Müller and J. Rafelski develop the following idea of a gluon sea:

Müller: So you think that the whole world is filled with a kind of solid body or liquid, which consists of such [gluons – author's note] lumps?

Rafelski: Yes, that's the way it probably is. We now believe that this explains why the building blocks of the elementary particles, that is, the quarks, cannot move freely in the vacuum in which we live.

Müller: The particles that are similar to photons but attract each other and stick together are very appropriately called gluons, that is, adhesive particles. But tell me, Jan, why don't we normally notice anything of this sea of gluon clusters?

Rafelski: Well, we do not see the gluonic structure of the vacuum because the gluons do not interact with photons or electrons because they are not electrically charged. They also do not interact with the building blocks of the atomic nuclei, although these are made up of quarks that can interact with gluons.

Müller: Yes, just as air bubbles can move freely in water, the movement of the nucleons is not hindered by the gluonic vacuum.

(....)

Rafelski: Well, the bubble we call the proton is made up of building blocks, just as an ordinary bubble is made up of air. And we call these building blocks quarks. The quarks would very much like to be free particles and move individually in a vacuum. But unfortunately, this vacuum is filled with gluonic "water."

Müller: Quarks, therefore, behave similarly to air bubbles under water, which would like to expand or dissolve if they were not held together by the water pressure.   [Rafelski 1985, p. 84ff.]

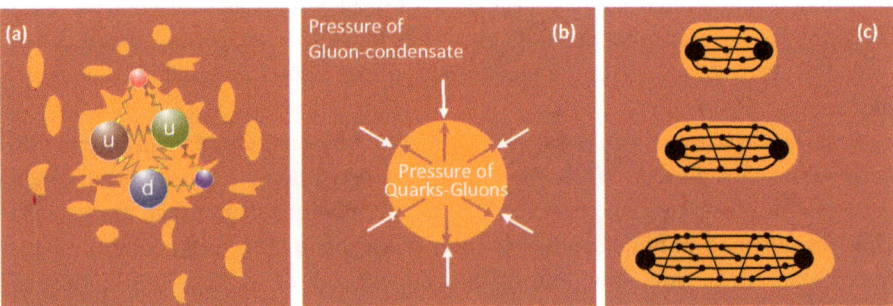

Figure 69: Quark structure of a proton and quark inclusion. (a) Bubble model of the proton consisting of a noncondensed "bubble" (orange) surrounded by the superfluid gluon condensate of the quantum æther (red). Inside the bubble, quarks and gluons are formed, whose color charges partially cancel each other out. (b) The stability of the proton is determined by the struggle and unity of the pressure of the gluon condensate and the counter-pressure of the free quarks and gluons in the bubble. (c) Quark–antiquark pairs (of any color charge) increase their attraction with increasing distance due to antishielding. Therefore, the generation of single free quarks is not possible.

While the concept of an "empty space" has been disproven by modern particle physics for a long time, the concept of "vacuum" is still being adhered to under the ideological dictate of positivism. Inside the bubble of nucleons or mesons, the gluon pressure is reduced. The presence of quarks is just another description of a local change of the structure of the quantum æther. In the veiled language of positivism, this structural change is called the transition from the "true" to the "false vacuum state" in the nucleon bubble. Both types of states of the quantum æther inside and outside the bubble form different phases. An approximately correct conception is that the gluons outside the nucleon bubble exist in a superfluid condensate, which is normally fluid inside the bubble.

Color-neutral ensembles of quarks in nucleons and mesons move like free particles at small distances in the bubble. According to the current state of experimental knowledge, it is not possible to separate a single quark from the ensemble by a high-energy impact with other particles. The binding energy between the color charges increases strongly due to the strong self-interaction of the gluons and the resulting anti-shielding effect. Above a critical energy level, new quark-antiquark pairs are formed from the sea of unstable gluons. The attempt to separate a single quark as a free particle results in the constant transformation of the superfluid gluon sea into new quark–antiquark pairs. To date, no single free quarks have been detected experimentally.

The phase transition into quark-gluon plasma

However, at a sufficiently high energy and density of nuclear matter, the inclusion of the quarks in the nucleon bubble ceases, and a phase transition to the quark–gluon plasma occurs (Figure 70). This is found in extremely high-energy collisions of atomic nuclei in particle accelerators. The indications come from characteristic particle jets in collisions of heavy nuclei in particle accelerators. They are generated by a modified excitation of the quark–antiquark pairs and gluons at energy densities above about 1 GeV/fm^3. This corresponds to a temperature of about 10^{12} Kelvin or a collision energy of 200 MeV [Boyanovski 2006]. Whether the quarks and gluons actually form a plasma state or instead a quark-gluon fluid has not yet been determined.

The condensed state of the gluons in the quantum æther outside the bubbles of nuclear matter is similar to a "superconducting" condensate of the gluons, as far as we know today [Greiner 1985]. At the phase transition, at high energy densities, it changes into a new normal-fluid phase. In this phase, quarks and gluons can move freely in a region that is large by comparison to the size of the nucleon. Such a state is suspected in the interior of neutron stars.

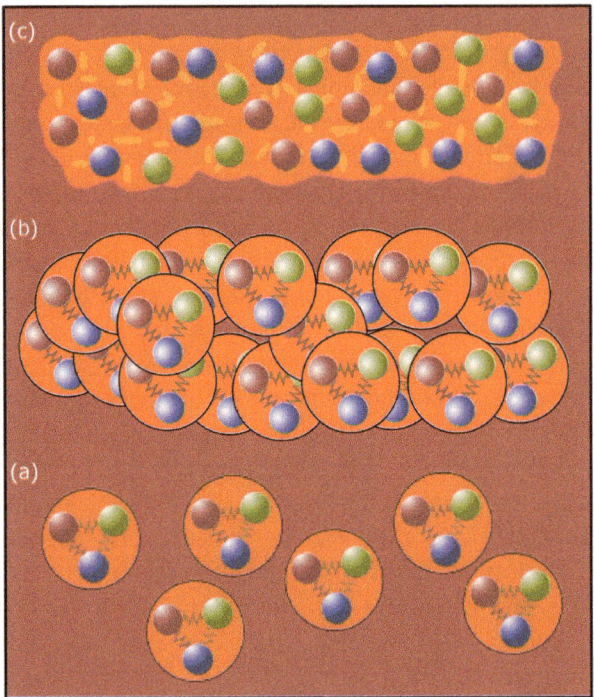

Figure 70: Phase transition in quantum æther. It comprises a state with normal nucleon matter (a), which is characterized by quark–gluon confinement, (b) an excited state with overlapping nucleon bubbles and the transition to the quark–gluon plasma (c) as the energy density/temperature increases. In the quark–gluon plasma, quarks and gluons form a free state. In the process, the superfluid gluon condensate (red), which causes the inclusion of the quarks in the nucleon bubbles, is converted into a normal-fluid phase (orange).

6.4 The electroweak phase transition in the quantum æther

According to current knowledge, the two "elementary particle" classes of the six leptons and six quarks differ fundamentally in their properties of color charge. Leptons have no color charge and are therefore not subject to interaction with gluons or the strong nuclear force. However, both particle classes are subject to electromagnetic and weak interaction. This indicates an inner connection of both classes of fermions. The question of the unity of leptons and quarks and their differentiation with regard to the color charge and the interaction with gluons is one of the great unsolved problems of our understanding of the microcosm.

Characteristics of the weak interaction

A characteristic of the weak interaction is that its electrically charged carriers, the W^+ and W^- bosons, only take part in reactions in which the basic building blocks move in left-handed helices and their antiparticles right-handed helices. This helical motion is caused by precession of the particle spin around the direction of the particle velocity. This property is known as helicity. The experimental observation that only particles with left-handed helical motion participate in the weak interaction is called parity violation – the reaction is not invariant under the symmetry operation of spatial mirroring. In 1957, the Chinese physicist Wu clearly proved the parity violation for the first time in the directional dependence of radioactive decay [Wu 1957]:

$$^{60}_{27}co \rightarrow ^{60}_{28} Ni + e^- + \bar{v}_e + 2\gamma$$

The electrons were always emitted in the opposite direction to the spin of the coatom. The violation of symmetry with respect to left-handed or right-handed rotation processes also occurs during the propagation of light in molecules with a helical structure, such as amino acids. Only light that is levorotarily polarized can propagate in molecules with levorotary structure, and dextrorotary polarized light is absorbed. The material cause lies in the structure of the molecule. The observation of parity violation in the weak interaction is an indication of the complex inner structure of the quantum æther, which probably has an inner angular momentum structure of one or more superfluid components.

The weak charge is inherent in all known basic building blocks in the microcosm; even the neutrinos, which carry neither electric nor color charge, are characterized by their weak charge.

The proof that W^+, W^-, and Z^0 bosons are excitations of the weak nuclear force field and mediate the radioactive transmutations of atomic nuclei is a great triumph of modern physics. They are not only involved in the nuclear transmutations (neutron decay to protons) but also in the transmutation of mesons (pion decay) and the transmutation between different leptons (muon decay to electrons). Figure 66c shows a process of transmuting leptons into hadrons by means of the Z boson, which played a decisive role in the precise determination of the mass of the Z boson (Figure 71).

The unity of the weakons with the photon

The properties of these three weakons have been investigated very precisely in experiments in particle accelerators. In particular, it was possible to precisely determine their mass at 80.4 GeV ($W^\pm$) and 91.2 GeV (Z^0), which is about one hundred times the

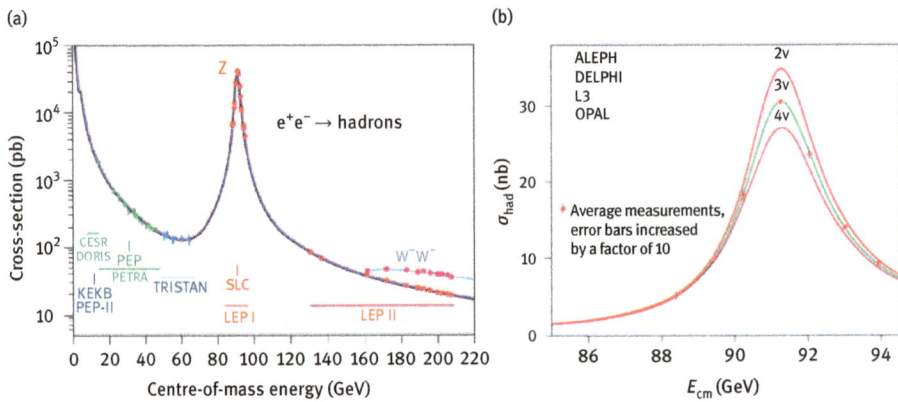

Figure 71: Determination of the weakon mass and the number of neutrinos in electron–positron annihilation experiments in the energy range of 100 GeV. (a) A sharp maximum occurs in the transmutation cross-section of electron–positron pairs into hadrons at 91.2 GeV, the mass of the Z boson. (b) From the line width of the maximum, the number of possible transmutation reactions and thus the number of neutrinos can be deduced [CERN 2004, CERN 2005].

mass of the proton. Just one year after the successful experiments at CERN, Carlo Rubbia and Simon van der Meer were awarded the Nobel Prize in Physics in 1984. Exchange particles of an interaction with such a gigantic mass have only a very short lifetime and range.

The weak interaction is, therefore, an expression of processes in the quantum æther on length scales below 10^{-18} m and thus on scales smaller than the diameter of nuclear building blocks.

The existence and mass of the three weakons was already be predicted before their discovery by means of the theory of electroweak unification. This theory was developed by Glashow, Salam, and Weinberg in the 1960s. It treats the three weakons and the photon as different aspects of uniform electroweak processes. According to their theory, at high energies, in addition to the transition into a quark–gluon plasma, another phase transition occurs in the quantum æther, called the electroweak phase transition. In the high-energy phase, both photons and weakons are massless and thus have an infinite range. Below the electroweak phase transition, the uniform electroweak force disintegrates into two components due to a structural change in the quantum æther: The weakons are strongly shielded and thus enclosed on extremely small length scales. The photons remain massless as a "residual interaction." The masses of the weakons are determined in the unified theory of electroweak interaction by "mixing parameters" between the photon and weakon states, known from scattering experiments. The correct prediction of the masses of the W and Z bosons is an excellent support for the real existence of such a phase transition in the quantum æther. It is expected to occur at an energy above the masses of the weakons in the range of 200–250 GeV (temperature of 10^{15} K) [Boyanovsky 2006].

The electroweak phase transition and the discovery of the Higgs boson

The shielding of the W and Z bosons below the electroweak phase transition is explained by the occurrence of a superfluid Higgs condensate in the quantum æther [Kibble 1964, Higgs 1964]. This superfluid condensate is named after Peter Higgs, who introduced it in the 1960s to explain the differentiation of the weak and electromagnetic interactions from a single excitation. This development was strongly influenced by a deeper understanding of superconductivity in solids. In such systems, the superconducting condensate of the electrons shields the electromagnetic interaction and gives the photons a mass (see Chapter 3). The shielding is caused by the phase rigidity of the superconducting condensate. Photons cause collective movements of the superconducting condensate (plasma oscillations) and their propagation is thus damped. The photon mass corresponds to the emergence of an energy gap in the excitation spectrum of superconductors. Because of the formation of a uniform phase of matter waves of electrons in the superconducting state, the possibility of a new quasi-particle excitation arises: It is a collective oscillation in the electron phases of the superfluid. It can be optically detected as a plasmon and is a "Higgs-boson fingerprint" for the existence of the superconducting condensate.

The quantum field theory of the unified electroweak interaction is based on the same mechanism: The occurrence of the Higgs condensate below the electroweak phase transition damps the propagation of the weakons and gives them a mass. The unshielded remnant creates the electromagnetic interaction. Above the electroweak phase transition, the superfluid Higgs condensate disappears or changes into a "normally-fluid phase." There, it is no longer possible to distinguish between electromagnetic and weak interaction. Above the electroweak phase transition, both fundamental interactions possess uniform massless carrier particles (photons and weakons). Part of Peter Higgs' theory was the prediction of a new quasi-particle as an excitation of the superfluid Higgs condensate: the Higgs boson. On July 4, 2012, CERN published the discovery of a new particle with a mass of about 125 GeV that is consistent with the predicted Higgs boson (see Figure 72). Peter Higgs and François Englert were awarded the Nobel Prize in Physics in 2013 for their prediction of this particle and their contributions to understanding the origin of particle masses.

The Higgs field and the formation of particle masses

According to the Standard Model, the Higgs condensate not only gives the $W^\pm$ and Z^0 bosons their masses of 80.4 and 91.2 GeV but also represents the general cause of the formation of a rest mass for all particles. The rest mass, and thus the Dirac energy gap of the particles, is directly proportional to the condensate density [Böhm 2003] (see also the connection between particle mass and Dirac energy gap in Chapter 5).

However, some physicists and popular descriptions present a very strange idea of "mass formation by the Higgs boson." It is as if the Higgs field "sticks" to the particles and gives them mass. The physicist H. Genz imagines the formation of mass as follows:

> The Higgs field forms a background field in space that is locally disturbed whenever there is a particle in it with which it interacts. The disturbance – the formation of bubbles – produces the observed effective mass of the particle. When we try to set such a particle in motion, we also pull on the condensation that surrounds it. It, and not the particle in its midst, resists acceleration
>
> [Genz 1994, p. 283].

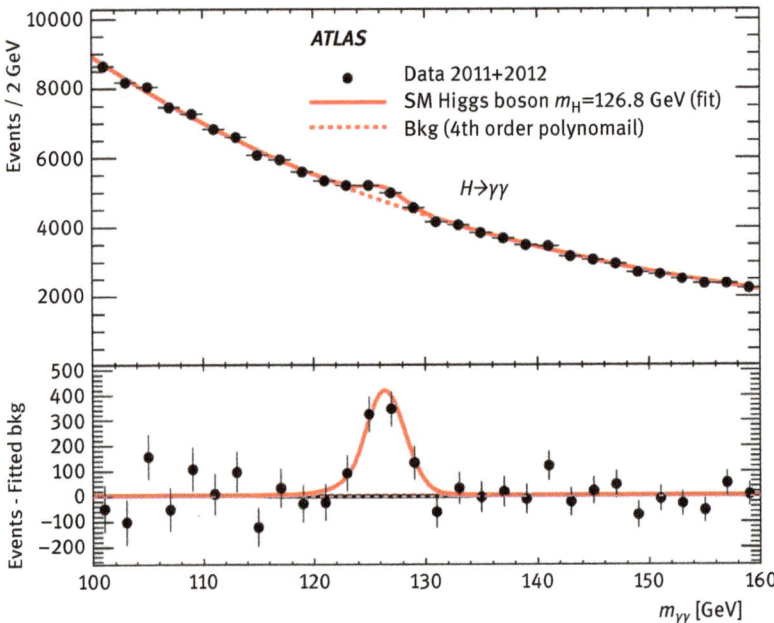

Figure 72: Discovery of the Higgs particle as strong evidence for the existence of a superfluid condensate in the quantum æther. The Higgs proof of the decay of the Higgs boson into two photon pairs is shown here. Further proof was provided by the decay in two Z bosons. ATLAS experiment at CERN [CERN 2013].

Here the Higgs field is treated as an *"external addition"* to the particle. Some even call it an "adhesive particle," a "ghost particle," or a "God particle." The idea of mass formation by "adhesive particles" conceals the progressive aspect of the Higgs condensate theory, namely the recognition of the real existence of superfluid components in the quantum æther.

But even in the Glashow–Salam–Weinberg model of unified electroweak interaction, the occurrence of the Higgs condensate remains, so to speak, an artificial addition to the quantum field theory of electromagnetic or weak interaction. The combination of the two quantum field theories about the formation of a Higgs condensate in the quantum æther in one phase transition represents only a first step forward in the uniform understanding of the differentiation of the various particles. There are also contradictions to experimental observations: these include the nonvanishing neutrino masses observed in experiments [King 2015]. According to the Standard Model, both left-handed and right-handed neutrinos should occur for finite neutrino masses, in contradiction to all experiments in which only left-handed neutrinos are observed [Murayama 2002]. A theory of particle masses needs a deeper foundation beyond quantum field theories, by understanding the formation of "elementary particles" and their masses as a self-organization process in a material medium.

6.5 The search for the unity of the forces of nature

The endeavor to understand the differentiation of the various particles and forces of nature in the microcosm uniformly as an expression of different forms of movement and of structure formation of the quantum æther is an important progressive impetus of natural-science research. To this end, systems of condensed matter, such as the quantum æther, must be studied at ever smaller length scales. Since the excitation quanta of interactions at higher energies have ever smaller de-Broglie wavelengths, smaller length scales can be investigated by higher excitation energies. The low-energy excitations of a condensed system are usually more diverse than high-energy excitations. They are determined by the orderings of the system and the resulting broken symmetries.

Let us first discuss this question briefly using the example of solid-state excitations. For example, the magnetic exchange forces leading to the alignment of the magnetic moments in a ferromagnet can be studied by optical excitation of magnons. Of course, this is only possible on such low energy levels that the ferromagnetic order is not destroyed, that is, for excitation energies of a few dozen meV. This energy level is connected to a length scale of the magnon excitations of 10–100 nm, at which of course nothing can be learned about the atomic structure of the ferromagnet on the 0.1 nm scale.

To analyse the arrangement and type of atoms, high-energy experiments with X-ray quanta, with excitation energies in the range of 10 keV, are necessary, for example. But does this lead to an understanding of the unity of all interactions in the ferromagnetic solid? Certainly not in the sense of "grand unification theories" with their arbitrary introduction of symmetricized new particles. Applied to the solid-state system, such a method would attempt to make all the different collective low-energy excitations of the system (phonons, magnons, plasmons, etc.) emerge from a supersymmetry. The elastic bonding forces between atoms, the magnetic exchange forces between spins, and the Coulomb repulsion between electrons, do not become identical even at high energies, however. Their differentiation from the electromagnetic interaction cannot be separated from the atomic and electronic structure of the underlying solid. The differentiation of different types of forces is "emergent"; it is a product of the development of the underlying structure of matter.

In high-energy particle physics, attempts are made to find the unity of electromagnetic, weak and strong nuclear force at high energies or small length scales. Figure 73c shows the coupling strengths over distance extrapolated from the quantum field theories of electromagnetic, weak and strong nuclear force, as well as the prediction from the grand unified theory (GUT) using SU(5) gauge symmetry.

The changes in the coupling strengths have only been experimentally tested up to the 100-GeV range and thus on length scales down to about 10^{-18} m (Figures 73a and b). The material cause for a distance or energy dependence of the natural forces in the microcosm is the shielding of the respective charge by the unstable excitations of the quantum æther.

While electromagnetic coupling decreases as a function of the distance from the "bare" electrical charge due to shielding, that does not apply to the weak and strong nuclear force: They are amplified by anti-shielding, which is why their coupling strength increases with the distance from their respective charge. Anti-shielding occurs when the exchange quanta of these forces, the gluons or W and Z bosons themselves have a charge, that is the color charge of the gluons and the weak charge of the weakons.

From the extrapolation of the coupling strengths of quantum field theories to large energies or small lengths in Figure 73, it appears that the strengths of all 3 forces of nature approach each other. However, they do not become identical but cross, instead. Only on the basis of supersymmetry, such as SU(5), would all curves merge into each other at an unimaginably small distance of 10^{-31} m. Do the 3 basic interactions become identical at these distances? The trends of coupling strengths at smaller lengths indicate that the differentiation into different forces, and thus different forms of motion mediating them, is a low-energy phenomenon of the quantum æther. However, an extrapolation of the forces by more than ten orders of magnitude on the length scale is highly questionable. At length scales of 10^{-18} m that correspond to an energy of the electroweak phase transition (200 GeV), a sudden change in the

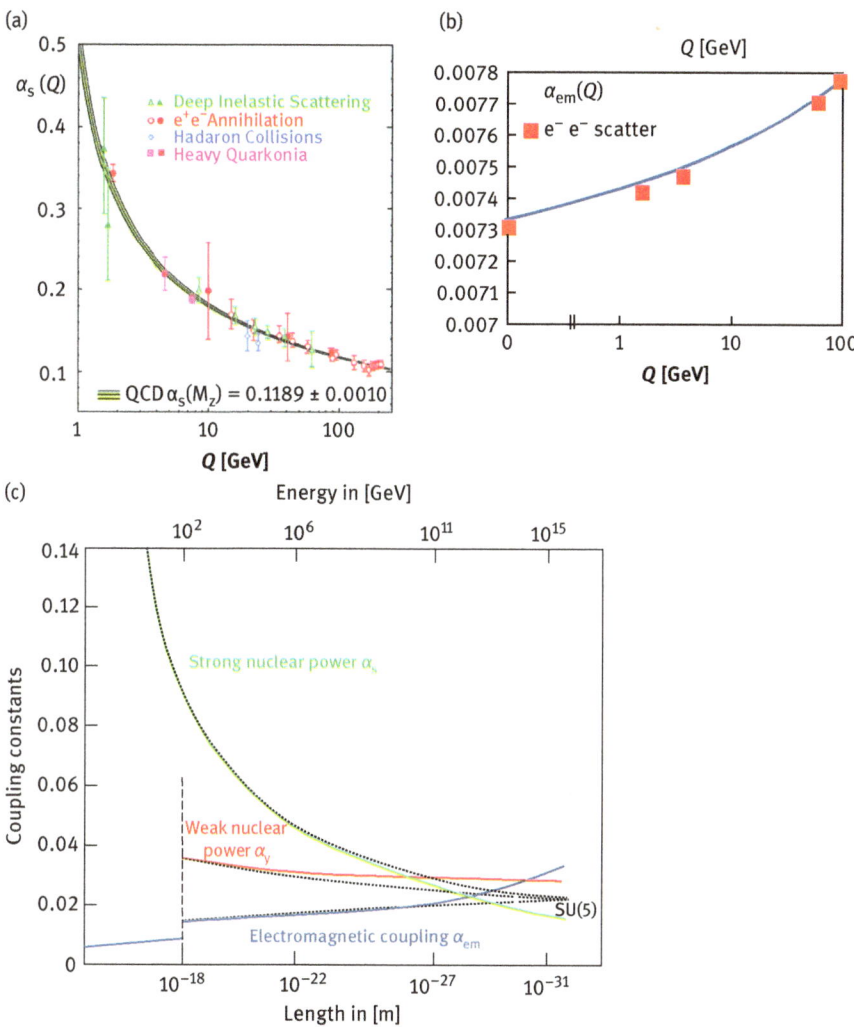

Figure 73: Distance-dependent coupling constant of the 3 fundamental interactions by partial shielding. (a) Weakening of the coupling strength of the strong nuclear force α_s as a function of the energy transfer Q [Bethke 2007]. (b) Amplification of the coupling strength of the electromagnetic force α_{em} as a function of the energy transfer Q [Achard 2005, Burkhardt 2011]. (a) and (b) show experiment and theory, respectively. (c) Extrapolation of the electromagnetic (blue curve), weak (red curve), and strong force (green line) from quantum field theory, as well as a prediction by the grand unified theory based on a SU(5) gauge symmetry (black dotted curves) [Georgi 1986].

strength of the electromagnetic coupling is suspected; but measurements have not been made yet. The experience of the organization of matter in structural levels on different length scales suggests that on length scales of more than ten orders of magnitude downwards a great variety of structures and forms of motion of the quantum æther will occur. It is likely that further new structural levels of matter will be discovered in the microcosm.

What does the inner structure of the electron really look like? Is there a core region with a different phase of the quantum æther, as the study of particle formation as topological defects suggests? The use of photons of higher and higher energy for the deep-structure elucidation of the electron in photon–electron scattering has so far led to nothing other than the discovery of the increasing contributions of higher-energy photon transmutation processes in leptons, quarks, and more complex hadrons in their interaction with the electron [Berger 2015].

One should not forget that all these interactions are approximated as point interactions within the framework of renormalizable quantum field theories. The resulting infinities are eliminated by renormalization techniques. While these are effective low-energy theories of the quantum æther, extrapolation to higher energies will inevitably encounter contradictions.

The dead end of the grand unified theories

However, all grand unified theories (GUT) ignore these limitations and strive to find a common origin for all "elementary particles" and forces by finding a "higher gauge symmetry." This state of higher gauge symmetry is supposed to prevail at extremely high temperatures in the Big Bang and is associated with the prediction of new particles as excitations of the *"symmetrical high-energy structure of the vacuum."*

One of the first grand unification theories is the model of Georgi and Glashow from 1974, which aims at a uniform description of the electromagnetic, weak and color forces. All three interactions are characterized by the fact that the transformations caused by them are accompanied by characteristic phase jumps of the matter waves of the particles involved. This connection is expressed by means of gauge symmetries. Thus, the electromagnetic interaction is invariant under global phase shifts of the matter waves of the particles with an electric charge generating it. As we have already discussed, however, a local phase jump of the matter wave is identical to the emission or absorption of a photon. This is called gauge symmetry $U(1)$ of the electromagnetic interaction. Likewise, the weak interaction by W and Z bosons is subject to an $SU(2)$ gauge symmetry and the strong interaction by gluons to an $SU(3)$ gauge symmetry. These gauge symmetries reflect the structure of the weak charge with two components, and of the color charge with three.

In Georgi's and Glashow's GUT theory, however, a supersymmetry is constructed to unify the electroweak and the gluonic forces, which has an $SU(5)$ symmetry as a mathematical product space $U(1) \times SU(2) \times SU(3)$. In addition to the twelve known exchange quanta of fields (one photon, three weakons, eight gluons), it predicts twelve further exchange bosons (the X bosons) with which quarks and leptons can transform into one another. The resulting prediction of proton decay has been experimentally refuted, and none of these X bosons were found in the expected energy range.

Another spectacularly failed theory of symmetrization was devised in the 1970s by the Soviet physicist Michail Schifmann: According to him, each of the known elementary particles had a partner in the fires of the Big Bang. After the transformation of socialism into a revisionist system, the debate about a dialectical-materialist interpretation of the development of matter in the Soviet Union suffered a defeat, and the idealistic world-view and method spread among physicists there as well. The beauty of the supersymmetry (SUSY) theory was elevated to a criterion of its truth: "*Supersymmetry has such a beautiful structure, and in physics we allow ourselves to be guided by the degree of beauty and aesthetic quality of a theory, in the search for truth*" [Greene in Wolchover 2012]. A world full of diversity and complexity is thus considered "ugly," while the primitive symmetrical world of simple equations is considered "beautiful."

SUSY's predictions have been experimentally refuted by the high-energy experiments that led to the discovery of the Higgs boson: None of the heavy particles it predicted were found [Wolchover 2012]. Schifmann also must now admit the deep crisis of SUSY and other symmetrifications of particles [Schifmann 2012]. In view of the lack of new discoveries in the high-energy accelerators costing billions, the big-bang theorist Steven Weinberg even speaks of a "crisis of large-scale research," and fears that no further money will flow [Weinberg 2012]. The unity and contradiction of the various interactions and particles cannot, however, be resolved by the mathematical construction of supersymmetries, but only by the concrete analysis and synthesis of the forms of motion and structures of new levels of matter to which the accelerator experiments of recent decades have made an important contribution.

The fundamental error in the theoretical approaches to the unification of the 3 forces of nature through gauge symmetries, such as SU(5), consists in viewing the exchange quanta of the forces of nature as "fundamental particles" in empty space instead of as collective forms of motion of a quantum fluid continuum [Klinkhamer 2005]. As a result, the physical cause of the distance/energy dependence of the couplings by the shielding effects of a real medium, as shown in Figure 73, is replaced by the idealistic construction of a supersymmetry. Materialistically, however, the gauge symmetries of the three interactions are an expression of the underlying material structure of the quantum æther. The convergence of the strength of the three natural forces reflects the fact that at higher energies, all forms of excitation increasingly contribute to the different natural forces so they become more similar. However, based on the findings about other systems of condensed matter, there is no reason to assume that the three forces of nature should become identical at all. One reason is the expected qualitative changes at higher energy due to phase transitions.

The subordination of particle physics to the Big Bang Theory

The theoretical idea that all natural forces should become identical at extremely high energies or temperatures serves primarily to subordinate particle physics to the big-bang theory. According to this theory, such a high-temperature state prevailed shortly after the Big Bang. From this state of extreme radiation, all current forms of matter are said to have developed after cooling.

However, the crisis of high-energy symmetrification of all particles has not yet led to a fundamental questioning of this subordination of particle research to the big-bang theory. These fantastic constructions replace a materialistic development theory of particles with a linear one-way street: 10^{-36} seconds after the Big Bang, with a radius of the universe of 10 cm, one of the supersymmetries is said to have broken, depending on the variant of the theory. In the SUSY scenario, the invented 24 SUSY twin particles acquire gigantic masses, thus becoming invisible, and the quarks and leptons known today are formed. In contrast to reality, there would now be as many particles as antiparticles for every kind of particle. However, since the universe known today contains hardly any antiparticles, the law of conservation of baryons and leptons must be broken to explain the dominance of matter over antimatter. In fact, there are experimental indications how a dominance of matter over antimatter is related to the chiral structure of the quantum æther (see Section 6.6), and in this context, an asymmetry of left-handed versus right-handed particles occurs. In the big-bang model, however, this asymmetry is chosen *ad hoc* in just such a way that the ratio of baryons and photons corresponds to the value observed today. Only with this *ad-hoc* assumption can the Big Bang theory correctly calculate the ratio of the light elements (hydrogen, deuterium, and helium), which is celebrated by its followers as a great success.

Another trick in describing this early phase of the universe is inflation. In a "normal big-bang scenario," a homogenization of matter and radiation would occur at most in an area of expansion that could be crossed at the speed of light in the period after the big bang. So, if the GUT phase transition begins 10^{-36} seconds after the Big Bang, then at that time, areas that interact with each other could be only 3×10^{-26} cm large. During the expansion of the universe, these would have expanded to just 30 cm today [Hoyle, Burbidge, Narlikar 2000]. Larger areas would thus be completely decoupled according to this absurd scenario.

As a fantastic way out, A. Guth [Guth 1981] in 1981 introduced the model of inflation, which in addition to the big-bang theory itself contradicts all known laws of physics. Through the effect of an "inflationary field," and driven by the GUT phase transition, the universe was supposed to have gone through a phase of faster-than-light expansion in which it increased its size by a factor of 10^{100}. In order for the universe to have only roughly the properties observable today after inflation, namely a flat space and the associated mean baryon density, an extreme fine-tuning of the model is necessary: The entire inflationary phase should not have lasted longer than 10^{-33} seconds, and the matter density of the state before the phase transition must have corresponded to the critical density for a flat universe to an accuracy of 1 in 10^{50}. Otherwise, the universe would either have collapsed to a point again or suddenly expanded to infinity, in a time of the order of 10^{-36} s [Hoyle 2000, p. 179]. Such a conglomeration of absurd ad-hoc assumptions is proclaimed by Big Bang prophets as one of the greatest achievements of twentieth-century science! [Blanchard 2006].

In an interview with *Der Spiegel* on the misbelief in a world formula, the Nobel Laureate in physics B. Laughlin called the big-bang theory "marketing" and "quasi-religious nonsense" [Laughlin 2008]. The attempt to understand the various particles and forces of the microcosm in a uniform way is turned into its opposite by the search for a "world formula" that encompasses all forms of matter and forces. The idea that there is any kind of world formula at all denies the qualitative and quantitative infinity of the universe and matter.

6.6 Self-organization of "elementary particles" in the quantum æther

Just as in the course of the nineteenth century the concept of the "atom" had to be dialectically negated by research into the atomic structure, from the structureless "mass point" to complex composite structures, so in the twentieth century, a vast amount of observational evidence accumulated that even "elementary particles" such as electrons, quarks, or photons are not structureless. That it is not possible to explain the subatomic structure of matter by a hierarchical system of ever smaller particles, sub-particles and sub-sub-particles, but that something qualitatively new occurs on a certain structural level of matter in the microcosm: The formation of particles out of a material continuum [Dickhut 1987]. This is proven by the observation that mass defects occur.

Mass defects in composite systems

In the transition from individual particles to a composite system, the mass of the composite bound system is not identical to the sum of the masses of the individual components:

- The mass of a hydrogen atom, consisting of an electron and a proton, is slightly smaller than the masses of free electrons and protons. When the H atom is formed, the binding energy is emitted in the form of a photon with an energy of $E = 13.6$ eV. Consequently, the mass defect associated with this is very small, a hundred millionth of the sum of the rest masses of the 2 particles: The binding energy of 13.6 eV is low compared to the rest masses of the atomic nucleus (938.7 MeV/c^2) and the electron (0.5 MeV/c^2).

- In the formation of atomic nuclei from protons and neutrons, the mass defects are considerable and can amount to several percent of the rest mass of the nucleus. The fusion of 2 H atoms to form a deuterium atom releases energy of 0.4 MeV, while the fusion of deuterium with a further hydrogen atom to form helium-3 releases 5.5 MeV. The reduction of the nuclear mass m in comparison to the free particles is an expression of the energy gain E of the bound system in comparison to the separated state. Since the nuclear binding forces are much stronger than electromagnetic forces, the mass reduction in the bound system is also much greater.
- At the transition to the next deeper structural level of matter, the mass defects become huge and change their sign. The proton consists of three valence quarks and a fluctuating system of unstable quarks and gluons. The proton has a mass of 938.28 MeV/c^2. The masses of the proton constituents, the individual quarks of 1.5–4.5 MeV/c^2 (up) and 5.0–8.5 MeV/c^2 (down) make up only a small fraction of the mass of the composite system. A large part of the rest mass of the proton comes from the kinetic energy of the valence quarks (some 100 MeV) as well as from the masses of the fluctuating gluons and unstable quark–antiquark pairs. In order to excite a proton in its internal degrees of freedom, energy comparable to its rest mass is needed. At the same time, it is not possible to produce single free quarks under normal conditions due to the antishielding effect. The energy of such excited states and thus their rest mass would be much greater than the energy of the proton's composite (color-neutral) system.

In the nineteenth century, the property of bodies to possess mass and to be weighable was equated with matter *per se*. Matter without mass was unthinkable in the mechanistic view of the world. The property "mass" was regarded as a fixed property of a substance, as an expression of its existence at all. When at the beginning of the twentieth century, with the discovery of radioactivity, it was found that the mass of particles can change, that mass can convert into kinetic energy, this was equated with the *"disappearance of matter."* In fact this is an idealistic interpretation of the relation between change of energy E and mass given by Einstein's famous formula $E = mc^2$.

"Mass" is a dynamic property of certain stages of matter as an expression of their structure formation in the quantum æther and not a particle property *per se*. The "mass defects," which become larger and larger at ever smaller length scales, indicate at the latest with the formation of hadrons from quarks that an image of the microcosm in which particles are composed of ever smaller particles, reaches a limit and even leads to absurdity. That is why W. Dickhut coined the term "continuous matter," with particles as products of development, in his study of dialectical materialism [Dickhut 1987].

In fact, a substantial part of the mass of the nuclear building blocks is formed due to the local phase transition from the superfluid to the normal phase in the gluon condensate within the nucleon bubble. Indeed, not all forms of energy contribute to mass, as analyzed in the critical study of G. Kalies [Kalies 2019]. Nevertheless, mass is an emergent property related to the energy of formation of topological defects in the quantum æther. It thus can also be assumed that the electron mass is essentially caused by a local phase transition of the quantum æther into the electroweak phase in the innermost core of the electron but strongly renormalized by shielding effects. The masses of the different particles correspond to different Dirac energy gaps in the excitation spectrum of the quantum æther, comparable to the band structure of the excitations of a solid body; see Figure 74. The size of the energy gap and

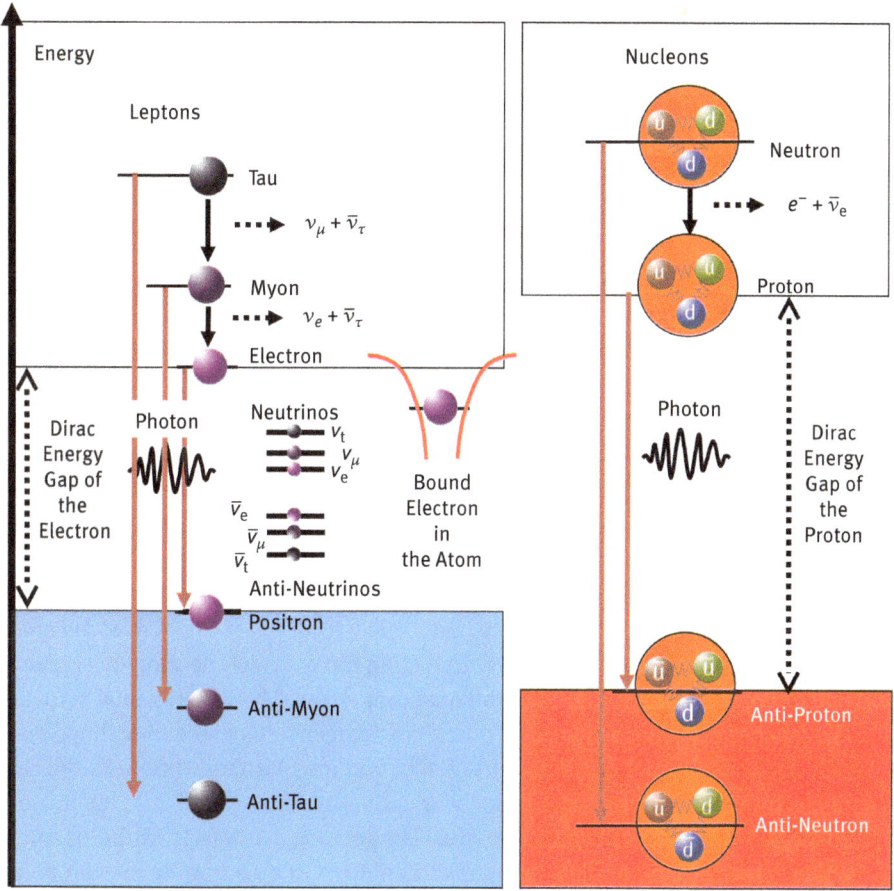

Figure 74: Dirac energy model of the mass spectrum of leptons and nucleons and of some low-energy transmutations via the weak or electromagnetic interaction (not to scale). The mass spectrum of "elementary particles" known today ranges from about 0.02 eV/c² (neutrinos) to 170 GeV/c² for t quarks. The illustration shows a sketch of the Dirac energy gaps of the two stable fermions, electron and proton, as well as of their unstable variants. Their two Dirac energy gaps differ due to the differing internal structure of the particles.

thus the mass of the particles depend on its aggregate state, its phase. The low-energy phase with "normal" mass spectrum of the particles is called the Dirac phase of the quantum æther in the following, in honor of the work of Paul A.M. Dirac.

Dialectic unity of excitations and structure of the quantum æther

The microscopic composition of the quantum æther and its material structure in the ground state are not known today. However, the recognition of the mutual determination of types of excitation by the ground state and the ground state by the type of excitations (see Section 2.6) in many-particle systems allows us to draw a number of conclusions.

At least the following classes of qualitatively different excitations occur in the quantum æther:

1. The zero-point excitations of the quantum æther. They consist of unstable photons. Higher-energy unstable excitations also occur at smaller length and time scales, such as particle–antiparticle pairs of all the "elementary building blocks," as well as fluctuations of weakons and gluons. In quantum fluids, this corresponds to fluctuating quasi-particle excitations at a finite temperature. There they are determined by the struggle and unity of ordered movements of the superfluid condensate and collective disordered forms of movement, which form the quasi-particle excitations.

2. The stable collective low-energy excitations of the quantum æther, where quantity transforms into new quality, governed by the law of action quantization. The formed quanta, photons, gluons, and weakons represent the exchange particles of the various forces. High-energy scattering experiments show that with increasing energy, photons also bear parts of weakons, lepton–antilepton pairs, and even quark–antiquark pairs [Berger 2015].

3. The topological structures as stable particle-like local phase-changes of the quantum æther. They are created by contradicting forces which develop by shielding macroscopic motion patterns in the quantum æther. The experimental evidence is overwhelming for the hadrons: quark confinement, color shielding, huge mass defects, and unusual combinations of spin and angular momentum. In the case of electrons, an experimental look "behind the shielding cloud" into the nucleus suspected has not yet been successful. The topological defects in the quantum æther have certain parallels to topological defects in quantum fluids with an internal rotational structure.

4. The global phase transitions in the quantum æther. The transition into the quark–gluon plasma (or fluid) has been proven experimentally. At about 200 MeV, the high thermal energy of disordered motion patterns overcomes the quark confinement of the surrounding superfluid gluon condensate. It is predicted that at energies above 200 GeV, the electroweak phase transition develops as

an order–disorder transition in the superfluid Higgs condensate. The mass ratio of the W and Z bosons and the (probable) observation of the Higgs boson as a collective excitation of this condensate are strong experimental evidences for the existence of this condensate and of a finite transition temperature.

This variety of excitations and forms of motion are an expression of competition and cooperation of different internal forms of motion of the quantum æther.

Conservation laws as invariants of the internal forms of motion of the quantum æther

These internal forms of motion are not arbitrary. Electromagnetic, weak, and gluonic forms of motion, as well as their respective transitions into one another, are subject to a series of conservation laws as invariants of the motion.

- The formation of stable (partially shielded) fields of photons, weakons, and gluons is bound to the existence of quantized charges of the respective type as topological invariants of the respective particles as "topological structures."
- The electric charge Q, the weak charge Y, and the color charge C are subject to conservation laws, at least in the energy range of particle transformations experimentally accessible today. For example, in the transmutation process from leptons to quarks, $e^-e^+ \rightarrow Z^0 \rightarrow q\bar{q}$, the electric charge, the weak charge, and the color charge are conserved. Here, the sum of the electrical charges for an electron–positron pair (e^-e^+) and for a quark–antiquark pair ($q\bar{q}$) is zero in both cases. The Z^0 boson is electrically neutral. The equivalent applies to the other types of charges.
- The law of conservation of the baryon number B. Baryons are particles consisting of 3 quarks. Baryons are all unstable except for the proton. Proton decay has not yet been observed. The law of conservation of the baryon number states that in any transmutation process, the number of baryons minus the number of antibaryons is invariant. For example, the Λ particle ($B = 1$) decays into a proton and a pi meson $\Lambda \rightarrow \pi^- + p^+$ while conserving the baryon number and all three charges.
- The law of conservation of the lepton number L. For example, a muon decays into an electron, a muon neutrino and an electron antineutrino $\mu^- \rightarrow e^- + \bar{v}_e + v_\mu$. The lepton number, the number of leptons minus the number of antileptons, is conserved.

The baryon and lepton conservation laws are an expression of the stability of all forms of matter in the cosmos under ordinary conditions, i.e. phase stability of the Dirac phase of the quantum æther. If they were violated, atoms would not be stable either. For example, the hydrogen atom could be transformed into a positron and 2 photons by a hypothetical proton decay $p^+ \rightarrow e^+ + \pi^0 \rightarrow e^+ + 2\gamma$. Various current experiments have not yet discovered a single proton decay, and prove a lifetime of the proton of more than 6×10^{33} years [Nishino 2009, Abe 2014]. Conversely, the laws of conservation of the baryon and lepton numbers raise the question of the origination of baryons and leptons and thus of their origin in the overall evolutionary process of matter.

The average electrical neutrality of matter is remarkable: As far as is known today, the number of positively charged protons and of negatively charged electrons in the cosmos are more-or-less identical. It is true that during the formation or decay of neutrons, an electron and a proton can be formed or fused in pairs. In the formation of heavy nuclei, for example, electrically neutral neutrons are formed through the transmutation of protons and electrons by nuclear fusion processes in the stars via $p^+ + e^- + \bar{v}_e \rightarrow n$. However, no electron can arise within the quantum æther without its positive "twin" particle being formed. Since the antimatter counterpart to the electron, the positron, as well as the antimatter counterpart to the proton, the antiproton, are unstable, it follows from the observation of the overall electrical neutrality of matter that there must be a process in which the formation of electrons is linked to the formation of protons, despite the great differences in the nature of the two particles. Chapter 7 shows that this process is probably associated with the electroweak phase transition at phase boundaries in galactic nuclei, where the conservation laws for the baryon and lepton numbers, respectively, are abolished.

The differentiation of the electron, the proton, and their fields

From the point of view of electromagnetic structure, the electron and the proton become identical as exact antipodes of the electrical charge, which however decays inside the proton to thirds of the charge of its quark constituents. Only left-handed electrons and quarks with weak charges −1 and +⅓ occur, while the right-handed particles are part of a condensate of the Dirac phase of the quantum æther. The structure of the electron and the proton also differ in their color charge and their gluon dynamics. This is due to local phase transition within the proton bubble. The decay of the Dirac phase into a color-charged phase inside the nucleon is largely shielded from the outside at energies that are not too high (Figure 75a, b).

In his ground-breaking book *the Universe in a Helium Droplet* [Volovik 2003] the Russian physicist Grigory Volovik compared excitations and particles of the Standard Model with the excitations and topological defects in superfluid helium-3. He developed the identity and contradiction of different forms of motion of this quantum fluid and the quantum æther (which he calls the "Planck medium"). Based on this comparison of the two quantum fluids, a structural model for electrons and protons is proposed below.

It turns out that the existence of two superfluid phases of helium-3, the isotropic B phase and the anisotropic A phase, with a node in the excitation spectrum of the

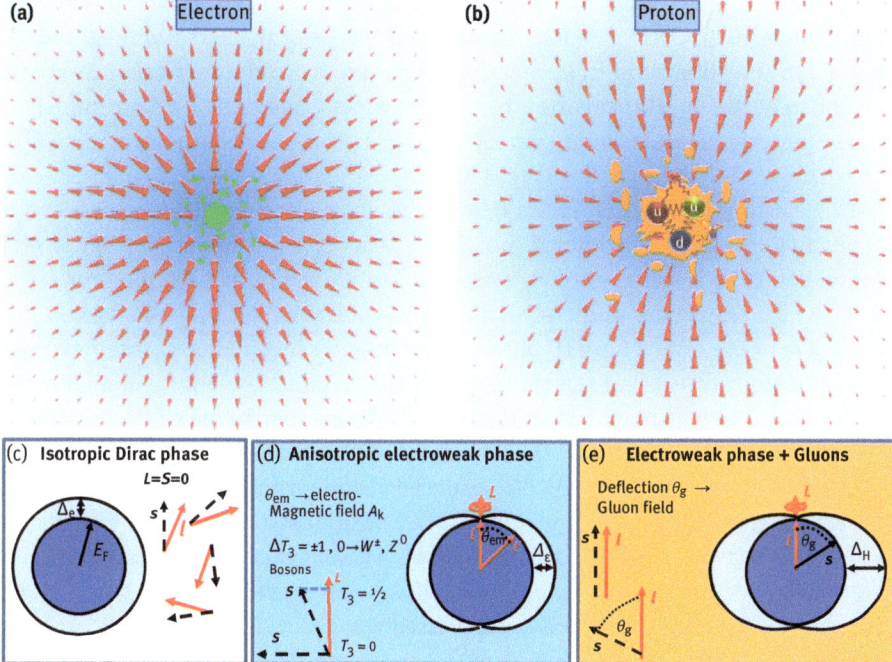

Figure 75: Model of the electron and the proton as topological defects in the quantum æther.
Top: Electrical field (red cones) of electron and proton with an electroweak core at electron (a) and color-charged core at proton (b). Bottom: Discussion of the phase transitions in the quantum æther based on the comparison with the isotropic (c) and anisotropic (c, e) superfluid phases of helium-3. See text.

quasi-particles, is a key to understanding the formation of particle-like topological defects in the quantum æther by self-organization. The isotropic B phase of the superfluid helium-3, with an energy gap in the excitation spectrum, is an approximate model for the structure and forms of motion of the quantum æther below the electroweak phase transition (Dirac phase). The anisotropic A phase of the superfluid helium-3 has a node in the excitation spectrum, which is determined by alignment (ordering) of the angular-momentum structure of the superfluid condensate. It is an approximate model for the anisotropic electroweak phase of the quantum æther above the electroweak phase transition.

Based on this quantum fluid model, the structure of the electron and proton (Figure 75) thus has the following characteristics:

Both have an electrical field of reversed sign (Figure 75a, b). The electrical field vector is identical to the hedgehog-shaped structure of the orbital angular momentum L of the condensate. It results from the topological defect in the isotropic Dirac phase.

In the core of the electron, the Dirac phase turns into the anisotropic electroweak phase, while in the proton bubble a modified electroweak phase with gluons is formed. In the superfluid helium-3, this corresponds to the formation of a hard core of helium-3-A in the vortex of a helium-3-B phase or of a soft split core which represents a skyrmion defect in the helium-3-B phase [Volovik 2003, p. 172ff].

The electromagnetic zero-point field occurs in the isotropic Dirac phase due to the fluctuations of the disordered orbital moments of the condensate (Figure 75c). The transition to a stable photon is identical to a dynamical 360° rotation of the l-field, which propagates at the superfluid speed of sound. Such a photon excitation can evolve in the Dirac phase (B phase of He) because it contains an internal angular momentum structure of the condensate, with microscopic orbital and spin moments l and s, which are, however, disordered. The electromagnetic vector potential $\mathbf{A}$ may be considered identical to short range dynamic partial ordering of l.

In the anisotropic electroweak phase, the microscopic orbital moments l of the condensate are aligned and form a uniform L axis (Figure 75d). With respect to this axis, the macroscopic spin S of the condensate has 3 possible values: $T_3 = +\frac{1}{2}$ (left-handed neutrinos), $T_3 = -\frac{1}{2}$ (left-handed electron and its unstable sisters m and t) and $T_3 = 0$ (right-handed electron, but which is suppressed and not included in Figure 75d). The $W^{\pm}$ and Z^0 bosons mediate a change of this characteristic, called the "isospin," and thus different transmutation reactions, which are associated with changes in the core of the topological defect. The isospin connects the electrical charge Q and the weak charge Y via

$$Q = T_3 + \frac{Y}{2}$$

In the electroweak phase, the uniform ordering of the individual spins s of the constituents of the condensate to form the total spin S is maintained The gluonic excitations correspond to a local deflection of the s spins with respect to the total angular momentum axis L (Figure 75e). The electroweak anisotropic phase inside the proton in the ^{3}He model would, therefore, exhibit gluon excitations by a local change of the spin-orbit coupling in the condensate. Such gluon-like excitations can evolve in helium-3 in the soft cores of the skyrmions, whose core decays into two or three "quarks." The formation of the color charge and its gluonic dynamics local phase inside the "proton bubble" (or as the analog inside the soft core of a skyrmion in ^{3}He) is shielded outside the phase boundary.

Based on this model, Figure 76 shows a three-dimensional representation of the fields associated with the electron as a topological defect. It gives an insight into the spin structure of the electron. The intrinsic angular momentum (spin) of the electron is quantized and has a value of half the angular momentum quantum $\hbar$. The model shows that the electron spin is neither identical with an inherent rotation of the core of the defect nor with the internal spin or orbital angular momentum structure of the

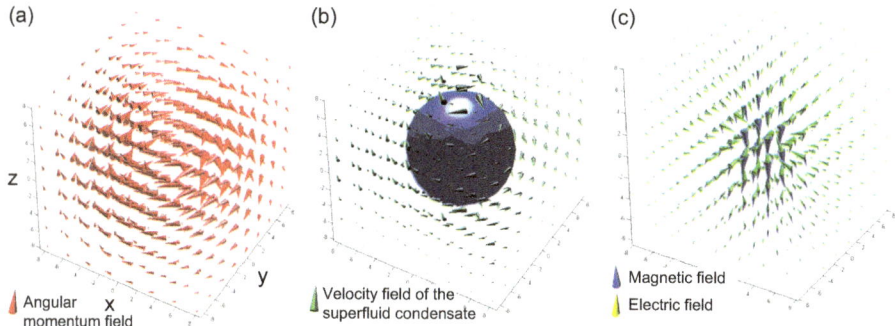

(a)

z

y

Angular   x
momentum field

(b)

Velocity field of the
superfluid condensate

(c)

Magnetic field

Electric field

Figure 76: Model of the electron as a topological defect in the quantum æther, based on a comparison with superfluid ^{3}He. (a) Angular momentum field L of the anisotropic phase induced around the core. (b) Velocity field of the superfluid vortex flow, projected onto a sphere in color, in addition to the arrowheads. (c) Electrical and magnetic field of the electron. Position coordinates in units of the extent of the noncondensed core. See also mathematical appendix in Section 9.5.

condensate. Rather, the L texture represents a quantized superfluid eddy current, which can be identified with an intrinsic angular momentum (spin) and is the cause of the magnetic moment.

Naturally, the proposed model of the proton and electron can only be a rough approximation of the structure of the electron. And the quantum æther and superfluid ^{3}He are two very different material systems with condensation energies and excitations on completely different energy levels. The comparison is nevertheless based on the current understanding of condensed systems with several components of quantum fluids, of their different phases with isotropic and anisotropic excitation spectra, and of their topological defects. The comparative method of studying the unity and contradiction of the two systems to form materialistic models provides a guideline for further experimental and theoretical research into the microscopic laws of motion of the quantum æther and, above all, for the establishment of a materialistic theory of development of the "elementary" basic building blocks of matter by self-organization. In the discovery of forms of motion and phase transitions of the quantum æther, twentieth-century physics has opened the door to a new, very diverse and complex part of the universe in the microcosm, which we are only now beginning to understand.

7 Dialectics of the evolutionary processes in the microcosm and macrocosm

The structure formation of the macrocosm is inextricably linked with the evolution of the forms of matter in the microcosm.

Inside stars, nuclear fusion processes take place under high pressure and high temperature as an energy source for the radiation of light. They are the result and driving force of stellar evolution and bring about the chemical evolution of elements from light hydrogen atoms to heavy elements such as metals. The state of stellar matter is determined to a large extent by the internal opposing forces of gravitational attraction and the radiation pressure of the radiation released in the nuclear reactions. This section presents arguments for the thesis that the evolutionary process of galaxies is also connected with evolutionary processes in the microcosm, especially with the formation of "elementary particles".

In the supermassive nuclei of the galaxies, the threshold for the electroweak phase transition in the quantum æther is crossed due to the extreme gravitational fields, so that processes of baryon and lepton generation begin. The resulting nuclear building blocks and electrons form hydrogen and other light atoms. While the nuclear fusion processes of the stars cause a transformation from light to heavy elements, at the level of galaxies, a reverse process of the transformation of heavy into light elements occurs, probably at active galaxy cores. Even at an yet larger scale, it is possible to identify a connection between evolutionary processes of the macrocosm and the microcosm. The formation of gigantic structures of galaxies in superclusters seems to be connected with the long-range change of gravity as a partially shielded interaction.

A dialectical-materialist theory of this evolution can only be developed through critical analysis of the diverse existing observational material. Particularly in the fields of high-energy particle physics and galaxy evolution, its theoretical processing is strongly influenced by the Big Bang theory. The qualitative picture of the evolution of the universe as a self-organized, coupled structure formation in the macrocosm and microcosm presented here is hypothetical at many points, and intended to provoke further scientific work.

A theory of the evolution of matter is only possible with the systematic inclusion of all known forms of motion and excitation of the quantum æther, including its phase transitions. This requires us to overcome the arbitrary positivistic method of misusing the "vacuum", sometimes considering it as "emptiness" and sometimes as "dark matter and energy", in order to save its world formulas.

https://doi.org/10.1515/9783110644203-007

7.1 Evolution of stars and chemical elements

Stars consist of a hot gas of atoms and nuclear building blocks, whose state is determined by the opposing actions of gravitational attraction and the gas and radiation pressure inside the star.

The gravitational attraction depends mainly on the stellar mass. The pressure acting against gravitational attraction comes from three sources, which contribute to different degrees, depending on the mass or temperature of a star:
1. The pressure of an ordinary gas that depends on the temperature and is determined by the impacts and interactions of the atoms.
2. The radiation pressure of the energy released by nuclear reactions, especially in the form of photons.
3. The Fermi pressure of electrons, protons and neutrons, which is important at very high gas pressures.

Depending on the mass, the composition and the stage of development of the star, different phases of matter form in the interior of the star and in the stellar atmosphere: they range from the plasma state, in which electrons detach from the shells of the atoms to phases of pure nuclear matter in the interior of extremely dense stars such as neutron stars. In these, the electron shell is pressed partially or completely into the atomic nucleus. Thus there is a close connection between the evolution of stars and the atoms and nuclei they contain, which undergo a chemical or nuclear evolution

The stars are classified according to the Hertzsprung-Russell (HR) diagram (Figure 77). It categorizes stars into different classes according to their surface temperature and their luminosity. The main-sequence stars, to which the Sun belongs, consist of stars with masses between about 0.1 and 100 solar masses. The higher the mass, the further to the upper left of the HR diagram is a young star.

When the Dane Ejnar Hertzsprung first drew up the HR diagram in 1913, based on data from the U.S. astronomer Henry Norris Russell, the various nuclear reactions were not known. They assumed that stars develop during their evolution only along this main series from states of high luminosity and temperature (blue color) towards the bottom right, into states of low luminosity and temperature (red), and gradually die. However, this is far too simple. The existence of other classes of stars is crucial for understanding the evolutionary process: at the right, above the main sequence, are stars of extreme luminosity and low surface temperature, called red giants and supergiants. To the left, below the main line, is the class of white dwarfs, stars with a mass of about 0.6 Solar masses, but an extremely high density.

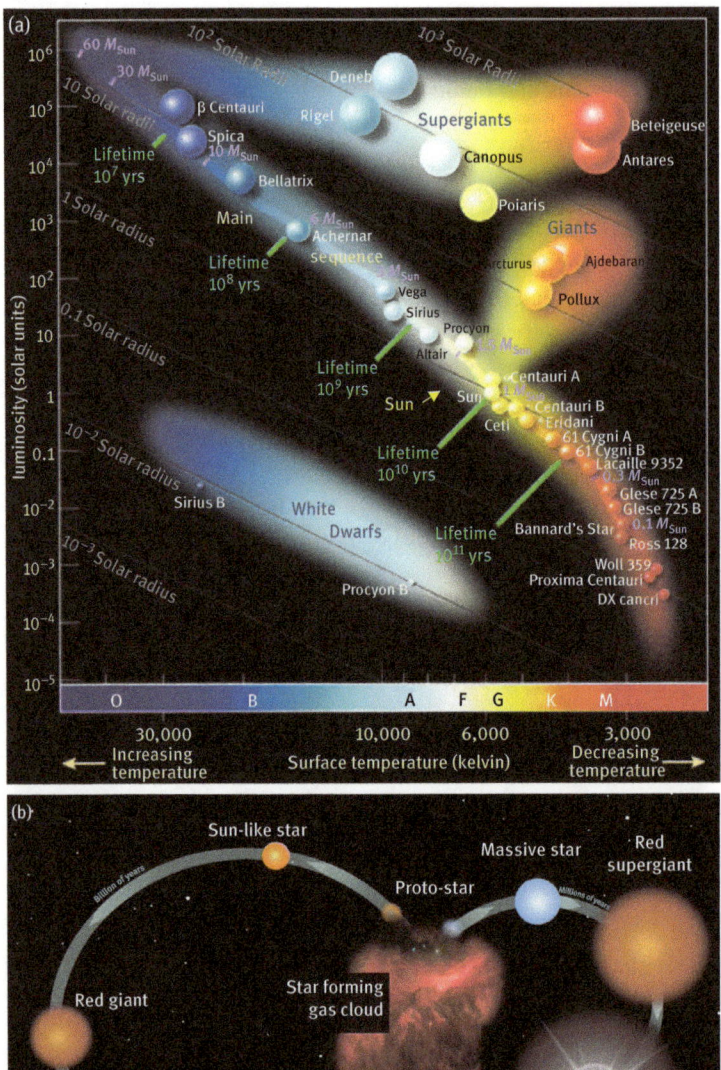

Figure 77: Evolution of stars. (a) The Hertzsprung-Russell diagram characterizes the evolution of the stars by changes in their brightness and surface temperature. The sun develops along the main sequence, and in about five billion years it will first enter the stage of red giants and then of white dwarfs. (b) Outline of the evolutionary cycles of Sol-type and massive stars. Images: (a) ESO, (b) NASA.

Chemical evolution through nuclear fusion in the main-sequence stars

The stars on the main series are characterized by a dynamic equilibrium of opposing forces that is stable over long periods of time, which is dependent on their mass. It is determined by the gas pressure, the radiation pressure of the released photons and the gravitational attraction of the distribution of masses. The energy of the radiation is fed by the fusion of light to heavy atomic nuclei (Figure 78a). This process goes through different stages of development, which depend on the mass at the time of star formation. Below about 0.08 solar masses, the pressure is not sufficient to ignite the nuclear fusion of heavy nuclei from hydrogen gas, and the star ends as a planet-like brown dwarf. Stars in the range of 0.1 to 0.26 solar masses can ignite hydrogen fusion at about 4 million degrees. A whole series of nuclear reactions then take place, including the fusion of two hydrogen atoms (^{1}H) to form a deuterium atom (^{2}D), releasing a positron, neutrinos and 0.42 MeV of kinetic energy per formed deuterium atom.

Deuterium can fuse with another hydrogen atom to form helium (^{3}He), releasing a photon with 5.49 MeV energy. The production of the isotope ^{4}He can take place via three different channels: The dominant process is the fusion of two ^{3}He atoms into ^{4}He, with the release of two hydrogen atoms and 12.86 MeV of kinetic energy. In the other two channels, lithium and boron are first created, which then disintegrate into ^{4}He through the release of neutrinos. The problem that the experimentally measured neutrino fluxes from the interior of the sun are smaller than theoretically expected was solved by the detection of conversion reactions between neutrinos ("neutrino oscillations") [Aharmim 2005]. These experiments indicate that even the ve neutrino with the smallest energy has a finite rest mass (see chapter 6.2).

Stars up to 0.26 solar masses are not able to ignite helium burning. When the supply of hydrogen is exhausted, the star contracts into a white dwarf and cools down. In contrast, stars in the mass range between 0.26 and 1.5 solar masses can ignite helium burning when the hydrogen supply reaches its end and the radiation pressure decreases. Then they leave the main series in the H-R diagram in the direction of the red giants, and the pressure and temperature in the interior continue to rise. Helium burning starts at about 100 million degrees. Essential nuclear reactions are the fusion of two ^{4}He nuclei into beryllium (^{8}Be) and the fusion of ^{4}He and ^{8}Be into a highly excited carbon nucleus (^{12}C). This decays again into mainly helium and beryllium, but some of the excited nuclei pass into the ^{12}C ground state, releasing photons with 7.65 MeV energy level.

At the stage of helium fusion, a whole series of complex accompanying reactions take place, which are not discussed here. In addition to helium fusion in the nucleus of such a star, hydrogen fusion continues to take place in the hydrogen shell surrounding the core. This energy production in two shells is a major reason why stars inflate at this stage and form red giants. When the Sun reaches such a stage in the future, it will expand to the orbit of Venus. Life will no longer be possible on Earth. During the red-giant phase, the star gradually discards its shell and ends up as a white dwarf.

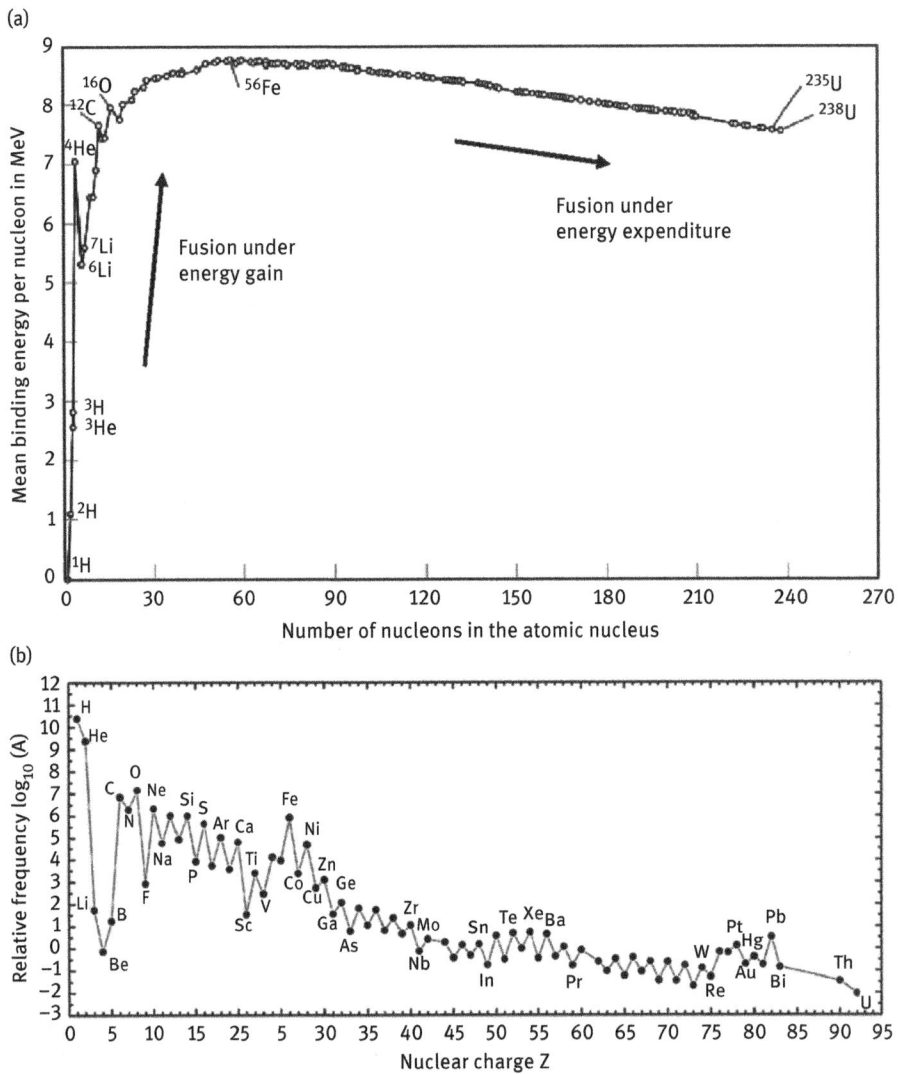

Figure 78: Evolution of the chemical elements in the nuclear fusion processes of stars. (a) Binding energy per nucleon of the different atomic nuclei. The binding energy initially increases from light elements such as hydrogen and helium until it reaches a flat maximum in iron. Iron is, therefore, the most stable element in terms of nuclear transmutations. (b) Relative frequency of the elements in the closer galactic surroundings of the Sun. The frequency of silicon was normalized to 10^6; logarithmic scale.

The role of supernova explosions in the formation of heavy elements

The nuclear fusion processes in stars explain how the distribution of chemical elements up to iron (Figure 78) is established. Stars with more than eight solar masses, but less than about ten times the solar mass, pass through the hydrogen and helium burning phases described much more quickly.

After the helium fusion and the formation of a carbon core, further fusion stages can be ignited. The evolution of nucleosynthesis then develops very quickly, much faster than the shell of the star can adapt to the changes in the nucleus as a red or blue supergiant:

- Above one billion degrees, carbon burning starts. It only takes about 600 years. Two carbon (^{12}C) atomic nuclei fuse to form neon (^{20}Ne), sodium (^{23}Na) or magnesium (^{23}Mg), releasing reaction products such as neutrons, protons and energy.
- At two billion degrees, oxygen burning begins, in which two oxygen nuclei are fused to form silicon, sulfur and other elements. Since the other stages of fusion also continue outwards in shells, such stars could develop an onion-like structure. However, this picture is idealized. In reality, convection currents also occur, that mix up the shells. Oxygen burning lasts only half a year.
- In the resulting silicon core, above about three billion degrees, silicon burning begins, in which various metals such as iron (Fe), calcium (Ca), nickel (Ni), titanium (Ti) etc. are produced. The process is dominated by the formation of the very stable iron isotope ^{56}Fe. At these high temperatures, there is an abundance of rapid nuclear reactions and back reactions, and thus an approximate nuclear thermal equilibrium. This is where the particularly stable nuclei are preferably formed.

If the fuel runs out during silicon burning, no further energy can be obtained by fusion of the next heavier atomic nuclei, since the binding energy decreases again above Fe as a function of the number of nucleons. The star can no longer withstand the gravitational pressure, and within milliseconds the core of the star collapses. The pressure is increased to such an extent that the electrons are pressed into the atomic nuclei and fuse with the protons. The collapse is associated with the formation of a shock wave which strips off the outer layers of the star explosively. In such a supernova explosion, enormous gravitational energy is suddenly converted into kinetic energy and nuclear reactions. An amount of energy of 10^{56} GeV can be released. 99% of this energy is transported away by neutrinos, only 1% goes into the shock wave and only 0.01% into the radiation of visible light. The ejected stellar atmosphere mixes with interstellar clouds of hydrogen and remnants of other stars. Shock waves from a supernova explosion can induce star-formation waves, in which dust and gas clouds form new stars and planetary systems.

The synthesis of nuclei above the threshold of the iron isotope ^{54}Fe is also possible on a small scale during different phases of the star's fusion processes. This is mainly done by neutron capture. Starting from ^{54}Fe and ^{56}Fe, the unstable iron isotopes up to ^{59}Fe are generated by neutron capture, for example. Due to the β decay in connection with the weak nuclear force, these are gradually transformed into nuclei with a higher number of nuclear charges. This "s process" (s for slow) also takes place outside supernova explosions. Nuclei with an ever higher nuclear mass up to the heavy element bismuth (^{209}Bi) are gradually formed along what is called the stability valley (see Figure 80). Nuclei with a higher nucleon number (mass number) are unstable and disintegrate by the emission of a ^{4}He nucleus.

During the supernova explosion (see Figure 79), the conditions for nucleosynthesis change. Such extreme temperatures and pressure conditions arise in a non-equilibrium state that the relative frequency, especially of nuclei with masses between that of oxygen and iron, changes. In addition to the s process, the synthesis of the heavy elements by the "r process" (r for rapid) now begins. The neutron density increases extremely so that in the r process unstable nuclei quickly build up on the neutron-rich side of the stability valley of nuclear matter. These neutron-rich nuclei then gradually decay into the stability range via the slower β decay. The "valley of stable nuclei" (marked by the s-process path in Figure 80) represents, in the sense of a phase diagram, the stability of the color-charged nuclear matter as a function of the protons or neutron number under the pressure and temperature conditions of stellar evolution.

The multitude of chemical elements of the periodic table which we find on Earth and which are the decisive prerequisite for the development of life on Earth is the result of the fusion of light nuclei into heavy nuclei in earlier stars. The existence and destruction of earlier stars is a prerequisite for the development of various chemical elements and complex molecular compounds based on them. So we see here a birth and death of star systems. However, this is not an eternal cycle, but a spiral of evolution: with each cycle a change occurs. The metal content in stars and solar systems increases in the course of the cycles of birth and death of solar systems: This allows a development from a simple to a more complex chemical composition of the systems.

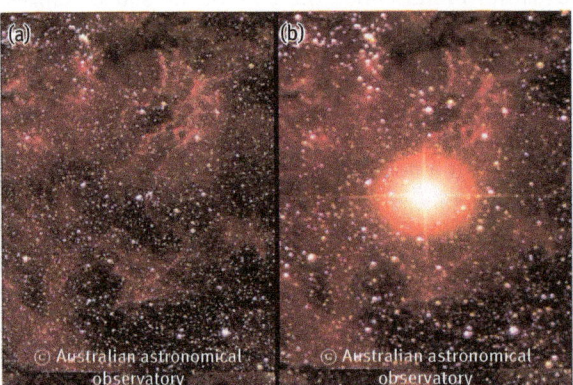

Figure 79: Supernova explosion of a massive star in the Large Magellanic Cloud. (a) Picture before and (b) after the eruption on February 23, 1987. (c) Image of the remains of the supernova taken by the Hubble Space Telescope. The ring structures show the ejected gas cloud of the precursor star of this supernova at the red-giant stage. The detail shows the expansion of the star's gas cloud in the years 1994–1996 after the supernova explosion. Pictures: (a + b) David Malin: Anglo-Australian Observatory (c) J. Pun (NASA/GSFC), R. Kirshner (CfA) and NASA.

Fermi fluids as a new state of matter in stellar corpses

In the stellar corpses, a new state of matter is formed. Studies of these illuminate further aspects of the dialectical unity of the state of matter in the macrocosm and evolutionary processes of subatomic building blocks in the microcosm.

White dwarfs are remnants of stars with a mass below the threshold at which a supernova explosion ends the life of a star. They have a surface temperature of up to 100,000 degrees Celsius, and are therefore hotter than the Sun, corresponding to their white color. They consist of the products of hydrogen fusion (mainly oxygen and carbon nuclei) and have a characteristic radius of about 7,000 km (slightly smaller than Earth) with a mass of about 0.6 solar masses, which corresponds to about 200,000 Earth masses. These observations show that there must be a form of matter unusual for terrestrial conditions in white dwarfs, with a very high mass density. The stability of the white dwarfs is primarily determined by the unity and struggle between gravitational attraction and Fermi pressure. The gas pressure of ionized atoms plays a subordinate role. So in the interior of a white dwarf, the state of matter is that of a Fermi fluid of interacting electrons and nuclear building blocks.

Inside the white dwarfs, the Fermi energy is much greater than the thermal energy (about 100,000 degrees in the interior). So despite the enormous temperature, it is what is called a degenerate Fermi fluid, which is in a state close to its ground state, similar to that of electrons in a metal. Such a "Fermi fluid drop" moving freely in space has unusual properties: its radius decreases when the mass increases – exactly the opposite of normal stars or normal gases [Hansen & Kawaler 1994]. The reason for this is the increase in mass of the fermions due to their high kinetic energy near the Fermi boundary. So when the mass increases, the gravitation increases disproportionately to the mass, which is why the volume decreases.

Stars with an initial mass above ten solar masses form neutron stars as stellar corpses. In addition to the gravitational pressure, the enormous pressure of the supernova explosion compresses the resulting iron atoms in such a way that their electron shells are pressed into the atomic nucleus. Under such conditions, the electrons combine with the protons of the nuclear building blocks to form neutrons and a neutron star is formed. In the star, the neutrons form a "superatom nucleus" with an enormous radius of about 10 km. This theoretical prediction of the transition from the degenerate Fermi gas of a white dwarf to macroscopic nuclear matter in a neutron star is based on the comparison of the Fermi energy of the electrons with the energy difference for the transformation of protons into neutrons of about 1 MeV. If the Fermi energy of the electrons exceeds this value, electrons from the electron Fermi fluid are fused with protons until the Fermi energy has sunk below the threshold for electron capture again. This produces nuclei on the neutron-rich side of the stability valley (see Figure 80). Under normal conditions, these nuclei would decay via the weak nuclear force. In the neutron star, however, this fails due to the occupied Fermi Sea of the electrons at high pressures. The neutrons, therefore, dominate the interior of stellar corpses above densities

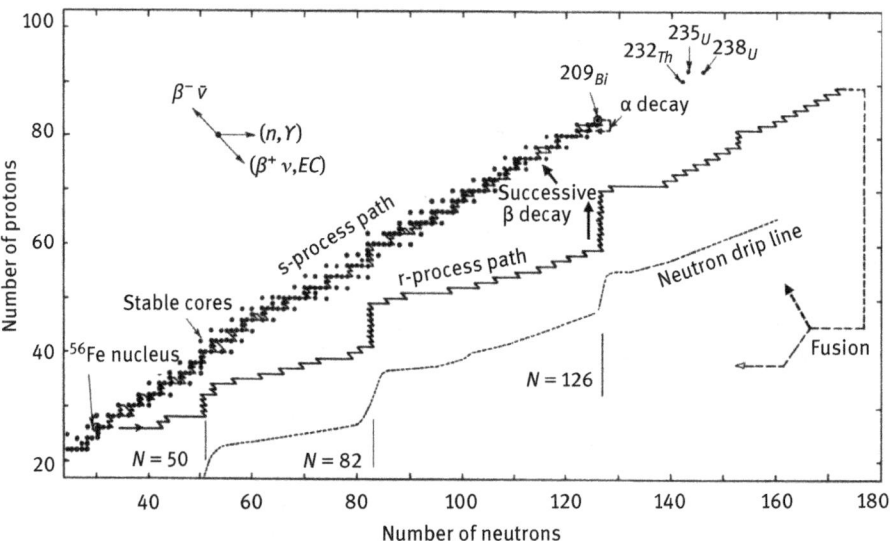

Figure 80: Evolution of the heavy elements above iron. There is an evolutionary path via gradual neutron capture (**s** process), as well as rapid neutron capture, for example during the supernova explosion (**r** process). The stable elements form a narrow "valley" in the field of possible nuclear structures of different numbers of protons and neutrons. On the neutron-rich side, nuclei decay by means of beta-decay (electron emission), and beyond the neutron-drip limit by neutron emission [Rolfs & Rodney 1988].

of 10^7 kg/m^3. Above 10^9 kg/m^3, the neutron gas itself forms a degenerate Fermi fluid, which is identical to the formation of a gigantic atomic nucleus of pure neutrons.

Astronomically, neutron stars are observed as pulsars (see Figure 81) that emit strong radio pulses at intervals of seconds. When a neutron star is formed by the collapse of a star, it is spinning at an incredibly rapid rate, on the order of one revolution per second, due to the preservation of angular momentum with a decreasing radius (pirouette effect). As a result, neutron stars have extremely strong magnetic fields of up to one million tesla. Radio waves are emitted along the poles of the magnetic field. They can be observed as periodic radio pulses.

Since the gravitational pressure increases from the surface into the interior of a neutron star, they must have a layered structure of different states of matter (Figure 81). Below a hot atmosphere that is only a few centimetres thick, an iron surface follows. This is followed by an outer shell in which iron reaches a thousand times its density on Earth. As the gravitational pressure increases further into the interior of the star, neutron-rich nuclei are formed, but the gravitational pressure is not yet sufficient to squeeze all electrons into the protons: A plasma of neutron-rich atomic nuclei and electrons is therefore formed at this level of the outer shell.

The inner shell probably consists of a solid layer of neutron-rich nuclei and free neutrons. The proportion of neutrons in this matter continues to increase inward.

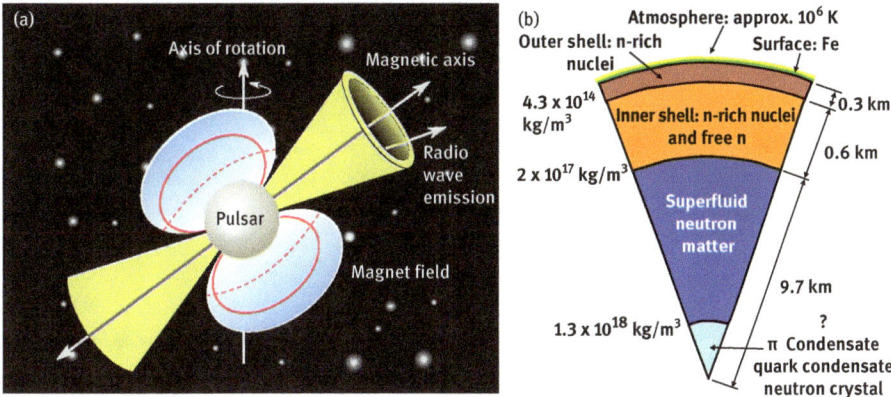

Figure 81: Properties of a neutron star with 1.4 times the solar mass and 10 km radius. (a) Due to its rotation and the alignment of the magnetic moments of the neutrons, a neutron star has a magnetic dipole field with enormous field strengths of ten thousand to one million tesla. By comparison, the terrestrial field is only 0.05 millitesla strong. Radio waves are emitted along the poles of the dipole. The axis of rotation of neutron stars often differs from the dipole axis. In this case, the neutron star is what is called a 'pulsar'; the radio waves sweep rapidly across particular directions in space. (b) Model of the possible layered structure of a neutron star [Ventura & Pines 2012].

From the laws of quantum fluids, it can be expected that the resulting degenerate Fermi gas from neutrons will condense into a superfluid state when the matter waves of the neutrons overlap sufficiently. A thimble of this superfluid neutron matter would have the unimaginable weight of 10^{10} tons. It can be expected that with a further increase in density towards the center of a neutron star, a phase transition to a neutron crystal will occur. When the neutrons touch each other, a phase can even occur in which the quark confinement is removed and macroscopically extended quark condensates similar to the quark-gluon plasma may be formed.

In fact, neutron stars have now been observed not only via radio signals from pulsars. Figure 82 shows the Cancer Nebula, which represents the remains of a supernova explosion at a distance of about 6,000 light years, seen in the year 1054 C.E. In the center, there is a pulsar with 30 pulses per second, which has been studied at different ranges of the spectrum, from the X-ray to the radio-wave range. The X-ray spectrum shows the formation of jets on opposite sides, with rings around the jet axis. These observations confirm that it is a rapidly rotating compact object, with the mass and size as expected for a neutron star. The X-ray image shows various ring structures around the pulsar that indicate how the "central rotating object" feeds the luminous gas clouds with energy. The presumed mechanism of energy transfer is based on particles (electrons and positrons) that are accelerated to nearly the speed of light by extreme electromagnetic fields of the rotating neutron star, and transfer the energy thus obtained to the surrounding luminous gas. In the visible spectrum (Figure 82b), the expanding gas cloud can be seen as a remnant of the supernova explosion.

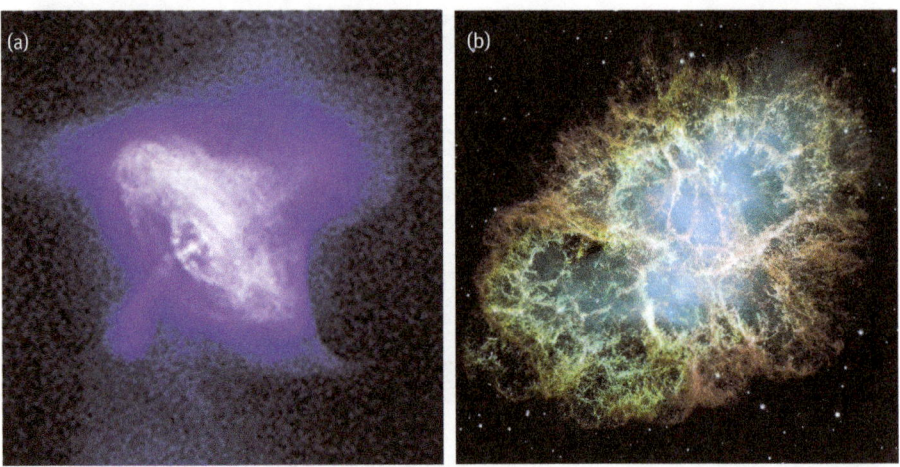

Figure 82: Neutron star in the Cancer Nebula. It emerged from a supernova explosion observed in 1054 CE. (a) Picture of the center of the nebula in X-ray light, showing a jet structure emanating from the invisible neutron star. (b) Picture of the ejected star shell in visible light. Images (a) NASA/CXC/SAO, (b) J. Hester & A. Loll, Arizona State Univ.

The stability limit of the Fermi fluid at higher pressures

The predicted emergence of new phases of nuclear matter in neutron stars with increasing gravitational pressure raises the question of how the equilibrium of forces between gravitational and Fermi pressure develops at even higher mass densities. The increase of the gravitational energy has to be compensated by an increasing degeneration pressure of the Fermi liquid. This increases their Fermi energy, which simply means an increase in the kinetic energy of the neutrons with a corresponding increase in mass. When the Fermi energy reaches the size of the Dirac energy gap of the neutrons, their velocity approaches the speed of light. Each further increase of the gravitational pressure causes the mass to increase strongly non-linearly. An exact theoretical analysis by the Indian-American astrophysicist and Nobel Laureate Subrahmanyan Chandrasekhar in 1930 showed that there is an upper limit to the mass, the Chandrasekhar mass limit. Above this threshold, an instability is created that leads via a "gravitational collapse" to a qualitatively new state. Above the Chandrasekhar mass limit, white dwarfs cannot exist in a stable state because their degeneration pressure can no longer withstand gravitational attraction. They collapse into neutron stars. At even higher mass densities, a further collapse occurs into a "black hole". These are the final stage of very massive suns. "Black holes" of even greater mass and qualitatively new properties occur in the center of galactic nuclei.

The possibility of the formation of "event horizons" in quantum fluids from which no quasi-particle excitations can escape was already discussed in Chapter 5.

Such material structures are not a "singularity" in space-time, but form a phase boundary surface to a new state of the quantum æther. The evolutionary process of the atomic nuclei in the qualitatively different evolutionary phases of the stars gives an indication for deriving a theory of this phase transition in the quantum æther. The stable atomic nuclei of various masses and nuclear charges form droplets of a color-charged phase of the quantum æther, in which the color charges of the quarks are rapidly shielded from the outside. The stable atomic nuclei with different proton and neutron numbers form a linear stability valley for the "normal" temperature and pressure conditions of our everyday world (Figure 80). We regard them as the valley of stable color-charged states of these topological defects in the gluon condensate of the Dirac phase of the quantum æther.

A change of the phase stability and thus of the valley of stable nuclei takes place as the gravitational pressure increases. At the level of the white dwarfs and neutron stars, the stability valley of the atomic nuclei begins to shift to the neutron-rich side by fusion of electrons and protons to neutrons. First, the equilibrium between the beta-decay of nuclei and neutron capture shifts. If, however, the beta-decay is suppressed by the Fermi Sea of electrons, a change in the electroweak interaction is to be expected. This enables the formation of macroscopic atomic nuclei in which the quark confinement is eliminated, and new collective quark states are formed. This would mean that, at such gravitational pressures, not only the electroweak but also the color interaction would change, so that their coupling strength decreases with increasing pressure. The phase boundary of the "black hole" is therefore probably determined by a pressure-dependent electroweak phase transition. Thus the Dirac energy gap collapses, and a macroscopically extended section of the quantum æther with the electroweak phase is formed. In other systems of condensed matter in solid-state physics, such pressure-dependent phase transitions are known as "quantum phase transitions". The hypothesis of a quantum phase transition in "black holes" is central to the understanding of fermiogenesis, the production of stable elementary particles from such states.

7.2 Evolutionary processes of galaxies

Stars do not appear randomly distributed in the universe but form a new structural level of matter: galaxies. Furthermore, they also form globular clusters as part of the halo of galaxies. Galaxies are relatively stable systems of a few million to a trillion (10^{12}) stars, which are shaped by the struggle and unity of gravitational attraction and repulsion by centrifugal or inertial forces. Different types of galaxies occur. They are divided into two main types: blue spiral, barred-spiral and irregular galaxies with high star-formation rates, and red elliptical galaxies, in which star formation has largely come to a standstill [Schneider 2006].

Our own galaxy, the Milky Way, is a spiral galaxy. It consists of a thin disk of stars, star clusters and interstellar gas arranged in the form of spiral arms around a central bulge with a massive core. The Sun is also located in the thin disk. The Milky Way has a total mass of about one trillion solar masses. Its largest part is in the form of stellar masses (about 80%) in the disk, which has a diameter of 100,000 light-years and a thickness of about 1,000 light-years. The central bulge contains about 20% of the stars of the spiral galaxy and has a diameter of about 10,000 light-years. The density of the stars increases by several orders of magnitude in the direction of the central bulge. Compared to the position of the Sun (28,000 light-years distant from the galactic nucleus), the star density is 50 times higher at a distance of 300 light years and 200,000 times higher at a distance of 3 light-years from the nucleus. In the vicinity of the nucleus, it reaches millions of times the star density in the surroundings of the Sun. The actual galactic nucleus consists of a very compact object with about 1 million solar masses and a diameter of a few light minutes, which corresponds to the diameter of our solar system out to Jupiter. Around the disk with the central bulge, there is a halo, which contains stars and globular clusters with roughly a hundred thousand to a million stars.

Hubble's classification of galaxies

Besides the spectrum of their emitted light, galaxies differ in their structure. The classification of galaxies into different structure types goes back to E. P. Hubble, who studied galaxies in detail in the 1920s and 1930s. He divided the galaxies into three basic types according to their visible structure of luminous matter: elliptical galaxies, spiral galaxies and barred spiral galaxies [Hubble 1936].

In the Hubble classification, these three main classes are refined further (Figure 83): spiral galaxies are divided into three subtypes: Sa galaxies have very strongly coiled spiral arms and a pronounced central bulge. Sb galaxies have more open spiral arms and the central bulge is less pronounced. Sc galaxies, which include the Milky Way, have very open spiral arms with many wave-like bulges (see also Figure 84d). The barred-spiral galaxies are also divided into three subtypes SBa, SBb and SBc, which have different degrees of compactness of the spiral arms. In general, the spiral arms contain many young blue stars, while the central bulge contains mostly older, red stars.

There is widespread agreement that both types of spiral galaxies are relatively fragile objects [Longair 1998], firstly because disc disks are unstable under the effect of their own gravity, and secondly because a stable spiral structure would require the orbital speed of the stars to increase with increasing radius of their orbit. Therefore, the spirals are interpreted as density waves of stars, that are accompanied by star formation waves.

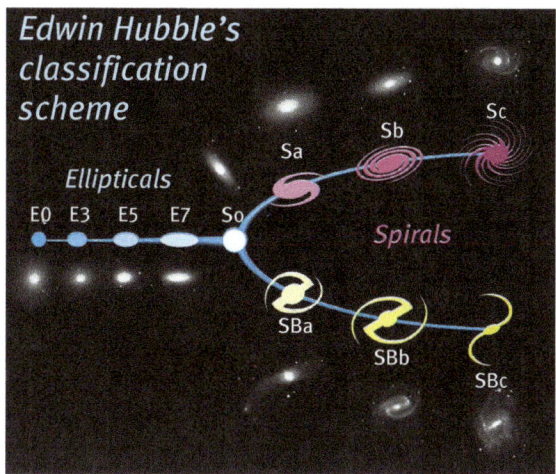

Figure 83: Classification scheme of different types of galaxies based on E. P. Hubble. A distinction is made between elliptical (E0–E7), lenticular (S0), spiral (Sa-Sc) and barred-spiral galaxies (SBa-SBc). Today, other types have to be added to this picture. These include the active galaxies, the irregular galaxies, the dwarf galaxies, and the intermediate forms between spiral and barred-spiral galaxies. Picture: NASA.

Elliptical galaxies have no distinct structure and belong to the brightest galaxies (Figure 84b). The subtypes E0 to E7 are distinguished according to the degree of ellipticity, with E0 having an almost spherical shape. Their stellar motion is irregular compared to the rotary motion in spiral and barred-spiral galaxies. Lenticular galaxies S0 (Figure 84c) are transitional types between elliptical and spiral galaxies. Furthermore, a whole series of irregular galaxy types occur, which led to extensions of Hubble's classification scheme [Vaucouleurs 1974]. One example are the Magellanic Clouds near the Milky Way. Such galaxies appear as companions of spiral galaxies and elliptical galaxies. Since remains of spiral structures can still be recognized in some irregular galaxies, these were classified as stages of the spiral and barred-spiral galaxies, which are called Sd, Sm and Im [Vaucouleurs 1974].

Controversy about the evolutionary processes of the galaxies

The analysis and classification of the different forms of a species necessarily leads to the question of their relation in its evolution. This method was followed, for example, by Darwin for the establishment of a theory of evolution of biological species. Hubble interpreted his classification as an evolutionary process of galaxies evolving from "early" (Sa, SBa) to "late" (Sc, SBc) types.

However, observations of galaxies in different regions of the spectrum allow different structural characteristics to appear in one and the same galaxy. For

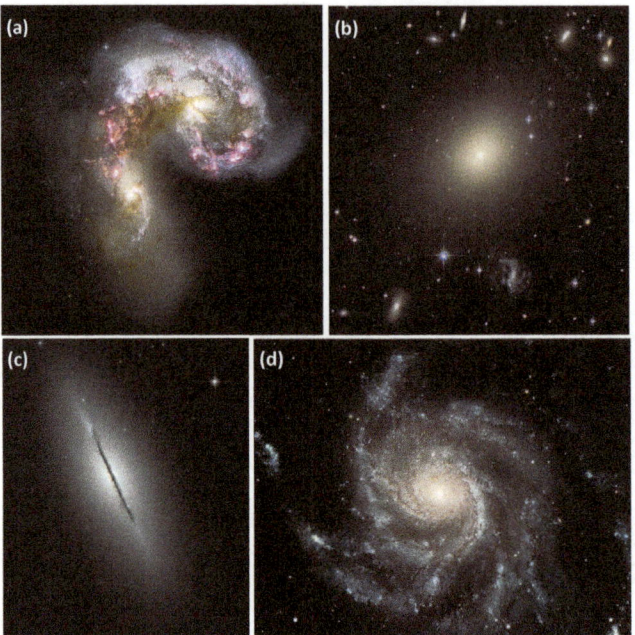

Figure 84: Different types of galaxies. (a) Antennae galaxy formed by the fusion of two galaxies. Such galaxies are also called "starburst galaxies" because they have an exceptionally high rate of star formation. (b) Giant elliptical galaxy in the Abell S740 cluster. (c) Side view of a lenticular galaxy (type S0). (d) The spiral galaxy Messier 101 of the Sc type. Photographs: NASA/ESA, (b) J. Blakeslee.

example, the galaxy NGC4303 appears as a spiral galaxy in the ultraviolet and as a barred-spiral galaxy in the infrared [Sheth 2003].

Such observations are considered an argument for completely rejecting Hubble's evolutionary theory. In a modern textbook on extragalactic astronomy it says: *"We explicitly emphasize here that this nomenclature is not a statement about the evolutionary stage of the objects, but only a nomenclature of historical origin"* [Schneider 2006, p. 88]. Even though Hubble's evolutionary outline represents an over-simplification, the core of it, an evolutionary process of galaxies in different stages, is correct. In fact, different processes, such as collision and fusion of galaxies (Figure 84a), propagation of star-formation density waves in the spiral arms, the formation of active galactic nuclei with jet structures, up to the possible decay of galactic nuclei and the ejection of fragments, are established in playing a role in the evolution of galaxies. Even for simple galaxies, the complex interaction of gravitational attraction and centrifugal repulsion is still not well understood (see chapter 5.4).

In the book *Ratlos vor der Großen Mauer: Das Scheitern der Urknalltheorie* (*"Perplexed in the face of the Great Wall: The failure of the big-bang theory"*) [Lutz 1991] Josef Lutz proposes an extended model of evolution based on the work of the Soviet astronomer Ambarzumijan (Figure 85). The heart of this extension is that galaxies are not formed simply by condensation of a gas cloud or fusion of existing galaxies, but that galaxies in their active core stage also produce new galaxies by ejection of matter.

The French physicist F. Combes' working group is investigating how active galaxies and barred-spiral galaxies can transform into one another: In bar structures, there is a pronounced flow of matter towards the central bulge. Above a critical matter density of the central bulge, the bar collapses, forming a galaxy with an active galactic nucleus. The bar then forms again after the density of matter in the center has been reduced [Delgado-Serrano 2010]. Hammer et al. investigated in detail the conditions under which spiral arms, rings and bars form again due to instabilities, after the fusion of two galaxies [Hammer 2009]. Many details of the evolution of galaxies are not yet understood [Cheung 2013]. Josef Lutz's model in Figure 85 is also an approximation. Active Seyfert galaxies probably develop more from the fusion of two galaxies than from elliptical galaxies. The positive core idea of cyclical galaxy evolution, in which the stage of active galaxies evolves as a developed contradiction between increased gravitational attraction in the nucleus, and the conversion processes thus induced, is correct and is discussed in Section 7.3 in more detail.

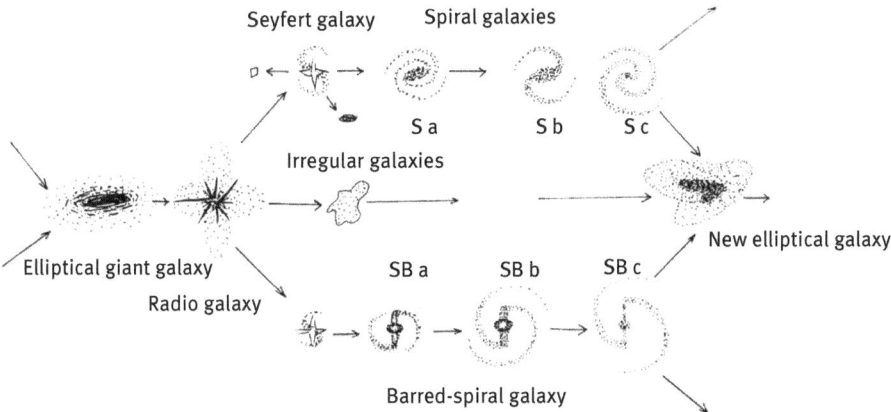

Figure 85: Proposed evolutionary process of galaxies according to Josef Lutz. Galaxy evolution is cyclic with phases of an active nucleus (radio and Seyfert galaxies), phases with high star-formation rates (spiral and barred-spiral galaxies) and quiet phases with many old stars (elliptical galaxies). Illustration: [Lutz 1991].

In contrast to an evolutionary cycle, the big bang theory claims that all galaxies were formed more or less simultaneously from the fluctuations of matter density at the transition from the radiation-dominated to the baryon-dominated universe. This theory is in crisis. It is not based on observations but must assume "dark matter" not only to explain the stability of galaxies (see chapter 5) but also for the nucleation of the formation of galaxies. For a non-expanding gas, James Jeans had already worked out in 1920 that structure formation in a plasma only begins when fluctuations in density occur at lengths greater than a characteristic length, known as the Jeans length. Only then is it possible for spiral structures to develop in a condensing ionized gas cloud in connection with plasma instabilities.

With the expanding gas of the Big Bang theory, however, such collapsing instabilities are suppressed extremely strongly. As a way out, Y. Zeldovisch, I. Novikov from Moscow and Peebles at Princeton in the 1960s presented the idea that extremely strong fluctuations in the decoupling of radiation and particles represent nuclei for later galaxy formation. However, this contradicts the extreme uniformity of cosmic background radiation [Longair 1998, p. 14]. To this day, the Big Bang theory can only model the structure formation of galaxies, galaxy clusters and superclusters if it is assumed that there is an invisible inhomogeneous background medium which causes the nucleation of galaxy formation, but is not perceptible in the radiation. The study of galactic evolutionary processes is thus subordinated to the Big Bang model. Individual development processes, for example through collisions and mergers of galaxies, are recognized, but the establishment of an overall evolutionary theory of galaxies is detached from experimental observations.

Metal content and the chemical evolution of galaxies

The generation of heavy elements by nuclear fusion processes in the stars leads to a complex chemical composition of galaxies.

An extensive study of more than 15,000 stars in the Milky Way shows no systematic increase of the metal content with the age of the stars, but rather a very strong variation at the same age [Nordstroem 2004]. However, due to the larger mass of the stars and the resulting shorter lifetime, the chemical evolution to higher metal contents in the central bulge should be more advanced than in the halo or in the disk. The very old globular clusters in the halo, on the other hand, should rather have a low metal content. Therefore the metal content in the galactic disk, the halo, and the central bulge should differ considerably. However, these simple assumptions only apply if galaxies do not alter their structure in the course of their evolution. If there is a cyclic galactic evolution in different stages, the chemical composition is mixed repeatedly. In fact, measurements of different galaxies show extremely contradictory results. While [Whitford & Rich, 1983] found a higher metal content of the stars in the central bulge, other researchers found a higher metal content in the disk [McWilliam 1997].

This refutes the naive assumption that there is a uniform chemical structure of all galaxies due to uniform formation after a "big bang". Wave-like star-formation processes play a central role in the chemical inhomogeneities [Pagel 1997, p. 287 ff]: zones and times of extremely intense star formation are triggered by collisions of galaxies, or by density waves from supernova explosions. Such processes favor mixing and homogenization of metal rich regions formed in earlier processes. This can explain that the metal content of spiral galaxies seems to be largely independent of the type Sa to Sc. A tendency towards metal content decreasing outwards is found in elliptical giant galaxies [Pagel 1997, p. 293], in which star formation has largely come to a standstill.

Globular clusters in the halo of galaxies are among the most beautiful celestial bodies visible through a telescope, because there is hardly any interstellar gas in them, and the stars appear particularly clear and without absorption. Based on our understanding of the laws of stellar evolution, it is estimated that globular clusters are up to 15–20 billion years old. High-resolution spectroscopic studies by the Hubble telescope have shown that galaxies have two types of globular clusters in their halo: metal-rich and metal-poor [Brodie & Strader 2006]. Metal-rich globular clusters up to an age of 14 billion years are observed [Spark & Gallagher, 2000, p. 72]. Indeed, all explanations by the collision of galaxies [Ashman & Zepf, 1992] assume that colliding galaxies already had metal-rich zones when the globular cluster was formed.

These observations contradict the thesis of the simultaneous formation of all galaxies according to the Big Bang scenario. Theoretically calculated times for the re-formation of galaxies after a merger are of the order of several billion, perhaps dozens of billion years. Thus some of them developed longer than the calculated age of the cosmos after the Big Bang. By investigating the spectrum of high-energy gamma rays, it has been demonstrated recently that even at a distance of 12.6 billion light-years, metal-enriched molecular clouds with 3% of the metal content of the Sun are found [Campana 2007]. But according to the Big Bang theory, a significant metal content of interstellar matter should only have arisen after the passing of a first generation of stars. Some measurements of gamma-ray bursts with increasing red-shift indicate a decrease of the metal content with greater distance [Savaglio 2006]. However, selection effects caused by the mechanisms of formation of gamma-ray bursts via collisions between neutron stars or between galactic nuclei might suggest such a dependence falsely.

An examination of the metal content, on the basis of spectra of objects with a high red-shift, shows that a high metal content can occur in quasars up to $z \sim 5$ and in stars up to $z \sim 6\text{--}8$ [Spark & Gallagher, 2000, p. 357]. Conversely, the evaluation of the metal content of 250,000 galaxies of the Sloan Digital Sky Survey found very near galaxies that have a very low metal content of only one twentieth of that of the Sun [Kniazev et al. 2003]. Therefore, there is no experimental evidence of an increase of the metal content over time, which would be expected according to the Big Bang theory. Even the plasma in the intergalactic space between the galaxies has a surprisingly high metal content, corresponding to 45% of the metal content of the Sun [Böhringer 1999] so that it must already have been part of the stellar evolution of earlier generations.

Overall, these observations are in agreement with a cyclical evolutionary process of galaxies and contradict the assumption of simultaneous formation after a Big Bang. A trend with distance or age does not seem to exist. Rather, large inhomogeneities of the metal content occur both within a galaxy and between galaxies. Has a dynamic chemical equilibrium of the elements in the cosmos developed over large time scales? This would only be possible if, in addition to the reaction to heavy elements in stars and supernovae, there were also a process of back-reaction from heavy to light elements.

7.3 Nuclei of Active Galaxies

Active galaxies have an active galactic nucleus. Such galaxies exhibit an extremely strong emission of electromagnetic radiation, which exceeds the emission of a galaxy of the Milky Way type by up to ten orders of magnitude. The spectrum of this electromagnetic radiation is non-thermal, and extends over fifteen orders of magnitude of frequency from radio waves to hard gamma rays. This distinguishes them from the emission spectrum of non-active galaxies, which have approximately a thermal emission spectrum, which extends over only two orders of magnitude of frequency, from near infrared to near ultraviolet. Active galactic nuclei are the source of the ejection of large quantities of matter in the form of jets, gas clouds, and possibly also parts of galactic nuclei. This makes active nuclei the brightest and most energetic objects found in the universe to date. They dominate the radiation of the entire parent galaxy. In the course of the discovery of different objects with extremely large energy radiation, a whole series of type designations has been introduced in the past, such as radio galaxies, Seyfert galaxies, blazars, BL Lacertae objects and quasars. It has become clear in recent years that they all possess active galactic nuclei that are in different stages of evolution, or are being observed from different directions.

Characteristic of active galaxies are the features of the active galaxy 3C 31, a dominant active galaxy in a galaxy chain, as shown in Figure 86. The common feature is a jet structure that can extend from several thousand to hundreds of thousands of light-years, and usually ends in lobe-like structures emitting radio waves. Furthermore, around active galactic nuclei, one or more disk or torus-like structures of gas or plasma can be seen, which revolve around an extremely massive central object with 10^6–10^9 solar masses, at sometimes very high speeds.

Types of active galaxies and galactic nuclei

A distinction is made between active galactic nuclei according to the strength or temporal variability of the emission in the radio wave and X-ray gamma range [Schneider 2006]:
- Among the strongly radio-emitting galaxies are the **radio galaxies**, elliptical galaxies with active nuclei. Their radio radiation comes mainly from lobe-like matter structures, which usually occur symmetrically on opposite sides of the

center of the galaxy. These lobes are fed by matter jets that emerge from the active galactic nucleus. One example is Centaurus A in Figure 87a.

- Also among the strongly radio-emitting objects are **blazars**, which are also called **BL Lacertae objects**. They have radiation in the optical, radio and X-ray ranges that varies greatly over time. It is assumed that here the jets and radio lobes extend in the direction of the observer. This is indicated by the high degree of polarization of the light in the optical range and the great variation over time of the gamma emission at energies greater than 100 MeV. Probably, these are also elliptical radio galaxies with an active galactic nucleus.

- **Quasars** (quasi-stellar objects) are extremely bright galactic nuclei with a large red-shift (typically $z \geq 2$), and very strong X-ray and gamma emissions. An example is 3C 273 in Figure 91. They occur in the active stage of irregular and elliptical galaxies. If their large red-shift were directly interpreted as a great distance, according to the Hubble effect, they would have to have an extremely high yield of energy, corresponding to the combustion of a mass of ten suns per year, and in the largest of up to a thousand suns. The prevailing opinion is that some active radio galaxies also contain quasars, but these are only visible from certain angles. In addition to the strongly radio-emitting active galaxies or quasars, which are associated with the formation of gigantic jet structures, there are also what are called "radio-quiet quasars".

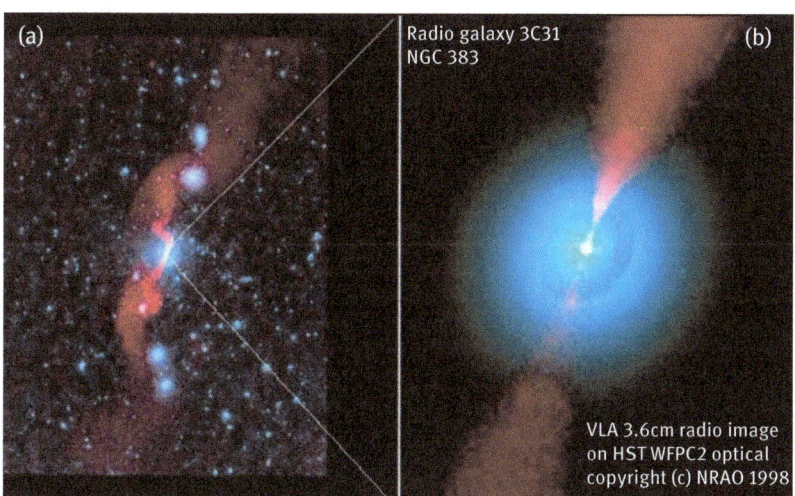

Figure 86: Active radio galaxy 3C 31 (NGC 383). Combination of an optical (blue) and a radio wave image (red). Clearly visible is a jet structure (blue) ending in a radio lobe (red). The jet with a lobe extends over 980,000 light-years. The radio emission mainly originates from the jet structure, the lobe, and the inner torus of the galaxy. The radio waves have a continuous spectrum. It is produced by the acceleration of a plasma of electrons and positrons to velocities close to the speed of light. Pictures from NRAO/AUI.

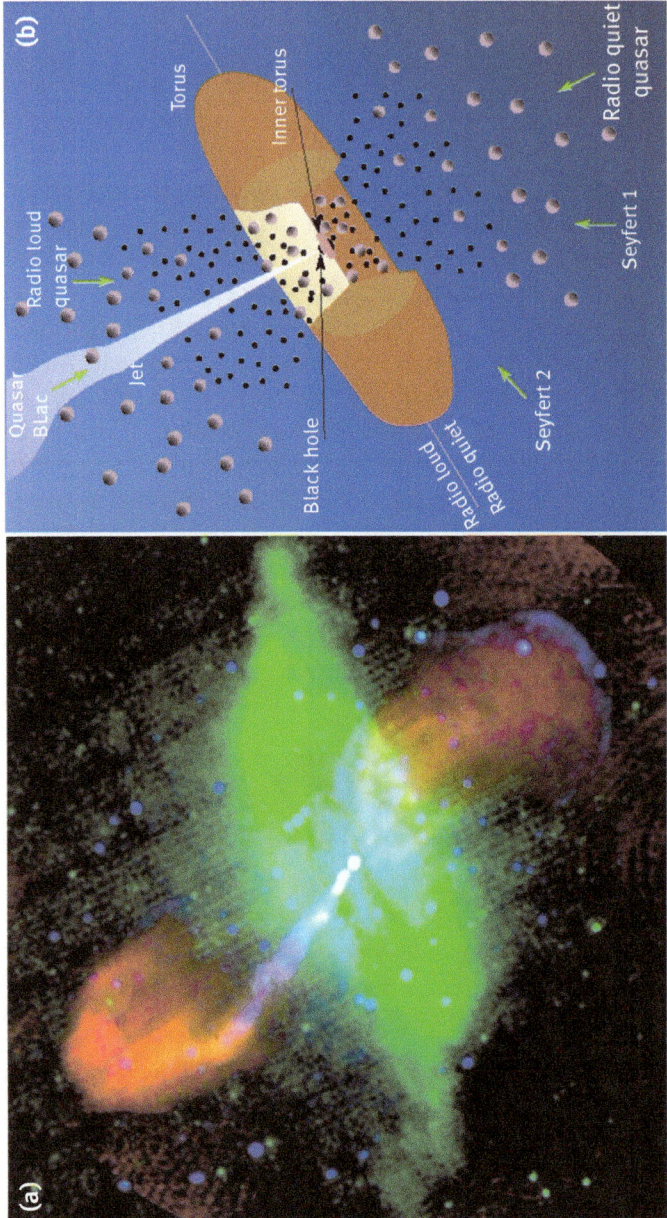

Figure 87: (a) False color representation of the radio galaxy Centaurus A in the range of radio waves (red), infrared radiation at 24 μm wavelength (green), and in the X-ray range at energies of 0.5–5 keV (blue). Radio waves are emitted especially by the lobes at the end of the jet. The jet itself emits strongly in all three spectral ranges. X-ray emission requires the acceleration of electrons to energies in the MeV range. The infrared emission is due to thermal radiation of the gas and dust distributed in the galaxy. (b) Unified model of active galactic nuclei having evolutionary phases with and without a jet. The visible features also depend on the direction of observation. Figure (b) after J. J. Condon & S. M. Ransom.

- Like quasars, the active galactic nuclei of **Seyfert galaxies** also exhibit a significantly weaker emission in the radio range, but a strong X-ray emission. They are
  part of spiral galaxies or irregular galaxies. Seyfert galaxies are the earliest class
  of galaxies with an active galactic nucleus discovered: they were discovered in
  1943 by C. K. Seyfert. Their X-rays resemble the synchrotron radiation of a rapidly
  rotating accretion disk. Seyfert galaxies are categorized as Type I and Type II.
  Type I Seyfert galaxies have greatly broadened emission lines of hydrogen, helium, neon and oxygen atoms. They are caused by the fast rotation, at speeds of
  500 to 4,000 km/s, of the accretion disk around the massive center of the galactic
  nucleus. Seyfert type I galaxies and quasars are very similar. Type II galaxies have
  only narrow emission lines, so that accretion disks are either absent or hidden.

A "unified model" of the different types of active galactic nuclei is shown in
Figure 87b [Urri & Padovani 1995]. It assumes that they represent evolutionary phases
with and without developed jet activity in active galaxies, leading to radio-loud or
radio-quiet active galaxies. Furthermore, it assumes that the angle of view in relation
to the direction of the jet has a considerable influence on whether one sees a quasar
or blazar, the dominance of X-ray or radio emission, narrow or widened spectral lines.
Strong radio emission occurs particularly if a jet is active. Extremely high-energy
X-ray radiation is seen particularly when viewed along the jet axis. In BL Lacertae
objects (BL Lacertides), the jet in the direction of view varies over time. Seyfert galaxies
are evolutionary stages without active jets. In quasars, the galactic nucleus is particularly bright; some quasars may even represent naked galactic nuclei.

The Seyfert galaxy Circinus

An example of a closely studied Seyfert type II galaxy is the Circinus galaxy [Wilson
2000]. It has a chaotic disk-shaped structure which is probably due to the interaction and fusion of two galaxies. At right angles to its galactic disk, there is a sharply
limited V-shaped outflow at about 150–200 km/s, which contains hydrogen and oxygen among other things. The center of the galaxy contains a ring of hydrogen with
a diameter of about 1300 light-years and a very compact hydrogen ring at a distance
of 130 light-years. These rings contain zones with extremely high star formation
rates. No hot stars are observed at a distance of fewer than 30 light-years from the
center of the galaxy.

A summarizing evaluation of Seyfert galaxies [Mathur 2000] comes to the conclusion that Seyfert galaxies usually originate from the fusion of galaxies and have
an an active galactic nucleus at early stage of evolution. According to [Mathur
2000], galaxies in the Seyfert stage undergo a "rejuvenation process" due to extremely high star-formation rates. Some of them have been shown to possess a double nucleus (for example, Arp220). In the course of the fusion of two nuclei, Seyfert

galaxies could enter the quasar stage, in which the radiation of the now merging nucleus increases still further.

Nuclei of active elliptical giant galaxies: Cygnus A and M87

The late, somewhat quieter stage of such an active galaxy formed from a merger is the giant elliptical galaxy. The orbit of the stars around the center becomes very chaotic during and after the fusion process, radial rather than rotary components of velocity occur increasingly, and an elliptical shape without a disk is formed. These galaxies are dominated more by old stars, and there are no high rates of star formation. As long as their galactic nucleus is still active, they form strong radio galaxies. One example is the brightest radio source in the sky, Cygnus A (Figure 88).

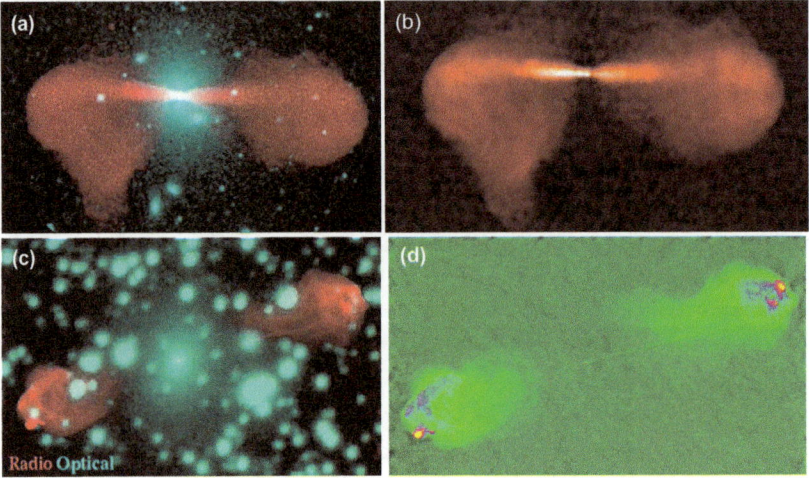

Figure 88: Jet structure and plasma lobes of the active elliptical giant galaxies 3C 296 (above) and Cygnus A (below). (a) and (c) each show a superposition of visible (white) and radio-wave regions of the spectrum (red). (b) and (d) show the radio-wave emission in a false-color image. While in 3C 296 the jet structure is extremely active all the way to the center of the galaxy, in Cygnus A the activity is weakly developed or greatly decreased. This reflects the different activity stages of the active galactic nucleus. The jets are about 500,000 light years long. Images: (a) 3C 296 NRAO/AUI 1999 (b) ATLAS of DRAGNs Project (c) W. C. Keel (d) VLA, NRAO.

Cygnus A has a symmetrical double jet with a diameter of about 500,000 light years, which is larger than the entire galaxy. The jets consist of plasma, probably mainly protons and electrons, with velocities close to the speed of light [Carilli 1996]. They end in a shock wave where lobular structures of ionized gas are formed. These are the main sources for their radio wave emission. However, the actual active nucleus of the galaxy has not yet been resolved, because it is covered by gas and dust. But there is a complex inner structure and evidence that the nucleus of Cygnus A resembles a quasar [Carilli 1996, p. 40]. Spectroscopic measurements show the presence of hydrogen inside the center, and high velocities of up to 500 km/s, indicating a torus around a massive central object.

Another example of an active nucleus in an elliptical giant galaxy is the M87 system. It forms a galaxy of about 300 billion solar masses in the center of the Virgo galaxy cluster. There are two huge jets with a length of about 5,000 light-years (Figure 89). Astronomers speculated that the high X-ray activity of the beam originated from matter falling into the nucleus of the galaxy, until the images of the Very Large Array Radio Telescope (VLA) in 1999 proved, by detecting node-like ejections, that the energy for the X-rays is pumped from the massive center into the jet (Figures 89c and d).

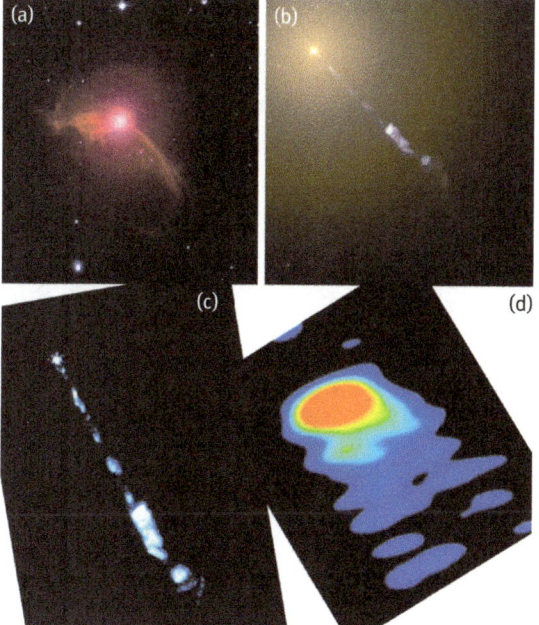

Figure 89: Active elliptical giant galaxy M87 made visible in different spectral ranges. (a) Overview with superposition of X-ray (red) and visible spectra (blue). The image shows the giant elliptical galaxy and the jet structures. (b) Hubble telescope photograph in the visible spectrum, showing the center of the galaxy and one of the two jets. (c) A high-resolution image of the jet in the ultraviolet spectrum. An interrupted nodular structure of the jet is clearly visible. (d) An extremely high-resolution image of the active galactic nucleus in the radio-wave spectrum. The red central region from which the jet comes corresponds to a size of 0.03 light years, i.e. about 20 times that of our solar system. Images: (a) NASA & DSS, (b) Hubble & NASA/ESA, (c) F. D. Macchetto/NASA/ESA) (d) NRAO.

Accretion disks and jets of active galactic nuclei

The compact central object of active galactic nuclei cannot be made directly visible, because it is surrounded by luminous dust, gas and plasmas. Only recently, scientists claimed to image the "shadow of a black hole" in M87 at a wavelength of 1.3 mm by assembly of several telescopes [Akiyama 2019]. They confirm that many galactic nuclei are surrounded by what are called 'accretion disks', which consist of gas and dust particles rotating rapidly around the nucleus in the form of a ring. Figure 90 shows a good example.

Figure 90: Center of the active galaxy NGC 4261, an elliptical giant galaxy with an active galactic nucleus. An extremely massive compact object with 4×10^8 solar masses is located there. The accretion disk on the right has a diameter of only 800 light years. Photograph: HST/NASA/ESA.

The investigation of the speed of rotation serves to determine the mass of the compact central objects. However, stable orbits of the particles in accretion disks do not exist very close to the compact central object. For a non-rotating "black hole", the innermost stable orbit lies at about three Schwarzschild radii R_s. For the compact central object of the active galactic nucleus of M87, with a mass of about three billion suns, R_s would be about 0.001 light years.

The spatial resolution of radio telescopes today makes it possible to obtain information about the origin of the jet down to distances of about 30 R_s from the compact central object. The investigations at 43 GHz of the active galactic nucleus of M87 (Figure 89d) show that the jet has a relatively wide aperture angle at a distance of about 30 Schwarzschild radii R_s and only focuses to a slender structure at a distance of about 100 to 150 R_s [Junor 1999].

This observation supported the theory that the jet is not ejected directly from the compact object, but is created by the rotating accretion disk of electrically charged particles. This disc generates extremely strong magnetic fields, at whose north or south pole the jets could be formed by a magneto-hydrodynamic self-focusing effect of the plasma [Koide 2006]. However, recent high-resolution radio-wave studies at 86 GHz show that the jet extends closer to the central object. At distances of 30 R_s, it is smaller and radiates more strongly than would be expected by magnetic self-focusing from an accretion disk [Krichbaum 2007]. Although it is not possible to draw final conclusions about its origin from high resolution observations [Akiyama 2015], the observation of jet bending [Kim 2020] points against a direct origin from the accretion disc.

Jet formation from a plasma instability of the accretion disk contradicts the pulsed jet structure, with a strong temporal variation of the jet within less than a year, however. A series of images from the Hubble telescope from 1994 to 1998 show that the speed of optically luminous ejected jet structures is seemingly up to six times the speed of light [Biretta 1999]. The acceleration of the particles necessary for such nodular structures, can hardly be supplied by electromagnetic processes at the accretion disk, since the velocities of the particles in the jets are many orders of magnitude (300,000 km/s) higher than those of the accretion disk (typically 500 km/s). Therefore, the major part of the acceleration would have to be supplied during the self-focusing phase, when the jet matter has to overcome the gravitational potential of the massive core at the same time. So all models based on the conversion of electromagnetic energy into kinetic energy of jet matter cannot explain the enormous speed of the particles, on the order of the speed of light [Harris 2006].

Like other active galactic nuclei, M87 also exhibits time-variable gamma radiation in the TeV energy range [Aharonian 2006]. This is far above the energy ranges of less than 200 GeV available today in the largest particle accelerators. The high energy of the jets and gamma radiation indicates that the energy scale of the electroweak phase transition in the quantum æther is reached or exceeded in the processes in active galactic nuclei.

Temporal variation of the jet structure of quasars

Figure 91 shows one of the brightest quasars (3C 273). It has a red-shift of z = 0.16, which according to Hubble's Law would correspond to a distance of 2.4 billion light

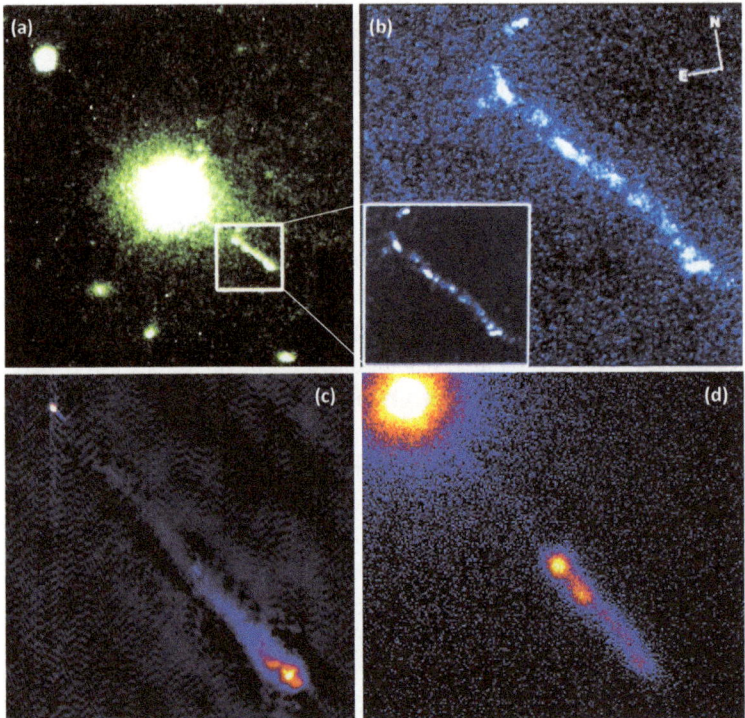

Figure 91: The Quasar 3C 273, shown in different regions of the spectrum. (a) Optical image from the Hubble Space Telescope. (b) Jet structure with higher resolution in the visible spectrum. (c) Radio-wave image of the quasar with jet, with a visible connection between jet and quasar. (d) X-ray image showing gamma radiation in the GeV to TeV range at the nucleus and in some nodes of the jet. Images: (a) NASA/ STScI, (b) R.C. Thomson, C.D. Mackay, A.E. Wright, (c) MERLIN (d) NASA/CXC/SAO/H. Marshall.

years. Its structure has been studied at different regions of the spectrum. It has a mass of about 900 million solar masses and radiates a hundred times as much energy as typical radio galaxies. At a diameter of 7000 light years, its active nucleus is as large as the entire width of the jet structure of the radio galaxy M87 (about 5,000 light years). The visible part of the jet of 3C 273 has a length of 200,000 light years; and the total length of the partially interrupted structure, starting from the quasar, is 550,000 light years. Such interruptions are typical for quasar jets. They are an expression of the variable activity of the nucleus. Because of the variable emission in the X-ray range over time scales from days to years, 3C 273 was also classified as a BL Lacertae object. The processed image shown as an insert in Figure 91b shows that the nodes are particularly prominent in polarized light. The polarization of the light confirms that it is a synchrotron emission of light due to the motion of very energetic electrons and ions.

Quasars are extremely active stages of active galactic nuclei in which the energy radiation even exceeds that of radio galaxies [Biretta 1999].

In general, the extent of the jets in quasars is about ten to a hundred times larger than in radio galaxies [Harris 2006], and the jet structure is often disconnected from the active nucleus. Using the example of 3C 273, it was shown recently that the temporal variability of the emissions in the X-ray and radio-wave ranges are strongly correlated [McHardy 2006], and therefore the activity of the jet and that of the "central machine" are interconnected. The quasar 3C 273 would already be as bright as the Sun in the sky at a distance of 36 light years. To produce such energy radiation, it would have to burn the mass of about a thousand Suns per year. Such energy and length scales in quasars are clearly incompatible with purely electromagnetic conversion processes in the plasmas of accretion disks. This statement remains valid even if the actual distance of the quasars is less than assumed by Hubble's Law. Thus a part of the red-shift of the quasars is surely intrinsic, caused by their gravitational red-shift. A reduction of the distance of the quasar 3C 273 to half would still correspond to energy radiation of 250 solar masses per year.

From the considerations of the enormous energy output, the jet energies and extents that occur, as well as the occurrence of gamma quanta in the TeV range, it follows that, in the quasar state of the galactic nucleus, a qualitatively new state of matter of the compact central object has established itself. This is interpreted in the following as a local electroweak phase transition in the quantum æther.

Chemical composition of quasars and jets

Spectral studies of the jets of quasars and BL Lacertae objects show that they are dominated by extremely high-energy electrons, positrons and protons, i.e. building blocks of the light elements [Kataoka 2007, Harris 2006]. It may be assumed that the gamma radiation of the jets constantly forms electron-positron pairs and, in the case of radiation up to the TeV range, other particle-antiparticle pairs, such as protons and antiprotons, as well. The experimentally observed composition of the jets is consistent with a fermiogenesis of protons and electrons in the electroweak phase transition [Jooss & Lutz 2006].

The quasar stage would thus be the climax of the growth phase of the most massive active galaxies with the most massive nuclei [Hamann 2007], which probably lasts only about a hundred million years. According to current knowledge, the quasar stage is already preceded by an active phase of the galaxy with an active nucleus and extremely high star formation, for example by the fusion of two galaxies.

The quasar with the largest red-shift known today ($z \sim 6.5$) was formed by a fusion of several galaxies [Li 2006]. So that would have been only a few hundred million years after the Big Bang. Yet this galaxy must have gone through a rich evolution, with the formation of stars with 10^{12} solar masses, even before the quasar stage. This result alone would already be enough to refute the Big Bang model, according to Popper's "falsification principle".

Controversy about the red-shift of quasars

Serious critics of the Big-Bang hypothesis among astronomers [Arp 1998, Hoyle, Burbidge & Narlikar 2000] express doubts about the idea of the great distance of quasars with high red-shifts. According to them, there are additional intrinsic red-shifts in quasars, which modify the distance determined according to Hubble's Law. This would make these objects closer and less energetic. For a whole series of quasars at small [Bahcall 1997], medium [Kirhakos 1999] and very large distances [Walter 2004], the red-shift of the parent galaxy was measured independently. It corresponds to that of the quasars. Also, along the path of the light from the quasar to the observer, absorption occurs in the intergalactic gas (the "Lyman-alpha forest"), which corresponds to the expected distances according to Hubble's Law [Keel 2003]. By contrast, [Arp 1997] finds many examples where pairs of quasars with a high red-shift are arranged around a parent galaxy with a low red-shift, and deduces the existence of intrinsic red-shifts. Red-shifted atomic line spectra would be expected, for example, if atomic building blocks such as electrons in quasars had a reduced Dirac energy gap. Possibly the main emission sources of atomic line spectra in quasars are not affected by such an effect because they are far away from the electroweak phase boundary, and thus from regions of reduced mass.

From the arrangement of quasars with respect to large parent galaxies, H. Arp [Arp 1998a] developed the thesis that some quasars are ejections from active galactic nuclei, possibly even "naked fragments of a galactic nucleus". Evidence of material connections between quasars of different red-shifts and a parent galaxy [Lopez-Corrredoira & Gutierrez 2002] supports such a hypothesis. The ejection hypothesis would shed new light on the evolutionary processes of galaxies and the formation of galaxy groups.

7.4 Fermiogenesis in active galactic nuclei

The experimental observation of the ejection of large quantities of electrons, protons, photons, and probably other subatomic particles, in the jets of active galactic nuclei raises the question of their possible origin in the instabilities of active galactic nuclei. Galaxies have a nucleus with an extreme force of gravity at their center, corresponding to a mass of 10^6 to 10^{10} solar masses [Longair 1998]. Their Schwarzschild radius at 10^6 solar masses would be comparable to the size of the Sun, and at 10^{10} solar masses to five times the distance from the Sun to Pluto. According to the theory of "black holes", no light would escape from such a region. As yet, the resolution of our telescopes is not able to perceive such small regions, however. The enormous increase in the orbital velocity of stars in the immediate vicinity of galactic nuclei, at distances of less than one light year, is proof of the extremely high mass densities [Schneider 2006]. A theoretical description as a singularity in general relativity theory only expresses the failure of the geometric theory at the boundary to extremely high mass and energy densities

(Chapter 5.5), however. The model of a space-time singularity is not adequate for describing qualitatively new states of matter under extreme conditions.

The state of the quantum æther in galactic nuclei

Like the Chandrasekhar mass limit for the gravitational collapse of a white dwarf to a neutron star, which results from the balance between gravitational pressure and Fermi pressure, another critical mass density exists for the collapse of a neutron star to a "black hole". This critical threshold is reached when the Fermi energy of the neutrons reaches the size of their Dirac energy gap as the particle density increases. However, the Dirac energy gap does not represent an absolute limit for the development of the balance between positive gravitational pressure and negative degeneration pressure. Instead of collapsing into a "space-time singularity", a new aggregation state of the quantum æther will develop. I suspect that inside the "black hole" the condition for the electroweak phase transition in the sense of a quantum phase transition is reached, in which the Dirac energy gap collapses. This may result in a phase with mass-less particles, shown as a gray core in Figure 92. This state therefore counters its cause, the extremely high gravitational pressure caused by the enormously concentrated mass in the galactic nucleus. To what extent the phase induced by gravitational pressure differs from the electroweak high-energy phase induced in the thermal phase transition is not discussed here for simplicity's sake.

Figure 92 extends the quantum-fluid model of the "black hole" from Chapter 5.5 to rotating states of ultra-dense objects, which are called "Kerr black holes". In addition to the event horizon (exceeding the Landau velocity perpendicularly to the phase boundary of the topological defect), the intrinsic rotation of the objects also results in a zone called the ergosphere: It encloses a region for which the velocity of the superfluid condensate in the tangential direction is greater than the Landau velocity. In the ergosphere, the excitation spectrum for bosons or fermions tilts in such a way that in the direction of exceeding the Landau velocity, filled states of negative energy now assume positive energy. However since particles can escape from this region in a tangential direction, in contrast to the event horizon, an external observer would regard such a region as highly excited. These highly excited states are annihilated in the various permitted decay channels, for example by photons or fermions.

Such an effect was first predicted by Zel'dovich and Starobinskii in the 1970s [Zel'dovich 1971, Starobinskii 1973], and the radiation was called superradiation. The ergosphere thus forms a complex transitional region between the phase of the quantum æther with a Dirac energy gap outside the phase boundary, and the region of the electroweak phase without Dirac energy gap inside this. At the poles, the ergosurface and event horizon become identical. According to theoretical studies by [Volovik 2003, p. 325], the possibility of excitations at superluminal velocity arises

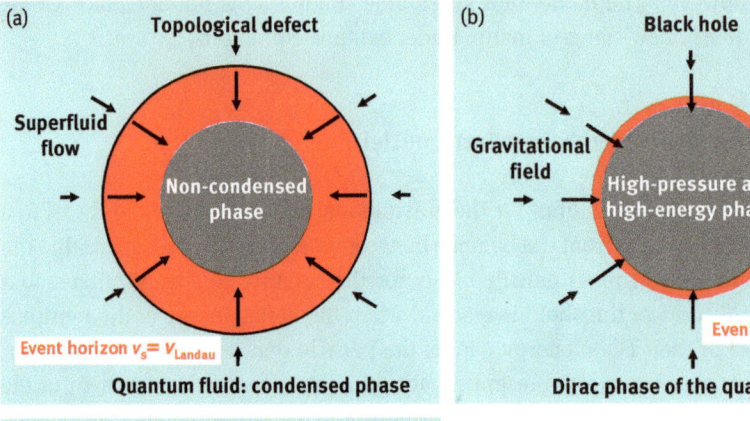

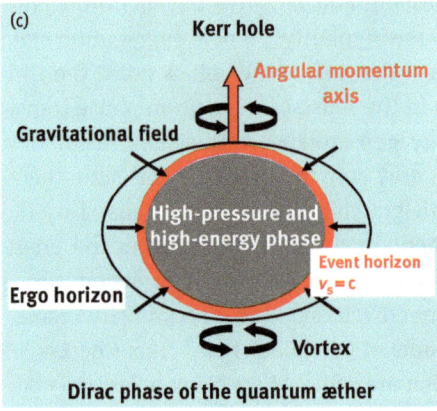

Figure 92: Superfluid model of "black holes" in quantum fluids (a) and in the quantum æther (b, c). (a) An event horizon is formed in the superfluid state of a quantum fluid when the Landau velocity of the condensate is exceeded. (b) In the quantum æther, the event horizon of objects with extreme gravitation occurs at the Schwarzschild radius. There the Dirac energy gap disappears. (c) A rotating "black hole", also called a "Kerr black hole". Between the event horizon and ergosurface, an ergosphere emerges, in which states of negative energy are raised to positive energy and radiate away. According to [Volovik 2003].

there, which can cross the phase boundary and leave the "black hole". We may assume that these poles are therefore the starting point of the gigantic jet structures observed, which may have superluminal velocity in part. They occur when a critical speed of rotation or mass density of the compact central object is exceeded, at the active state of the galactic nucleus.

Electroweak phase transition and fermiogenesis

A net generation of fermions, i.e. electrons and protons, and thus hydrogen, requires breaking the symmetry of the paired particle-antiparticle generation. In the Dirac phase of the quantum æther, electron-positron pairs or proton-antiproton pairs are formed by excitations of sufficient energy. However, the net production of particles is zero, since for each particle an antiparticle is formed, which can then annihilate themselves in pairs.

The decisive indication of the mechanism of a net production of fermions comes from the parity violation of the weak interaction. Below the electroweak phase transition, only a minor violation of the symmetry between particles and antiparticles occurs: particles or antiparticles with mass, in which spin and momentum have left- or right-handed helicity differ in their weak interaction. Right-handed particles and left-handed antiparticles are not subject to the weak interaction at all. And according to theoretical arguments, the asymmetry between left- and right-handed particles could be eliminated above the electroweak phase transition. This would also eliminate the differentiation between weakons and photons, and they would form what is called a hyper-electromagnetic field of unique weak and electromagnetic interaction.

Under certain conditions, such as non-equilibrium or rotation of the Kerr black holes, an excess of levorotary electrons can occur in the electroweak phase, characterized by their hypercharge. When cooled to below the electroweak phase transition, the Dirac energy gap develops. The electrons with right-handed helicity condense into the occupied states of the Dirac Sea. Thus the hyper-electromagnetic field decays into massive weakons and mass-less photons. The mass of the weakons as residues of the electroweak interaction is due to their partial shielding by the superfluid Higgs condensate. The left-handed electron acquires a mass, and the hypercharge develops into an electric charge.

As early as 1967, the Soviet physicist Sakharov outlined the general preconditions for the process of forming a surplus of fermions over antifermions [Sakharov 1967]: (1) Conditions for a violation of the law of conservation of the number of baryons or leptons. (2) Interactions that are asymmetric between particles and antiparticles. (3) Deviations from equilibrium to minimize backreactions of the excess formed. These conditions can be realized in the electroweak phase transition as follows [Volovik 2003]

– It represents a first-order phase transition between the electroweak high-energy phase without an energy gap and the low-energy Dirac phase, so that the baryon and lepton numbers can change at the phase boundary surface.

– The assumed anisotropic structure of the electroweak phase (node in energy gap) in which the symmetry between left- and right-handed particles or antiparticles is broken by a field or dynamic process, so that different densities of particles and antiparticles can form.

- A sufficient non-equilibrium to prevent the difference in density of particles and antiparticles formed in the electroweak phase from being equalized again during the phase transition to the low-energy Dirac phase.

The generation of a strong hyper-magnetic field, such as by the intrinsic rotation of the "Kerr black holes", seems to be the key to increasing the asymmetry between the density of particles and antiparticles in the electroweak phase.

It shifts the chemical potential for fermions and antifermions in opposite directions, thus promoting the production of fermions, and reinforces the character of the electroweak phase transition as a first-order transition [Krusius 1998, Piccinelli 2004], in which left-handed antifermions and right-handed fermions form superfluid condensates below the Dirac energy gap. Indeed, [Adler 1969, Bell & Jackiw 1969] could already show theoretically that the production rate of leptons is proportional to the energy flux density of the hyper-electromagnetic field.

The properties of the jets of active galactic nuclei suggest that ideal conditions for the generation of fermions (leptons and baryons) prevail precisely at their exit points from the ergosphere: as shown in the model in Figure 93, a strongly directed hyper-electromagnetic field would be generated at the poles of the rotating high-density Kerr black hole. It is fed by the decay of the antifermions excited to positive energies the ergosphere. When crossing from the phase boundary into the Dirac phase of the quantum æther, the hyper-electromagnetic field decays in a non-equilibrium process into an oriented magnetic field, into photons of a broad frequency range, and into electrons and protons. The velocities of the jets close to (or possibly above) the speed of light, their composition and TeV radiation are indications of such a mechanism. A more detailed experimental study of jets and improved modeling could provide evidence for such a scenario.

Can we understand that for every left-handed electron produced, a proton is formed, and thus the matter in the cosmos is electrically neutral on average? The theoretical conceptions of this are still very vague: probably the condensation of right-handed electrons in the Dirac Sea is in fact the formation of skyrmion-like topological defect (proton) due to the angular moment structure of the involved condensates. The skrymion core separates into a soft split skyrmion nucleus (quark substructure), while retaining the electrical and color charge, and thus produce the bubble structure of the nuclear building blocks in the quantum æther with their quarks (see Chapter 6.6).

The investigation of electroweak phase boundaries, their instabilities, and the processes of fermiogenesis require extensive further experimental and theoretical studies. The basic prerequisite for this is to break away from the idealistic constructs of "black holes" as singularities of space-time and "fermiogenesis during the Big Bang" and to consider real observations in astrophysics and particle physics without bias.

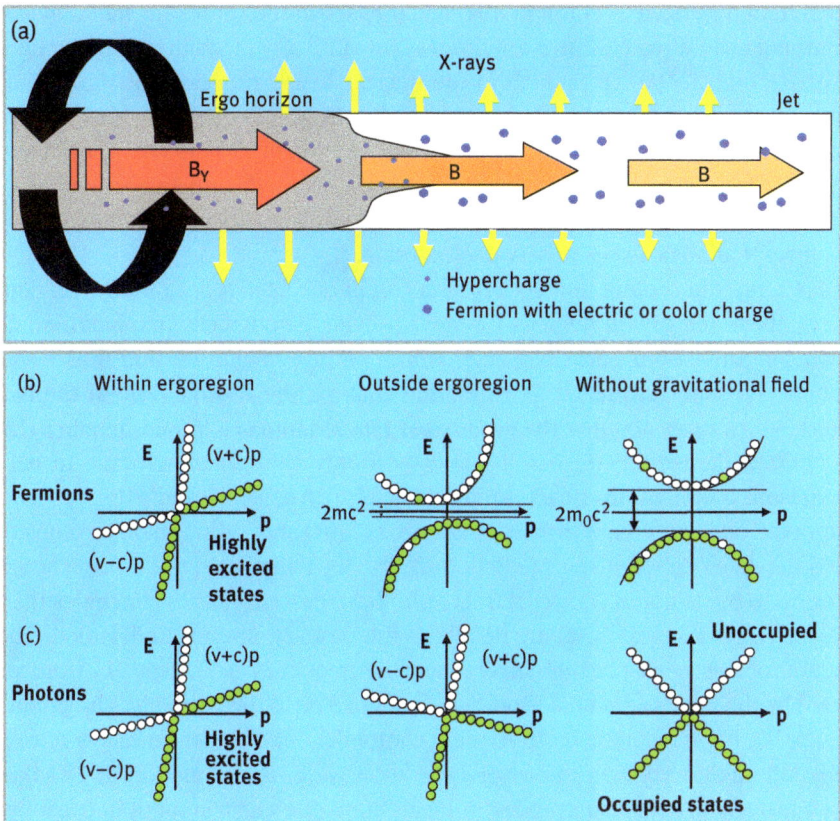

Figure 93: Model of fermiogenesis in the jet of an active galactic nucleus. (a) Condensation of fermions from hyper-electromagnetic fields. The hyper-magnetic field B_γ decays outside the ergosphere of the jet into a magnetic field B. Thus, the highly excited hyper-charges of the electroweak phase develop into the fermions of the Dirac phase with their characteristic electrical charges, color charges, and mass. At the phase boundary, high-energy X-rays and synchrotron radiation are produced. Evolution of the excitation spectra of fermions (b) and photons (c) as they leave the jet: In the ergosphere, the occupied sea of negative states is elevated to positive energies, and it begins to radiate. The velocity v of a condensate of the Dirac phase close to the electroweak phase boundary just stands here for the impact of a strong gravitation field on the speed of light c. The mass m of the resulting fermions increases with the distance from the ergosphere, and finally reaches the rest mass m_0 in the Dirac phase.

Composition of the light elements – a success of the Big Bang theory?

In equilibrium, the reaction rate between photons and particle-antiparticle pairs would produce a ratio of baryons to photons of 10^{-18}, contrary to the roughly 10^{-10} actually observed. This contradiction and the alleged explanation of the high ratio of baryons to photons became an important argument for the proponents of a hot Big Bang. In addition, it was impossible to understand the observed ratio of the light elements from hydrogen to boron in interstellar space in the context of stellar evolution. Burbidge had already drawn attention to the difficulty of synthesizing the light nuclides D, ^{6}Li, ^{7}Li, ^{9}Be ^{10}B and ^{11}B in the nuclear fusion processes of stars, since they represent very fragile objects due to their low binding energy, and would rapidly transform into more stable nuclei under the conditions close to equilibrium and high temperatures in stars [Burbidge et al. 1957]. Later, Hoyle and Tayler (1964) recognized that stellar nucleosynthesis is not able to explain the observed abundance of ^{4}He, more than 25%, in the observable universe.

Indeed, rapidly cooling non-equilibrium conditions from a hot phase with temperatures above ten billion degrees (>1 MeV) provide a convincing explanation. At such temperatures, both particle species would be in equilibrium, and thus occur with about the same frequency. During rapid cooling, the relative ratio of the two nucleons, which came about at the decoupling temperature of 1 MeV, is frozen in (at $n/(n + p)$ ⁻0.21 [Longair 1998, p. 236], and then only changes slowly due to beta decay. Protons and neutrons now form light nuclei such as D, ^{3}He and ^{4}He.

However, a recent review article by Prantzos (2007) shows that there are considerable differences between the prediction of the Big Bang theory and observation with regard to the isotopes D, ^{3}He, ^{6}Li, ^{7}Li and ^{11}B, so that the author concludes that, despite 50 years of research, many details of the problem are not understood. The connection of the mean photon density with the frequency of ^{6}Li and ^{7}Li isotopes predicted by the Big Bang model is off by a factor of 3 or 10, respectively. The prediction of the Big Bang theory and observation contradict one another for heavy nuclei, as well. Although according to our present knowledge, the formation of Fe takes place exclusively in the fusion processes inside stars, the metal content in the intergalactic medium, especially the Fe content, is higher than the mean content in many galaxies [Renzini 1997]. It is at least five times higher than would be compatible with the evolutionary history of the elements according to the Big Bang theory. All these are clear indications that the formation of the light elements takes place mainly locally and inhomogeneously, e.g. in high-energy processes in active galactic nuclei far from equilibrium. They act as a "rejuvenating tonic" for atoms and nuclei and can thus lower the metal content in the galaxy locally. This would thus allow for an infinite circular process of generating light atoms out of the quantum æther, evolving them into heavier species in stellar nucleosynthesis and transforming them back into condensates and light elements at galaxy cores.

The phase diagram of the quantum æther

A qualitative phase diagram of the quantum æther (Figure 94) derives from the scientific synthesis developed in this book from the materialist findings of high-energy

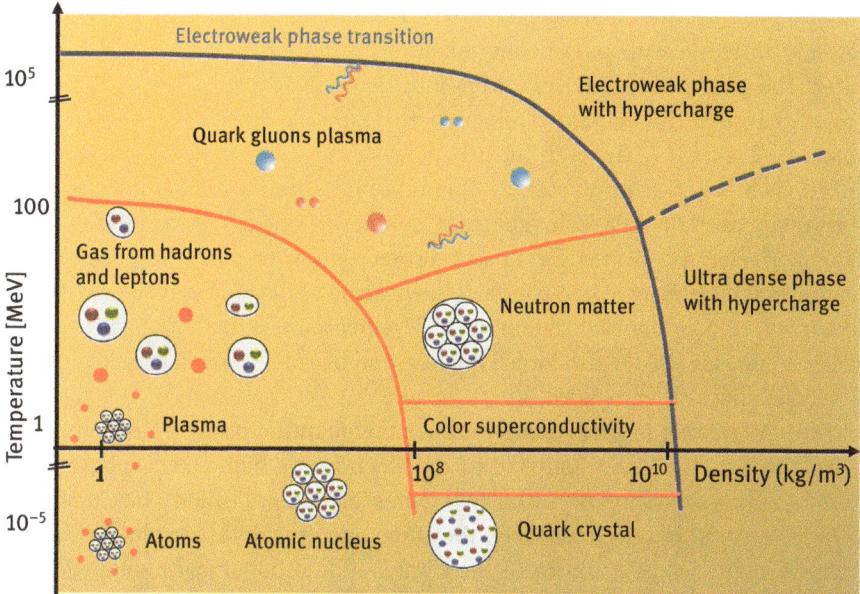

Figure 94: Schematic phase diagram of the quantum æther as a function of mass density and temperature. Neither the axis scales nor the size of the areas of the forms of matter shown is true to scale. The temperature and density figures are rough estimates.

particle physics and astrophysics. It reflects the phase stability of the Dirac phase of quantum æther with possible Higgs and gluon condensates and of the stability of the "elementary particles" as topological defects of the condensates:

- Under our ambient conditions, the quantum æther is present in the Dirac phase with an isotropic energy gap. Quarks form bound systems of nucleons, which in turn combine with electrons to form atoms. The shielding of the color charge in the nucleons and the electrical charge in the atoms enables the development of higher structural levels of matter. The weak charge is shielded and determines the decay of all high-energy leptons and hadrons into electrons, neutrinos, protons and neutrons. Free neutrons decay into protons.
- As the temperature increases, atoms ionize and enter the plasma state. A further increase in energy at low density leads to the formation of gases of leptons and hadrons. At even higher energy, the phase transition to the quark-gluon plasma occurs. There, the shielding of the color charge is removed by the elimination of the gluonic component of the superfluid condensate. Quarks and gluons can move freely. At even higher energies, the electroweak phase transition occurs, in which the Higgs component of the superfluid condensate and thus the shielding of the weak charge disappear. The quantum æther probably acquires an anisotropic, possibly chiral structure there, and the Dirac energy gaps of the leptons and quarks disappear.

- As the mass density increases in the Dirac phase, the qualitative leap from an atomic structure with nucleus and shell to the state of nuclear matter occurs first. The phase of nuclear matter is also determined by the shielding of the color charge in the nucleon bubbles that occupy the "valley of stable nuclear building blocks". At an even further increase of the mass density, such as in white dwarfs and neutron stars, a quantum phase transition occurs due to the beginning overlap of the color-charged "bubbles" of the nucleons. "Macroscopic atomic nuclei" are formed e.g. in star corpses. At low temperatures, such states can presumably change into coherent movements of the quarks (color superconductivity) or quark crystallization. At high temperatures, the high-density color-charged phases should also form a quark-gluon plasma in which the inclusion of the color charge is eliminated.
- A further increase in mass density causes the quantum phase transition into a phase that is similar or identical to the electroweak high-energy phase. This phase may appear in the ultradense cores of galaxy, called "black holes". Along with the Higgs condensate, the Dirac energy gap disappears. Thus, the mass of particles as a source of their gravitational potential is eliminated so that the concept of an increasing mass density in this phase is subject to a qualitative change, a negation.

The thesis of fermiogenesis in jets of rotating high-density galactic nuclei, starting from the electroweak phase boundary, takes on another meaning in view of the proposed phase diagram. If a quantum phase transition evolves above a critical mass density / gravitation pressure, "mass" as the source of the extreme gravitational field is eliminated in phases with zero Dirac energy gap. The structure formation of matter at the level of galaxies by the establishment of their mass distribution, and the development of the gravitational field, form a unity of contradictions. The self-consistent development of mass distribution and gravitation can reach a critical threshold, where both mass and gravitation can be cancelled. This creates the conditions for fermiogenesis at the phase boundary between different states of the quantum æther. Evolutionary processes on the structural level of the "elementary particles" and the structural level of galaxies thus connect reciprocally and inseparably.

Conclusion

The picture of the unity of evolutionary processes in the microcosm and macrocosm outlined in this chapter is based on the synthesis of insights from astrophysics, high-energy physics, condensed-matter physics and quantum fluids. The picture is well-founded and paves the way to a general dialectical-materialist theory of evolution of the forms of matter in the cosmos as a unity of contradictory evolutionary processes in the microcosm and macrocosm, and their mutual influence.

Gravity and masses have not always been there but develop in a process of self-organization. Instead of arbitrarily introducing "dark matter" to repair "fundamental" theories, there is much to be said for treating gravity as a partially shielded and thus size-dependent force. The quasi-religious fantasy of formation of baryons and leptons by a unique act of creation in the Big Bang is in reality an "everyday process" in active galactic nuclei. Thus a reverse process occurs, with respect to the generation of heavy atoms from light atoms in the nuclear fusion processes of stars. While in the evolutionary processes of stars, light elements are transformed into heavy elements by nuclear fusion, galactic nuclei collect different forms of matter like a galactic "trash can", and in their active stage transform them back into light elements such as hydrogen.

7.5 The Structure Level of the Galactic Superclusters

Galaxies are not arranged randomly in the cosmos, but form galactic clusters and superclusters. Galactic clusters consist of dozens of galaxies, and are millions of light-years wide. One example is the Local Group, to which our Milky Way with four billion solar masses, and the Andromeda Galaxy (M31) with 5.7 billion solar masses, the most massive members, belong. It also includes another spiral galaxy (M33) with 1.5 billion solar masses, a small elliptical galaxy (M32), and 33 dwarf galaxies. Galactic superclusters consist of thousands of galaxies and extend for hundreds of millions to a billion light years.

Galactic clusters are concentrated in such vast superclusters, forming the surfaces of gigantic bubbles and filamentous structures (see Figure 95). Isolated galaxies are an exception. In the superclusters and "great walls", there is also hot intergalactic gas, which is noticeable due to its X-ray emission and its effect on the microwave background radiation. The large-scale structure of the galaxies therefore resembles a foam-like structure, in which the galaxies are arranged on large surfaces and bridges around "cavities". However, the bubble-like "cavities" are not empty: the galaxy density is drastically reduced, but there are many different forms of less concentrated matter such as intergalactic gas, neutrinos and electromagnetic radiation.

Spatial arrangement of galaxies and their axes of rotation

In the Local Group, 11 dwarf galaxies are gravitationally bound to the Milky Way, and 10 dwarf galaxies to M31. All 11 dwarf galaxies of the Milky Way lie in a plane which is approximately perpendicular to the disk of the Milky Way, and includes the axis of rotation of the Milky Way [Spark & Gallagher 2000].

The Local Group of our Milky Way, and the galaxy group of the Virgo cluster, are part of the Virgo supercluster, which consists of at least 2,000 galaxies, about

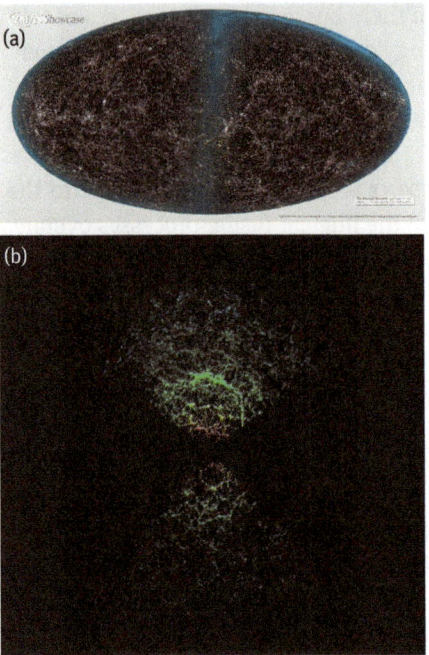

Figure 95: Structure formation of galaxies in galactic superclusters. (a) Polar representation of the distribution of galaxy superclusters in the nearby universe up to a red-shift of z = 0.6 (8 billion light years), based on one million galaxies detected. (b) Large-scale distribution of galaxies in two sections of the universe. The map shows approximately 250,000 galaxies up to a distance of 5 billion light years. Images: (a) T. H. Jarrett et al 2MASS (b) Sloan Digital Sky Survey.

90% of which are dwarf galaxies. The Virgo cluster consists of three sub-clusters, each grouped around a very bright elliptical giant galaxy (M49, M87 and M86) [Binggeli 1999]. The M49 sub-cluster consists mainly of spiral and irregular galaxies, while the M86 and M87 sub-clusters comprise a larger proportion of elliptical and S0 galaxies. This indicates different evolutionary periods. The jet axis of M87 and the galaxy distribution in the M87 and M86 subgroups are aligned.

The Swiss scientist B. Binggeli, in his detailed investigations of the Virgo cluster, comes to the conclusion that the coincidence of the jet and lobe axes with the large-scale structure of the cluster cannot be a coincidence. He writes: *"Not only M86 and M84 are lying "in the way"; there are also M59 and M60 on the other side, and there is a whole chain of elliptical tracing out the projected jet axis (...). This is suggesting some kind of causal connection between the orientation of the jet and the distribution of matter on an MPc scale ... "* [Binggeli 1999, p. 27]. [Note: MPc = megaparsec, approx. 3.2×10^6 light years]

The US astronomer Halton Arp, who contributed much to the investigation of the spatial correlation of active parent galaxies with jets and their surrounding galaxies, developed a model of the formation of galaxy clusters by the ejection of quasars from active galaxies, and their subsequent evolution into new galaxies [Arp 1998]. For example, he held the view that the Virgo cluster consists of three generations of galaxies, with the active (but probably decaying) giant galaxy M49 representing the origin of the cluster, from which M87 and the quasar 3C273 originated as the second generation. However, according to its red-shift, the latter ought to be far further away than the Virgo cluster. And M87 in turn represents the origin of a third generation of galaxies, which includes several elliptical galaxies on the jet axis, as well as a number of quasars as younger objects.

After almost twenty years of controversy as to whether there are alignments of the axes of rotation of galaxies at the level of superclusters as well, detailed investigations come to the conclusion that this is indeed the case [Hu et al. 2006]. This applies not only to our supercluster, within which the rotation axes are oriented to the Virgo cluster, but also to other filament-like arrangements of superclusters, as well as to bubble-like arrangements of galaxy clusters and superclusters around large "voids" with greatly reduced galaxy densities [Trujillo et al. 2005, Lee & Erdogdu 2007]. The authors find with 99% significance that the axes of rotation of galaxies are preferentially arranged in the plane of the bubble-like galactic superclusters, and that thus the galactic disks are arranged perpendicular to the line or plane of the supercluster (see artist's conception in Figure 96).

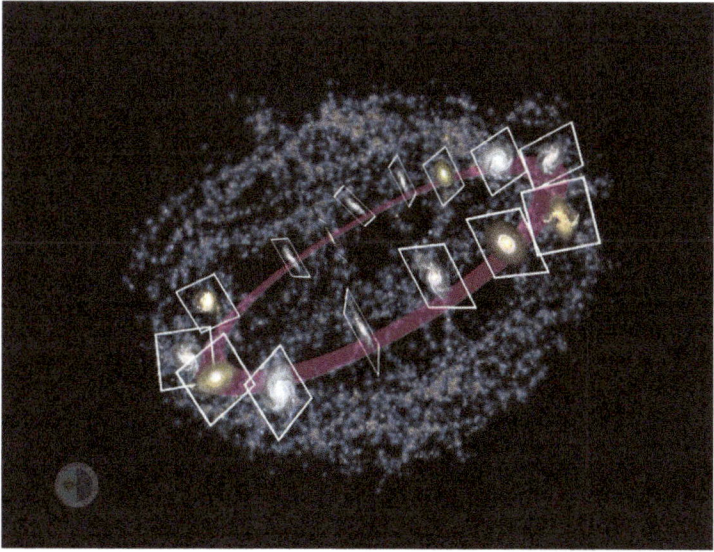

Figure 96: Orientation of the rotational axes of galaxies in large-scale structures. Artist's conception. Image: G. Pérez, IAC (SMM).

There are also exceptions where galaxies have a preferred orientation of their axes of rotation perpendicular to the main plane of the supercluster [Aryal & Saurer 2005]. However, the axes of rotation of the galaxies in the superclusters seems to be linked to their large-scale arrangement in a filament-like network structure. The observation of the ordering of galactic axes of rotation, together with their arrangement, puts the concept of a sole evolutionary process of matter in the universe from bottom to top in question. The notion that ordered large-scale structures were formed by the gravitational attraction of gases and plasmas alone up to galaxies, galactic clusters and superclusters is incompatible with the occurrence of an ordering of galactic axes of rotation on length scales of more than 500 million light years. The formation of large-scale network structures that is accompanied by alignment of the axes of rotation and jets evokes considerations of further connections of the evolutionary processes in the macrocosm.

Gravity as a partially shielded force and the stability of matter in the macrocosm

A central property of superfluid condensates is the incompatibility of the coherent motion of the condensate with macroscopic rotation (Chapter 3). The superfluid condensate must inevitably disintegrate into an ensemble of vortex filaments with quantized circulation and an aligned axis of circulation. The recognition of possible real existence of superfluid condensates at a high energy level in the quantum æther means that macroscopic rotary motions of the condensate, for example by rotary movements of matter in solar systems, galaxies or superclusters, should also be shielded. This may even result in structures with quantized rotation, which characterize stable states in the gravitational field. Franklin Potter, a student of Feynmann's, and his colleague Howard Preston, have developed a theory of partially shielded gravity [Preston & Potter 2006]. It predicts states with quantization of the angular momentum which alter the long range structure of the gravitational potential. Figure 97 shows the structure of the resulting gravitational potential using the example of the giant elliptical M87-type galaxy with 3×10^{11} solar masses in Schwarzschild approximation, i.e. the contribution of masses other than M87 to the gravitational field is disregarded. Here, the qualitative properties are discussed; details are given in the Appendix, Chapter 10.5.

Large parts of the potential follow the Newtonian law of gravity. At small distances from the Schwarzschild radius of the massive central object (10^{-2} light years), the potential flattens out. At the size of galaxies, the proposed quantization of the angular momentum increases the gravitational attraction. So Potter and Preston's theory can explain the flat rotation curves of the galaxies (compare with Figure 60).

A new length scale is set in this theory of partial shielded gravity by the galaxy superclusters, where the sign of the force of gravity changes, and the potential drops again. Gravitational shielding leads to a cosmic red-shift of photons. Potter and Preston's theory therefore explains Hubble's law of the increase in the red-shift

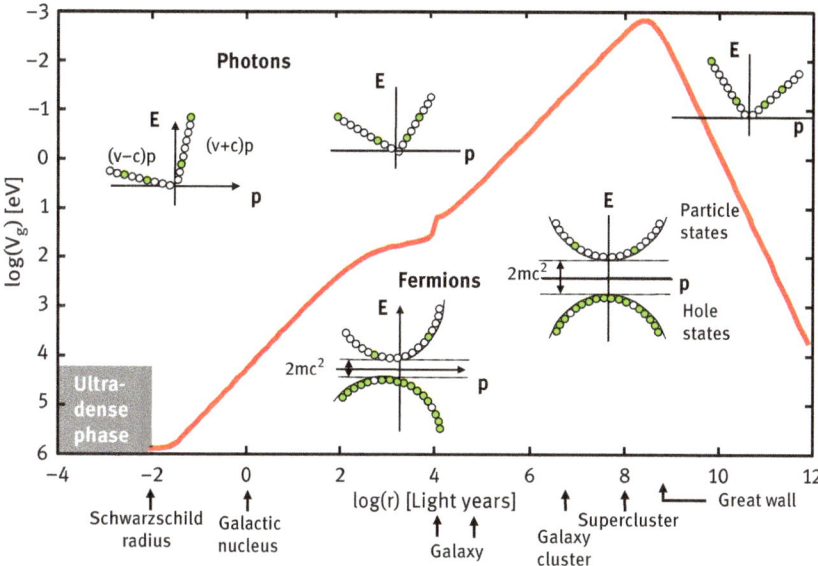

Figure 97: Relation between gravitation potential per electron mass V_g and distance r in light years (log scale) for the model of the partially shielded gravitational potential of an elliptic giant galaxy of the same size as M87, calculated according to [Preston & Potter 2006]. The energy-momentum relationship of photons and fermions distorted in the gravitational field is shown qualitatively. The decline of V_G at over 100 million light years leads to Hubble's red-shift law.

of incoming photons from greater distances, without cosmic expansion. For the connection between the shielding length of gravity and a related photon mass (from Hubble's law) see appendix.

Due to the repulsive character of partially shielded gravity on length scales of a billion light-years, superclusters can repel one another like vortex filaments in the superfluid state of a quantum fluid. In quantum fluids, vortex filaments form either regular crystalline or vitreous arrangements, due to their long-range repulsion (Figure 98). The arrangement of the galactic clusters in filaments and the long-range alignment of the galactic angular momentum resembles astonishingly a quantum-fluid model for large-scale structure formation via vortex filaments. In such a model, the phase transition from the Dirac phase to the non-condensed phase without the Dirac energy gap would occur only in the nodes and bulges of these filaments, and thus in the nuclei of large galaxies. The gravitational potential would thus be a potential partially shielded by macroscopic motions of the condensate, which depends both on the mass distribution and to a considerable extent on the proposed formation of large scale vortices of condensates in the quantum æther. Is it possible that the development of the matter distribution in the part of the cosmos visible to us is a result of the decay of a homogeneous superfluid quantum æther by shielding from movements on a much larger length scale?

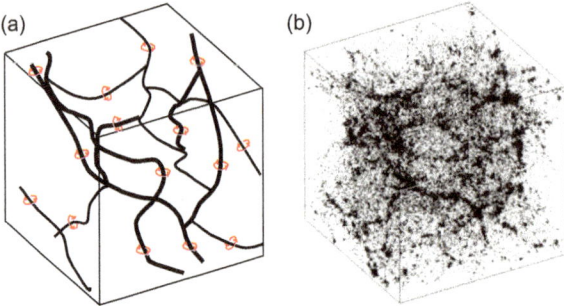

Figure 98: Filament arrangement of topological defects. (a) Schematic representation of the reticular arrangement of vortex filaments in a superconductor in the vitreous state of the vortex filaments. The superconducting eddy currents are indicated. (b) Large-scale reticular structure of galactic clusters and superclusters with formation of filaments (from a simulation). Nodes with an increased density of galaxies are formed when filaments meet. Image (b) from J. R. Bond et al. Nature 380 (1996) 603, with the permission of Macmillan Publishers Ltd.

7.6 Red-shift and microwave background

The explanation of the red-shift of the light of distant galaxies by a cosmic expansion and the existence of a microwave background radiation as residual radiation of the Big Bang are main arguments of the protagonists of the Big Bang theory. However, both effects have relatively simple other explanations and are by no means proof of the correctness of the Big Bang theory.

Different types of red-shifts

The increase in the red-shift of light with distance was discovered by E. Hubble in 1929. Hubble discussed the origin of this effect very cautiously [Lutz 1991]. The Big Bang theory that emerged in the late 1920s seized on this effect, explaining it by a Doppler shift of the photons due to a general expansion of the cosmos.

Alternative explanations assumed an energy loss by light on its way through the cosmos [Zwicky 1929, Dickhut 1987, Vigier 1990, Lutz 1991]. This explanation, however, was largely ignored in the prevailing doctrine. It was also rejected by means of the argument that energy changes can only be caused by the scattering of photons, which would also lead to a change of propagation direction. This would make cosmic objects appear increasingly blurred and faded as a function of their distance, but this is not observed. Four different causes of red-shifts are known today:

- The Doppler red-shift by movement of the source or the observer relative to the light field of the photons. The explanation of Hubble's law by the Doppler red-shift requires a homogeneous and uniform expansion of the cosmos. But the inhomogeneous distribution of matter in large-scale structures of galactic clusters would lead to measurable deviations from Hubble's law [Wiltshire 2007].

– The gravitational red-shift of photons propagating against a gravitational field. In order to explain Hubble's law, the disposition of the gravitational field on length scales greater than galactic superclusters must be known. Long-range repulsive aspects of gravity would be able to explain Hubble's law [Mannheim 2006, Preston and Potter 2006]; see Figure 99. But neither of these papers gives a clear interpretation of the gravitational repulsion of galactic clusters found, calling it "cosmic acceleration".

– The quantum red-shift arises from the interaction of photons with the unstable electron-positron pairs, which causes an intensity-dependent refractive index [Halpern 1933]. Today, it is a standard method of nonlinear optics to use crystals with an intensity-dependent refractive index for frequency conversion of photons. This parametric conversion is a coherent effect without scattering of the light. But an exact theoretical analysis of this effect within the framework of quantum electrodynamics [Adler et al. 1970, Jooss 2005] shows that the red-shift would be frequency-dependent, contrary to observation. Depending on the strength of the magnetic fields in intergalactic space, the effect is two to four orders of magnitude smaller than the effect of the cosmic red-shift observed.

– A fourth hypothetical type is the evolutionary red-shift in connection with a phase transition in the quantum æther, in which the value of the Dirac energy gap and thus the rest mass of atomic building blocks evolve inhomogeneously in space, or temporally by cooling from the high temperature state at the phase transition. This effect cannot explain Hubble's law, but it may contribute to the red-shifts of various objects such as active galactic nuclei and quasars.

A whole series of facts contradict an explanation of Hubble's law by a general expansion of the universe: first of all, there is not a single independent experiment that supports an expansion motion. The Tolman test of the distance behavior of surface brightness clearly speaks in favor of a non-expanding universe [Lerner 2006]. The claim that there is a time dilation in the luminosity behavior of distant supernovae, which would prove cosmic expansion at high velocities, has proved false [Jensen 2004]. In a cosmic expansion, the quantum æther would also have to expand. The reduction of its density over billions of years would have to cause drastic changes in its excitation spectrum, which is not observed.

The theoretical explanation of Hubble's law according to the theory of [Preston & Potter 2006] in Figure 99 is based on a repulsive aspect of the gravitational potential due to macroscopic angular momenta in the cosmos. It can be derived from the Hamilton-Jacobi equation of general relativity theory for the effect of a gravitational field [Landau & Lifshitz 1965], taking into account conservation of energy and momentum. From this theory, it follows that both shielding effects of gravity and

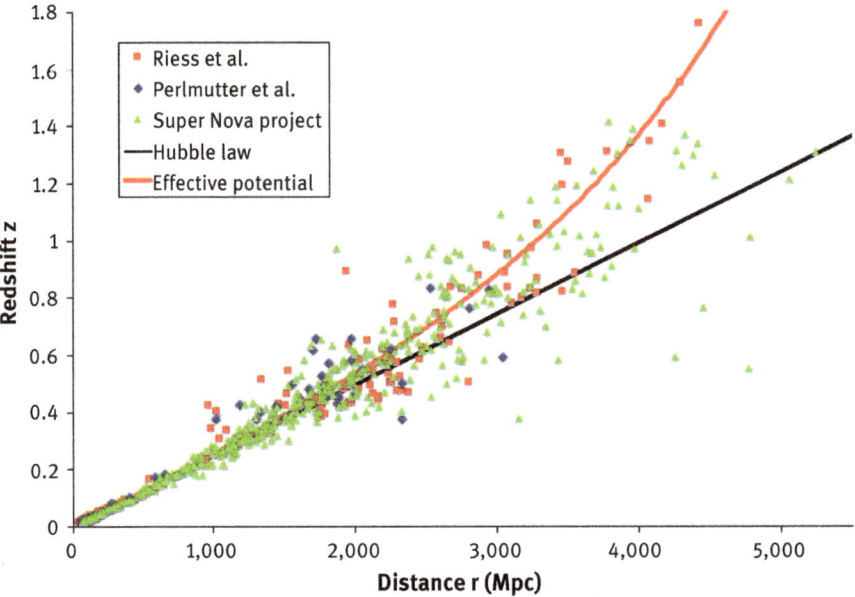

Figure 99: Interpretation of Hubble's law as a gravitational red-shift due to a repulsive gravitational potential according to [Preston & Potter 2006]. The black line is the conventional Hubble's law, while the red curve represents the result of a partially shielded gravitational potential. The experimentally measured red-shifts use supernova explosions as "standard candles" for determining distances, and are taken from [Perlmutter et al 1999, Riess et al 2004, Super Nova Project 2010]. Details in mathematical appendix 10.5 and 10.6.

macroscopic rotary motions of the quantum æther (or its superfluid components) can produce a cosmological red-shift.

The theory also explains the experimentally determined deviation from Hubble's law [Perlmutter 1999, Riess 2004] directly and without additional assumptions. In the Big Bang model, a fanciful "acceleration of expansion" due to "dark energy" must be assumed for this purpose. Even if Preston and Potter express the red-shift of photons by an "apparent escape velocity", they make it unmistakably clear that: " ... *The apparent escape velocities of the galaxies are a direct consequence of the solutions for the inner metric of general relativity by means of our wave equation and have their origin in gravity. There is no need for spatial expansion and its 'cosmological red-shift' or for modifications of the 'cosmological constant'.*" [Preston & Potter 2006].

A repulsive, very long-range behavior of the gravitational potential is a possible expression of the partial shielding of gravity due to condensates in the quantum æther. The theory of partial shielding of forces leads to a clear prediction: The shielding length is inversely proportional to the mass of the excitations of the gravitational field. Chapter 5 deals with the relationship between the gravitational field and inertial forces and the electromagnetic interaction. For a simple estimate, we assume that the partial shielding of the gravitational field corresponds to a photon mass. It can be determined from Hubble's law as $m_{ph} \approx 10^{-68}$ kg, which corresponds to a shielding length of four billion light-years (see mathematical appendix). This is only a bit larger than the size scale of the galaxy superclusters with sizes of up to one billion light years.

Cosmic microwave background radiation

When in 1965 Penzias and Wilson pointed their radio antennas into the sky and found microwave radiation at a wavelength of 7.3 cm, this was celebrated as "proof" of the Big Bang theory. It possesses a spectrum which is in large parts the radiation of a body in thermal equilibrium with its environment with a fixed temperature. So it was interpreted as thermal radiation which was supposed to have remained from immediately after the "decoupling" of particles and the radiation field during the cooling of the hot universe. With the further expansion and cooling of the universe, it was supposed to have reached the "current value" of 2.7 K. The predictions from the Big Bang theory by [Gamov 1961] were wrong at first, with radiation at a temperature of 50 K, and later approached the correct value. As Assis and Nevens have pointed out in their history of 2.7 K background radiation [Assis and Neves 1995], this radiation was already predicted in 1896 by the Swiss physicist Charles Édouard Guillaume [Guillaume 1896]. Based on the estimated total energy radiated by starlight, and the Stefan-Boltzmann law of black-body temperature radiation, Guillaume arrived at a temperature of interstellar space from starlight of about 5–6 K. A similar estimate was made by Eddington in 1926.

In fact, however, the energy density of starlight is only comparable within galaxies with that of the cosmic microwave background radiation of about 0.3 eV/cm³. As early as 1912, however, it was discovered that high-energy cosmic radiation exists, which contributes considerably to the energy density of the overall radiation. Taking it into account, Regener and Nernst obtained temperatures of 2.8 K [Regener 1933] and 0.75 K [Nernst 1938], respectively. These brief historical remarks already show that the idea that the universe is in a dynamic radiation equilibrium, in which an average temperature of intergalactic space with a thermal radiation results from the equilibration of radiation of hot sources, possessed significantly more predictive power than the Big Bang theory. Nernst went even further and showed in his remarkable work on the formation of a stationary state in the cosmos [Nernst 1937] that the sum of the energy loss of photons according to Hubble's law, the energy density of cosmic rays, and the energy density of starlight, corresponds approximately to the rate of formation of new fermions from the æther to maintain star formation, and establishes a dynamic equilibrium. Nernst assumed that the energy loss of photons was due to absorption in the æther, causing thermal radiation by the æther.

The current estimates of the energy densities of the various radiation fields in the cosmos do not differ significantly from those of Nernst and Regener [Spurio 2015]:

- The energy density of the photons from starlight in intergalactic space is about 0.04 eV/cm^3, and is thus significantly smaller than that of the photons from the microwave background of 0.3 eV/cm^3.
- The energy density of cosmic rays of 1 eV/cm^3 is, on the other hand, even somewhat higher than the microwave background. The fact that it is isotropic with respect to the galactic disk proves that its origin is extragalactic.
- The energy density of the magnetic field, of about 4 µG in intergalactic space, is comparable to that of cosmic rays, and amounts to 1 eV/cm^3.

This shows that, to a good approximation, the different components of the radiation spectra are in energy equilibrium with each other, and the microwave background is not at all an evidence of a big bang (see also [Fahr 2015]). The claim by the adherents of the Big Bang theory that the large ratio of photons to baryons in the cosmos of about 10^9 can only be explained by a hot Big Bang [Blanchard 2006] is not scientifically tenable. The main difficulty is to understand the process that leads to the thermalization of the radiation from stars and of cosmic rays. The spectrum of the microwave background shows a perfect curve of a Planck black-body radiator. The absorption of photons of different energies in the intergalactic plasma, their thermalization by plasma excitations, and subsequent emission in the form of heat radiation from the plasma, with a temperature of 2.7 K, is a plausible process [Ibison 2006]. Another possible contribution is the interaction of photons with Rydberg states in plasmas [Holmlid 2005].

The Big Bang theory, on the other hand, with its fantastic idea of inflation, must introduce arbitrary processes to decouple the structure of the galaxy distribution and the radiation field, in order to explain the extreme homogeneity of microwave radiation. In contrast, the thermalization of electromagnetic radiation via intergalactic plasma leads to a great uniformity of the radiation, with tiny fluctuations in intensity and temperature, due to fluctuations in plasma density. But a thermalization of photons via red-shift and plasma processes with a perfect Planck spectrum in the microwave range needs sufficient time. It indicates a stationary state of the cosmos, whose stability extends far beyond the age of stars and galaxies, and reaches periods of 500 billion years or more [Ibison 2006].

7.7 Self-organization versus fine-tuning

All forms of matter at the different structural levels form interacting systems of their discrete and continuous components. They are determined by the struggle and unity of different collective forms of motion of their components. All systems are subject in their evolution to relatively quiet stable phases, followed by instabilities with abrupt changes. Each system produces characteristic stable forms of motion, such as the planetary orbits in gravitationally bound solar systems, or the orbitals in the electron shells of atoms, bound by electromagnetic fields. They change at the transition from one system to another and can be attribute only partially to the laws of motion of other forms of matter. Newton's laws emerge from quantum mechanical laws of motion but cannot be reduced to them. The dialectic of self-organization considers the change of matter forms from system to system and the associated development of laws of nature on different structural levels as a unity of contradictions. Stability and instability, stationary states and catastrophic upheavals, are expressions of interacting and competing collective modes of motion and are only partially dependent on the microscopic structure and movements of individual components.

The Big Bang theory's fine-tuning "problem" with matter density

The fine-tuning problem of the Big Bang theory arises solely because it packs certain forms of matter taken out of context into a constructed system of laws of motion that stand above nature. To calculate the "cosmic expansion", the Friedmann model considers an equation of state similar to that of an ideal gas. Instead of grasping the total matter content of the cosmos, including the quantum æther, more and more generally, this view is limited to that part of matter which is present in the form of particles that form stars and galaxies. So in the end, the Big Bang theory considers nothing but the expansion of a gas whose temperature and density decrease with expansion. But in order to "explain" how the density of visible matter in the universe observable today comes about, which corresponds exactly to the critical density for the observed "flat universe", extreme "fine-tuning" is necessary. The density of matter actually observed at present is exactly $\Omega = 1 \pm 1\%$ [Spergel 2007], if normalized to the critical density $\rho_c = 8 \times 10^{-27}$ kg/m^3 for a "flat space" according to general relativity. For the universe not to have collapsed again immediately after the hypothetical Big Bang, or expanded so quickly that no stars and galaxies could form at all, the matter density must have had exactly the value $\Omega = 1 \pm 10^{-60}$, when extrapolated back to the Big Bang mathematically. This means that the density of radiation and particles during the actual Big Bang must be fine-tuned to a precision of 1 out of 10^{60} [Mannheim 2006] for a universe to be calculated at all.

The structural level of the quantum æther must be largely ignored for any attempt to calculate a "big bang". The compressibility and the expandability of all known forms of matter are very limited. Although there may be a certain expansion or compression of the quantum æther in connection with phase transitions, and nucleation and growth processes of condensed phases, extrapolating this to a singularity in the Big Bang is absolutely inadmissible.

The "fine-tuning" of the Big Bang theory is alien to a scientific description of a system, and already shows that there is something wrong with the entire approach. This applies not only to the "creation out of nothing in the Big Bang", in stark contradiction to the conservation of energy and momentum of matter. But also to the subsequent evolution of the cosmos. In the context of the theory of partially shielded gravity, the observation that the universe is "flat" simply means that even when using the geometrical approximation of general relativity theory, repulsive "negatively curved" and attracting "positively curved" areas balance out on average. The structure formation of the mass distribution in the cosmos visible to us, and the development of gravitation as a partially shielded residual interaction of the quantum æther, develops self-consistently as a unity of contradictions. They do not have to be specified by "external" fine-tuning.

The fine-tuning "problem" of "vacuum energy"

With the fantastic theory of inflationary expansion in the initial phase of the Big Bang, the fine-tuning of matter density was extended to a fine-tuning of "vacuum energy" [Mathiesen 2006, Mannheim 2006]. With this inflation, a superluminal extreme expansion was conceived, which was to "solve" not only the problem of "flatness", but also that of the uniformity of the background radiation. For this purpose, it was assumed that the expansion immediately after the Big Bang was driven by a "dark energy" similar to the "vacuum energy", which results in negative gravitation. This conception was such that during inflation, the density of matter became completely irrelevant. By complete transformation of the "vacuum energy" into matter immediately after the inflationary phase, a density of matter was supposed to be achieved that would correspond exactly to the critical density of matter for a flat universe. However, such a model could only be made to match flat space and the structure formation of galaxies by introducing large quantities of "dark matter". Here too, the arbitrariness of the thermodynamic approach becomes apparent. If one wanted to describe an expanding matter system seriously at all, one would have to include both ground state and excitation energy completely in the free energy. Instead, an "inflationary field" is conceived arbitrarily, which on the one hand is supposed to provide a large amount of "vacuum energy" in the inflationary phase, which drives the expansion, but on the other hand, is supposed to disappear below the phase transition. Fine tuning here means that the energy density of the "vacuum" of $\Lambda = 10^{111}$ J / cm^2 is said to have adjusted to the current value of 10^{-17} J/cm3 immediately after inflation, and not to have changed since [Steinhardt & Turek 2006].

With the "present value of vacuum energy", another blatant paradox of Big Bang theory arises, namely that the values of "vacuum energy", which determines "cosmic expansion and space curvature", and the vacuum energy" that follows from modern quantum field theories, differ by about 120 orders of magnitude (see for example [Steinhardt & Turek 2006, Vilenkin 2006]). A more recent turning point in this development is the discovery of deviations from Hubble's law for great distances [Perlmutter 1999, Riess 2004], which is interpreted in the context of the Big Bang theory as a re-emerging "accelerated expansion" of the universe. The "vacuum energies" that are supposed to drive this accelerated expansion are called "quintessence", "essence" or "mystery". These names alone already show how cosmology is degenerating into un-scientific mysticism. To "acceleration the expansion", the "vacuum energy" Λ now becomes an arbitrary adjustment parameter to save the interpretation of Hubble's law as "cosmic expansion". It is supposed to account for 70% of the total matter-energy content of the universe. Nevertheless, the stark contradiction already mentioned, the difference of 120 orders of magnitude from the value of "vacuum energy" from quantum field theories, remains. More and more complex ad-hoc assumptions are reminiscent of the epicycles to save the geocentric world view of the Middle Ages. They were the expression of a fundamentally false picture of nature, which was maintained for ideological reasons.

The total energy of a quantum fluid

In order to correctly determine the total energy of a quantum fluid, the potential and kinetic energy of all components must be taken into account. Any theory which arbitrarily picks out only certain forms of motion, such as the zero-point excitations of the system, or the condensation energy, will be dramatically wrong.

In a continuum field theory of a quantum fluid that only picks out the zero-point excitations, its energy would diverge if a cutoff wavelength were not introduced. This also remains arbitrary if it does not reflect a real structure of matter. For example, the "Planck length" of superfluid helium is given roughly by the inter-atomic distance of the helium atoms forming the condensate.

The determination of the energy of the ground state and the excitations of the system requires a microscopic theory. Only then is it possible to analyse the "vacuum energy" and the "visible matter content" correctly. In the case of superfluid helium, this is the theory of the interacting system of helium atoms. Depending on external conditions such as temperature, pressure and macroscopic fields, the unity and contradiction of its internal motions and interactions determine the state as a stable phase; it is self-stabilized. The formation of topological defects, particles and textures change the motion of the condensate, which must be taken into account when considering the total energy. The self-organized structure formation of particles and fields even lowers the total energy of the system when a macroscopic field is shielded. Only the consideration of the total system of ground state and excitations, quantum æther, and structured matter, including the energy contribution of all macroscopic fields makes sense in the energy balance.

The conception of a "contribution of vacuum energy" to gravity in expanding universe cosmologies breaks this connection. There is no contribution of "vacuum energy" to gravity. Rather, gravity and mass distribution, inhomogeneity of the "vacuum", and topological excitations occur self-consistently. They follow the tendency of minimizing the free energy of the overall system.

Self-organization versus fine-tuning

The structure formation in nature is self-stabilizing and self-organized. It does not require fine-tuning. The fine-tuning problem of the Big Bang theory shows that it is in every respect a construct placed artificially above reality. It is also no wonder that hordes of religious and esoteric elements are jumping at this model, and using the fine-tuning problem of the Big Bang as "proof of the effect of a supernatural force", for "intelligent design" and see a god.

According to such ideas, fine-tuning not only determines the density of matter and the "vacuum energy", but also determines favorable values of "natural constants". If, for example, the strong nuclear force of the proton were only a few percent greater, deuterium would be stable, and the nuclear fusion process in stars would be accelerated greatly. The question arises as to whether, under these conditions, chemical evolution would have produced such favorable distributions of the elements for life as we find today. As interesting as the question of the relationship between the chemical evolution of the elements and the structure of the quantum æther is, and whether this relationship might be modified in other regions of the cosmos, it becomes unscientific when the theory is detached from the structure of matter. A view of nature that regards natural constants as "above nature", no matter how subtle the arguments of fine-tuning and intelligent design, is ultimately identical in essence to religion. Whether the world was created by a god, by a supernatural idea, or by the anthropic act of retrospective observation, are only different versions of idealistic ideology.

Even the retreat to the weakest form of the anthropogenic principle is no basis for a scientifically based theory of the evolution of matter. This simply states that the natural constants have precisely those values that are favorable for life because otherwise life and us would not exist and we would not be able to observe nature. That is not wrong as an observation. Perhaps there might be many other regions in the cosmos where the natural constants have different values and no life is possible. But to use this weak form of the anthropogenic principle as an explanation means "intelligent design". Or it remains pure happenstance that such a complex universe could develop, in which life could also arise.

From the point of view of the dialectical theory of the development of matter, just such self-stabilizing manifold systems of matter are formed in the struggle and unity of contradictory forms of motion, because in this way, the "favorable" values of the forces and "natural constants" arise in the process of collective self-organization. "Natural constants" such as the "elementary charge" or the "quantum of action" are expressions of the forms of matter that emerge out of different types of motion of the quantum æther, and deeper structural levels of matter. Their collective forms of motion are reflected in corresponding laws of motion of electrodynamics, quantum mechanics, and other laws.

8 Self-organized development of matter systems in the cosmos

The scientific generalization of knowledge gained from individual systems is an important and necessary part of scientific work. This requires the conscious application of a suitable scientific method. The dialectical-materialist method combines analysis and synthesis, induction and deduction, and theory and practice by finding unity and contradiction as an ongoing process. It must itself be constantly further developed according to the progress of knowledge. In this chapter, insights into the common laws of development underlying the various systems of matter will be summarized and discussed as a self-organization process. The limits of previous models must be consciously recognized, creatively negated and, in part, set aright. The scientific-technical revolution of the twentieth and early twenty-first centuries creates the conditions for a more comprehensive theory of the development of matter. Natural science can thus reach a higher stage of development, which develops the laws of self-organization of systems from the accumulated knowledge.

The development of a theory of the evolution of matter is only possible by deliberately overcoming the predominant influence of the positivist and idealist ideology, which has caused a deep crisis in physics. It rejects the scientific generalization of observations by induction as an *"inadmissible mixture of empirical experience and metaphysics."* In the positivist literature, this is referred to as the "demarcation problem" [Wendel 2013, p. 41]. Scientific progress, however, is based on the identity of essence and appearance, natural law and what is directly given. Popper's *Logic of Scientific Discovery*, on the other hand, is directed directly against higher development of the natural sciences: *"For we can utter no scientific statement that does not go far beyond what can be known with certainty 'on the basis of immediate experience'."* [Popper 2005, p. 76]

The rejection of the inductive method justifies the "immersing in details," the eclecticism, and, quite contrary to its claimed "freedom from ideology" itself represents a worldview. Under the banner of the struggle against "mingling of science and worldview," the method and worldview of positivism itself opens the door wide to the influence of idealist ideology. This takes place in two steps: first, the rejection of the derivation of a theory of the evolution of matter forms from observation and theoretical generalization. Secondly, it goes hand in hand with degrading deduction to a one-way street of deriving theories from "basic propositions," which are elevated to quasi religious dogmas.

"Basic theorems" such as the principle of relativity, the second law of thermodynamics or principles of symmetry, which have a certain validity as a reflection of some aspects of nature, are dogmatically imposed on the development of the entire cosmos. What is needed is a carefully thought-out interaction of inductive and deductive methods.

https://doi.org/10.1515/9783110644203-008

The worldview of imposing "basic propositions" on nature has been strongly integrated into the way of thinking of today's natural sciences, and opposes the task of natural science of reflecting matter and its evolutionary processes ever more deeply and comprehensively. It will never be possible to establish an ultimate theory of the development of all the various forms of matter in the cosmos. Human knowledge gradually encounters ever-new development processes in the macrocosm or microcosm, their mutual connections, and their transitions. In the following, an attempt will be made to summarize the findings on the evolution of the cosmos through self-organization processes of matter dealt with in the previous sections, and to generalize them into a larger picture of the evolution of the cosmos.

8.1 Development from within through self-organization

The scientific significance of the critique of the theory of the Big Bang [Lutz 1991, Lerner 1992, Alfvén 1981, Fahr 1995, Arp 1998, Lerner CCC 2006] is by no means limited to the refutation of a "specific cosmology." The thesis of the "emergence from nothing," from an "initial singularity," is ultimately the thesis of the "external impulse," and thus the antithesis to the theory of the evolution of matter from itself through self-organization. This also applies to the less primitive forms of Big Bang theory, such as the idea of universes oscillating through time, in which Big Bang and "Big Crunch" replace one another. The same goes for the fantastic idea expressed in the context of superstring theory, of continuously occurring "big bangs" when membranes [or "branes"] collide and "parallel universes" are created. What all these models have in common is that they place "iron, eternal, unchangeable laws of nature," which would thus exist outside nature, at the beginning of the development and above matter.

The dialectical-materialist view of the self-organized development of matter from itself, on the other hand, investigates how different structural levels of matter with their laws of motion emerge, merge into one another and in their infinite interaction with one another produce ever new forms of matter. The scientific-technical revolution in the late twentieth and early twenty-first centuries enlarged our view of the microcosm and macrocosm explosively. The new stage of synthesis and processing of the gigantic mass of observations must therefore be systemic. It must, as far as possible today, consider the coupled development processes of the various structural levels of matter in the macrocosm and microcosm in their overall context. The explanation of the evolution of the cosmos requires neither external impulses, nor a God, neither a *"wound-up clock spring"* (entropy destruction) of the Big Bang, nor *"fundamental laws of nature"* standing above nature, nor the "fine tuning of natural constants," because the development occurs from itself.

8.2 Formation of systems, stability, and instability

Matter systems occur both discretely in the form of structures/particles and continuously in the form of condensates and fields. Both types form a unit: fields on a certain level emerging from collective movements of the particles, and particles forming in turn as excitations and forms of motion of the material continua. When investigating the formation of different structural levels of matter, both the structure and laws of motion of the individual discrete forms of matter/particles, and their collective laws of motion, must therefore be derived from reality at each stage of development.

Systems of particles occur in different stable phases, which are abruptly separated from each other by instabilities or phase transitions. They are determined by the struggle and unity of thermal random motion, that is, disorder and order through interactions between the particles. This applies regardless of whether the systems consist of subatomic building blocks, atoms, molecules, nucleons, electrons, vortex tubes, stars, or galaxies. Atoms form gaseous, liquid, and solid phases in which the unity of spatial order and forms of motion change qualitatively. For example, these produce a large part of the structure of matter on our planet Earth. Another phase, the plasma state of atoms, consisting of ionized atoms and electrons, characterizes large parts of the cosmos. They form intergalactic and interstellar plasmas of low density, and plasmas of higher density in the stars. A further important class of phases are superfluid states in quantum fluids. They are formed by ordering the motion of the particles and their matter waves forming them.

Superfluid phases can occur regardless of whether the multiparticle system is formed from atoms, electrons, nucleons, quarks, gluons, or the constituents of the Higgs condensate in the quantum æther. Under conditions in which their discrete composition can be ignored, many-particle systems appear as continua and their excitations as fields. Solids therefore form elastic continua, liquids viscous continua, and quantum fluids superfluidic continua. Even below the structural level of subatomic particles, there exists with the quantum æther a further level of matter, which like other systems of condensed matter, is a material continuum, which produces different forms of excitation and motion. The occurrence of different excitations (e.g., electromagnetic, electroweak, gluonic, and gravitational) and different phases/aggregate states proves that it is far more than a structureless continuum. However, its microscopic components down to infinitely small forms of matter in the microcosm are not yet known.

Two main forms of the evolution of new, higher matter systems are known: the formation of structures and particles as unstable and stable excitations of the continua, as well as the formation of new composite systems from the organization of particle collectives. Important mechanisms of structure formation in continua are topologically stable defects, structures, and patterns of motion. In them, the quantity of a disturbance or movement of a continuum transforms into the new quality of a stable "particle-like" structure. The change in the internal microscopic properties (of the defect) produces a macroscopic change (a field).

In this book, we have dealt with topological structures/defects in materials and quantum fluids. But electrons, protons, and neutrons also represent topological defects in quantum æthers. The formation of local microscopic properties, such as electrical charge or particle mass, is inevitably associated with the formation of long-range electrical or gravitational fields, as macroscopic forms of motion of the quantum æther. In addition to this, they interact and form stable composite systems such as atoms.

At higher pressures and energies, a whole series of other states of these particles occur: ionization in plasmas, compression in nuclear matter, or changes in their collective behavior and structure due to phase transitions in the quantum æther, such as quark–gluon plasmas or color superconductivity. The stability of these states and their transformation into other systems, is determined by the struggle and unity of opposing internal forms of motion and forces. Depending on the character of the contradiction, consolidation, stability, transformation, or decay and dissolution of the structure, are the results.

8.3 The formation of laws of motion and nature

Each system of matter possesses collective forms of motion with characteristic system behavior. We humans extract them from reality as laws of nature. Such laws have different levels of generalization. As with Kepler's laws of planetary orbits, they can directly describe the movements of the system and thus have an empirical character. In the further progress of knowledge, more general laws underlying these movements, such as Newton's gravitational law or even a dynamic field theory and Einstein's gravitational theory were induced, which cover the laws of a larger class of movements and mutual influences of different matter systems. Many laws can be expressed by mathematical relationships. These can often describe individual motions quantitatively with high precision, such as Newton's laws describing the motion of bodies subject to forces. But even these relatively simple laws have no mathematically exact solution for three bodies. Such laws always only deal with certain aspects of the system behavior, and therefore are always approximations to the system behavior.

In addition to deterministic processes, random motions and developments also occur. These two are interdependent. Thus, the random quantum fluctuations in the microcosm are caused by the regular behavior of the zero-point fields. According to our current view, the mutual influence of random zero-point excitations of the quantum æther and moving particles produces the matter waves. They not only deterministically govern the mean state of motion of particles, but also cause the statistical character of the quantum-mechanical laws of motion.

These stochastic laws express the fact that lasting changes in the state of motion occur only at the threshold of the quantum of action.

Chance and necessity are therefore not strictly separated processes, but interact and condition each other. The quantum-mechanical unit of matter wave and particle as a self-organization process is an expression of the fact that, in the constant absorption and emission of zero-point quanta, cause and effect constantly change sides and penetrate each other (see [Dickhut 1988, p. 65 et seq.]).

A progressive driving force of the natural sciences is the attempt to trace system behavior back to the structure and dynamics of the individual components. A fundamental error of the metaphysical method in physics, however, is to follow a one-way street of explaining the entire cosmos from a few microscopic building blocks. As justified as the analysis of the individual components of systems is as one side of the investigation, their absolutization in the form of the reductionist method and worldview is one-sided. The reductionist search for "original particles," which are subject to a "world formula" from which the whole cosmos can be understood, ultimately leads to metaphysical idealism.

The laws of atomic structure cannot be attributed one-sidedly to the structure of subatomic building blocks. It is only by the interaction of these building blocks, via their influence on the dynamics of quantum fields, that a self-organization process occurs. The resulting stable structure of the electron shell of the atom is determined by the creation of matter waves, which are formed by resonant emission and absorption of zero-point photons. In these stationary states, the motion of the electrons and the Fermi pressure of the matter waves act as a repelling force against the electrostatic attraction of the opposite electrical charges of electron and proton. But whether subatomic building blocks are organized into plasmas, atoms, or pure nuclear matter in neutron stars or quark condensates is not determined by the structure and motion of individual atomic building blocks, but by the collective state of the particles in phases. The interdependence of individual and collective behavior in systems is fundamental, with the individual behavior usually being subordinate to the collective system dynamics. Even if, for example, the individual movement of a single atom in a liquid is subject to random fluctuations, the amplitude of these fluctuations is determined by physical laws of the collective.

Collective behavior of complex multiparticle systems can also lead to "simple laws." Newton's law of inertial motion and the Lorentz invariance of motion are the results of the self-organization of moving particles that modify the zero-point fluctuations of the quantum æther by forming matter waves that produce a stable state of inertial motion. Lack of knowledge of these processes led to the idealistic notion of an undisturbed movement in "empty space."

Simple laws of force, such as the laws of inertia and gravitation, result from the interaction of masses with currents or inhomogeneities of the zero-point excitations. Neither Newton's laws nor quantum mechanics or relativity theories, are therefore "fundamental laws." They are emergent laws of one or more structural levels of matter that emerge through collective self-organization. They produce the permanent, the invariants of system behavior. Thus also natural constants, such as Newton's gravitational constant or Planck's quantum of action represent emergent properties, are materials constants of the underlying material structure.

8.4 Instabilities of systems at phase transitions

Qualitatively different phases suddenly and abruptly change into one another at phase transitions. This results in qualitatively new properties. At the transition from the solid to the liquid and the plasma state, the laws of elasticity change into those of viscous hydrodynamics and of plasma magneto-hydrodynamics. They are an expression of the change in the collective forms of motion of the systems. Phase transitions also occur in the quantum æther. At higher temperatures and pressures, new phases, such as the quark-gluon plasma and the electroweak phase, emerge from the Dirac phase present under our ambient conditions, in which only the electromagnetic and gravitational forms of motion produce long-range forces. There, the forms of motion of the color and weak charges change. Today, only effective low-energy approximations to these forms of motion exist in the form of the quantum field theories of the electromagnetic force and the weak and strong nuclear forces. The development of the microscopic theory of the quantum æther with its various phases is still in its infancy.

The stability of the phases has limits. It is caused by contradictory internal forces and patterns of motion. Breaking out of the "slavery" of the collective forms of motion and structures that make up the respective phase is only possible globally at phase transitions. At phase transitions, therefore, strong random fluctuations and movements occur on a wide range of length and time scales as an expression of the antagonism of the contradictory forces and the instability of the collective state of matter. The result is a struggle between different states of order. Among other things, close to the critical point, this results in long-range correlations of the random movements that break out of the existing phase. For example, the boiling of a liquid produces turbidity (critical opalescence). The cause is the refraction of light due to the chaotic formation of gas bubbles in the liquid, which is accompanied by a sudden change in density. Gaseous and liquid states co-exist and strive antagonistically to displace one another. Such instabilities can also lead to fascinating new structures. One example is the formation of polycrystalline microtextures during the solidification of melts, which assume the form of growth dendrites for rapid solidification. The instabilities during the electroweak phase transition in compact nuclei of active galaxies lead to gigantic jets, and are decisive basic conditions for fermiogenesis, the formation of subatomic particles from processes in the quantum æther.

The transition from the relative stability to the instability of any system is based on a change in the character of the internal contradictory forms of motion and forces. At the solid–liquid phase transition, the predominant side of the contradiction between atomic bonding forces and thermal motions changes from the dominance of bonding to the dominance of motion. As we approach the critical point of phase transition, gradual quantitative changes in motion turn into a new quality of system dynamics.

8.5 Phase interfaces – development of metamorphosis and complexity

Phase coexistence and phase competition create phase boundaries and thus a spatially abrupt transition from one state of matter to another. Solid, liquid, and gaseous phases of water co-exist on the Earth's surface. This is an expression of the richness of structure formation that goes hand in hand with phase coexistence and competition. The complex, diverse structure of the Earth's surface is characterized by competition and co-existence of solid, liquid, and gaseous phases of differing chemical composition and morphology. It develops under various influences such as gravity, convection currents in the Earth's mantle, solar radiation, atmospheric circulation systems due to the Earth's rotation, as well as air and water currents. The conflicting structures and motion patterns of matter, which meet at phase interfaces, produce a metamorphic process between different phases and substances. They therefore have the potential for higher complexity. Thus, the development of life and the biosphere is largely linked to the proximity to phase boundaries between the Earth's solid surface, the liquid water, and the gaseous atmosphere in lakes, rivers, and oceans. Through the exchange of substances and energy, they enable the formation of complex molecules up to metamorphic processes of life.

Other systems such as stars, subatomic particles, or compact galaxy nuclei also form phase boundaries. At these interfaces, the collective behavior of the respective system changes more or less abruptly. In addition to discontinuous phase boundaries, continuous changes within phases and phase mixtures also occur. One example is the interior of a star, where the ratio between the ionized fraction and the gaseous fraction of the atoms decreases continuously toward the outside. The plasma and gas states co-exist here as a mixture. At the photosphere, the star becomes transparent to electromagnetic radiation and the density of the matter drops drastically. This transition to the stellar atmosphere is discontinuous and occurs as a phase boundary, at which the coupling of the photons to the plasma changes abruptly. There, photons can be emitted into space. Phase boundaries between the electroweak and the Dirac phase of the quantum æther, and the metamorphosis between both phases via hypermagnetic fields probably play a decisive role for the production of subatomic particles and the formation of light elements.

8.6 The formation of topological structures by partial shielding of macroscopic fields

Today, we know of a whole zoo of topological defects/structures in different systems of condensed matter. They are therefore, on the one hand, the source of long-range fields and, on the other hand, are generated by partial shielding of such macroscopic fields. In solids, for example, dislocations and vacancies are created by shielding elastic stress fields. However, vacancies and interstitial atoms are also present as thermal excitations. In ferromagnets, long-range stray magnetic fields are partially shielded by the formation of magnetic domains which are separated by domain walls.

In quantum fluids, above a critical velocity threshold, macroscopic rotary movements are shielded by the formation of microscopic vortex filaments or complex topological defects, such as skyrmions. Skyrmions are an important class of topological structures in liquid crystals, quantum fluids, and magnetic systems with a chiral microscopic structure, that is, they occur in left-handed and right-handed forms. But "elementary particles" also form topological structures/defects in the quantum æther: The shielding of electroweak and color fields determines their structure.

What they all have in common is that under certain conditions the formation of topological structures reduces the free energy of the underlying material continuum. The energy of a macroscopic form of motion of continua (a field) can be reduced by partial shielding and local structure formation. Thus, the contradiction that builds up between the state of order of matter and the disturbance by the field is resolved by spatially concentrated structure formation, which often has sharply defined quantized properties, such as a "particle" in the quantum æther or a "defect" in a material. For example, the antagonistic contradiction between the vortex motion of the superfluid and the uniform ordering of the matter wave of the superfluid condensate is resolved by the formation of a normal-fluid core at the center of the vortex filament. In the balance of free energy gained through field relaxation, and free energy expenditure through the formation of the new phase in the core of the particle or defect, the total free energy is reduced. This determines the structure of topological structures. Their sharply quantized properties are reflected in topological charges, which are identical for each individual species of a defect type. Dislocations are characterized by their Burgers vector, point defects by their excess volume, vortex tubes and skyrmions by their number of turns, domain walls by their magnetic charge, and subatomic particles by their electric, weak and color charges.

Stable topological structures occur in all condensed phases of many-particle systems. The term "defect" correctly refers to the disturbance of the state of order of the phase from which it originates, but in conceptual terms, it is only a simple negation. The creative, double negation is the concept of a topological structure. It makes it clear that this forms a new structural level of matter with qualitatively new properties and laws of motion. For example, dislocations in the case of alternating deformation of material are organized in slip bands, a new structure of a material, in which they have a characteristic arrangement and direction of motion. The dislocation movement is also influenced by other defects such as foreign atoms and grain boundaries. The control of dislocation motion by the microstructure has been used for centuries in steel-making by forming, forging, and rolling to control mechanical properties. However, the transition from empirical knowledge to understanding at the microscopic level has only occurred in recent decades. An ensemble of vortex filaments in quantum fluids forms "vortex matter": they can assume liquid, glassy, or crystalline orderings. When vortex filaments move in the background medium of the quantum fluid, forces similar to the inertia and gravity of masses in the quantum æther are generated.

The formation of topological structures is, therefore, an important general mechanism of the structure formation of discrete systems out of continuous forms of matter in the cosmos. It is determined by the constraint of local concentration of forms of fields or motion that contradict the macroscopic order of a phase, so that a locally new state is formed. In a certain sense, a phase transition occurs locally in the core of a topological structure. This region is surrounded by a complexly structured phase boundary and spatially inhomogeneous partially shielded fields. The formation of topological structures creates a hierarchy of structural levels of matter: A larger number of "elementary particles" as topological structures in the quantum æther themselves form new multiparticle systems with new phases. They occur as nuclear matter, Fermi liquids, atoms, plasmas, etc., depending on their density and kinetic energy. Multiparticle systems of atoms, in turn, form new forms of condensed matter such as solids and liquids, in which unstable excitations and a zoo of topological defects such as vacancies, interstitial atoms, dislocations, grain boundaries etc., can arise. On the structural level of stars, "elementary particles" can agglomerate to macroscopic regions of nuclear matter – the local phase transition in the topological structure changes into a new macroscopically extended phase in the quantum æther.

The change in the concentration of topological defects, therefore, has repercussions on the respective forms of matter from which they originated, if critical quantities are exceeded. Critical dislocation densities cause atomic bonds to rupture up to fracture. If a critical density of vortex filaments is exceeded, the superfluid phase producing them is destroyed and it changes into the normal fluid state. The increase of the density of "elementary particles" with their mass as a source of gravity leads to mass defects in bound systems. Extreme concentrations of nuclear matter lead to phase transitions in the quantum æther, for example in neutron stars and galactic nuclei. At another critical threshold of density, a phase transition to an electroweak phase probably occurs, in which the particles and their property of mass as a source of the gravitational field may be eliminated in the interior of active galaxy nuclei, playing an important role in fermiogenesis.

8.7 Stationary states and cyclic processes

In nonequilibrium, matter systems can form stationary states with relatively stable patterns of motion, which approximately take the form of cyclic processes. By coupling of a chemical reaction with structural surface transformations of the catalyst involved, a chemical wave and spiral formation occurs. In Chapter 2, we discussed the example of the chemical conversion of carbon monoxide to carbon dioxide molecules on a platinum catalyst.

It is a cycle in which the chemical transformation of molecules creates a depletion of oxygen on the platinum surface, changing it and stopping the reaction. Only when the concentration of carbon monoxide rises again does it restart, creating a periodic pattern. Chemical pattern formation at the macroscopic level and surface modulation at the atomic level, are inextricably linked in an ordered process involving trillions of molecules and surface atoms.

Another example is the formation of convection cells in heat transport in gases or liquids. They occur as circulation cells in the atmosphere or inside stars. An unimaginable number of more than 10^{30} molecules can be involved in the organization of ordered circular motions, which maximize the heat transfer by the order of their motion. In stars, electromagnetic radiation is transported to the outside and hydrogen to the inside via ordered motion in convection cells.

But even the cycle of becoming and passing away of whole solar systems is approximately a circular process. The gravitational collapse of an interstellar gas or plasma produces stars with their planetary systems. Under the pressure and temperature of the star being formed, nuclear fusion processes are ignited. The process of forming heavy elements lowers the free energy of the nuclear matter and photons with high energy and low entropy are formed and emitted out of the stars.

The light of the Sun drives a huge variety of coupled cyclic processes in the biosphere of the Earth. The matter thrown into interstellar space after the death of a star becomes in turn a starting point for the formation of new generations of stars. The death of stars in the form of supernova explosions can trigger entire waves of star formation in the closer neighborhood of the spiral arms of galaxies. But macroscopic processes, such as the collision of galaxies and the subsequent transformation to an elliptical or spiral galaxy, can also cause long-lasting phases of waves of star formation. The cycle of star formation and death is driven by the contradiction between gravitational attraction and the counter-pressure of internal motions of matter. These include radiation pressure, gas pressure, and Fermi pressure. So gravitational attraction drives microscopic processes of nuclear fusion and particle transformation until the supply of light elements is exhausted. With a sufficiently large stellar mass, the further gravitational collapse leads to a supernova explosion in which the stellar matter is redistributed into the interstellar medium with a changed composition, with more heavy atoms. The cycle of stellar evolution thus interacts with processes of galaxy evolution, such as plasma filamentation, and waves of star formation in the spiral arms.

Cyclic processes by coupling between processes at different structural levels are common in the cosmos (see Figure 100). The concept of cyclic processes is always to be understood only as an approximate description of the repeated passage of qualitatively relatively similar states, such as the periodic course of the seasons. A cycle is not the same as the previous one, nor is it the same as the next. The repetition is

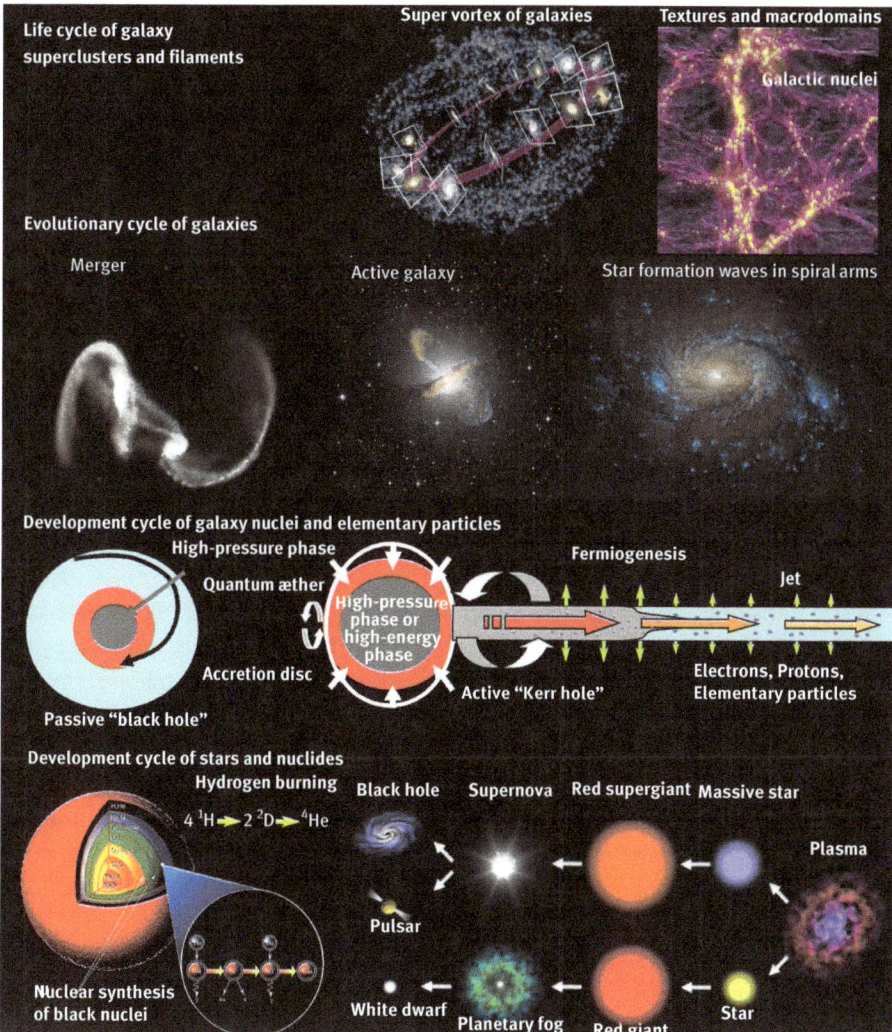

Figure 100: Overview on coupling the evolution of matter systems in the macrocosm and microcosm. The mutual interaction of the development of different structural levels causes tendencies to equilibrium on one structural level to be reversed by evolutionary processes at other structural levels. This results in an infinite process of emergence and decay of complex systems on different time and length scales.

relative; the change is absolute. The system behavior can vary in the course of its evolution, either due to an external influence (chance), or as a result of the changes which are caused by the cycle itself. The chemical pattern formation in catalysis comes to a standstill when the concentration of reactants falls below a critical threshold. Star formation processes subside when hydrogen is increasingly used up as an energy source. The material cycle in galaxies is subject to a qualitative change with the transition into active stages when a critical mass density of its core is exceeded. Because processes on different structural levels of matter interact with each other, the evolutionary process of the system components does not come to a standstill when the cyclic processes of a system are terminated. It is replaced by a new system behavior or displaced by another existing system.

8.8 Transitions between stationary states

Quantitative changes in stationary states or cyclic processes can change abruptly into qualitative changes. The transition of the heat transport from diffusion to convection occurs abruptly at a critical limit of the heat flow. Billions of molecules change from a disordered thermal motion to an ordered movement in convection cells. By means of the random disordered motion as a carrier of the thermal energy, however, the heat is transported into the ordered convection flow in the first place. The random and ordered motion form a unique process, where the random thermal motion of the molecules must subordinate to the ordered motion of the entire convection cell.

The dialectic of chance and necessity becomes even more evident at the critical points of transition from disorderly to orderly motion, or when jumping between different types of cyclic processes. Thus, the convection flow can be organized in cylindrical forms with clockwise or anticlockwise movements. The nonequilibrium phase transition from diffusion to convection is characterized by what is called bifurcation, a branching of the evolution of the system in the future [Haken 1990]. The necessity to organize the thermal transport in ordered movements in order to maximize the heat transport leaves room for random processes, which, for example, determine the direction of rotation of the convection cells. In Figure 12, the direction of rotation of the convection cells is random. On closer inspection, however, it can be influenced by the shape of the vessel or the rotation of the Earth, for example.

The stationary states of electrons in the shell of atoms are also determined by constantly repeating processes of emission and absorption of zero-point photons. This results in a stationary matter-wave field, in which electron movement and Fermi pressure counteract the electrostatic attraction between atomic nucleus and electrons. Changes of the stationary state are only possible above the threshold of the quantum of action, by absorption and emission of photons. The exact time and place of the transition from one stationary state to another are subject to random fluctuations. The effect of the zero-point photons on the process of absorption and emission is below the threshold of current observation abilities, and is therefore covered by stochastic laws, although each individual interaction process underlies causal laws.

In the cosmos, randomness constantly occurs in the transitions from one dynamics to another. Coupling and mutual influences of different evolution processes and forms of matter is creating an ever increasing richness and complexity of all developments. The evolution of the cosmos is open and not completely determined by its present state. The evolutionary tendency of individual systems can be predicted on the basis of the recognized laws. However, chance often plays a major role in the transitions from one self-organized state to another.

8.9 Infinity of evolutionary processes in a system of structural levels

In the history of natural science, new discoveries have challenged every assumption of the finiteness of systems. This also applies to the history of the study of the macrocosm and microcosm. Atoms or electrons are not the smallest, and galaxies or clusters of galaxies are not the largest systems of matter. Therefore, the assumption of an infinite system of structural levels of matter is the only sensible idea. In epistemological terms, the coming into being and passing away of all forms of matter corresponds to the unity of continuous and discrete matter on all development stages. The existence of an infinite system of structural levels is consistent with the idea that evolutionary processes also progress without end.

Any process that would eventually lead to equilibrium, and thus to the end of evolution on an isolated structural level, can be modified, unbalanced, or even canceled by interaction with other structural levels.

An isolated consideration of the repetitive processes of star formation and decay and the associated chemical evolution would lead to the conclusion that such a process would inevitably end due to the enrichment in metals. However, the evolutionary process of the stars is modified by the evolutionary process of the galaxies. This influences both macroscopic and microscopic processes. Thus, the formation of active galactic nuclei with their jets of ejected hydrogen and light elements affects both star formation in the spiral arms, and the chemical composition of the interstellar matter: the process of formation of heavy elements in the nuclear-fusion processes of stars is canceled out (in a dialectical sense) by the process of formation of light elements in the evolutionary process of active galactic nuclei: Ultimately, formed heavy elements, star corpses and other forms of matter are transformed in the "thrash cans" of galactic nuclei to ultra compact matter states which then are "recycled" into light elements during the electroweak phase transition in the galaxy nucleus. The coupled system of stellar evolution–nuclear fusion–galactic evolution–fermiogenesis forms a metamorphic process at a higher level, which couples several cyclic/spiral processes to one another.

Since the time scales for stellar and galactic evolution are different, this coupling produces a temporally and spatially oscillating pattern: galaxies that repeatedly pass through different stages. Phases with active nuclei and jets alternate with phases of intense star formation and formation of new spiral arms. This development is influenced by a number of other factors, such as collision and fusion of whole galaxies or decay and ejection of galaxy nuclei. Thus, a large variety of different types of galaxies develop.

The mutual influence of the evolutionary processes not only affects two structural levels that build on each other, but also continues over many structural levels of matter and is systemic. Thus, during nuclear fusion in the evolutionary processes of the stars, not only the composition of the atoms is changed, but also the structure of the quantum æther at a certain stage of development. Initially, these changes are only gradual, due to the formation of heavy nuclei in the interior of the star along the "valley of stable nuclides," and affect the structure of the nuclei themselves. When heavy stars die, this gradual evolution turns into phase transitions of nuclear matter within star corpses. Depending on density and temperature, even quantum fluid states of this nuclear matter may arise in the form of color superconductivity in neutron stars. Furthermore, at extremely high pressure, an electroweak phase transition may occur in supermassive objects that are called "black holes." The macroscopic evolution of the stars is thus not only coupled with the chemical evolution of the elements, but also in the long term with the evolution of various other stages of matter.

8.10 Galaxy supercluster: agglomeration or self-organization?

The conventional theory of the formation of galaxies and galactic superclusters, which still prevails today, assumes that these were formed solely by gravitational attraction from fluctuations in particle density in an agglomeration process. According to this theory, the law of gravity is independent of the formation of the matter structure of the cosmos and stands "above and beyond the matter structures."

The qualitative model of the formation of galactic superclusters in unity with the partial shielding of the gravitational field, as discussed in this book, is based on the dialectic that the formation of a structure at ever greater dimensions is accompanied by microscopic structure formation, via the formation of topological structures. Gravitational fields caused by heavy masses interfere universally with other forms of matter, in the microscopic movements of atomic building blocks, in the propagation of light, and also influence the propagation of matter waves and thus the zero-point fields. Consequently, the question of the interplay between the distribution of mass, the gravitational fields, and the state of the quantum æther is natural.

The recognition of the real existence of superfluid states in the quantum æther, such as color and Higgs condensates, naturally raises the question of their state of motion on a macroscopic scale, and their influence on gravity and the formation of masses. It is a law of the superfluid state that it is incompatible with a macroscopically inhomogeneous or rotary motion. It decays into vortices forming an uncondensed phase in the center. This could be a mechanism for the formation of masses by partial shielding of macroscopic condensate movements. In Chapter 7, the observations were listed which hint at the formation of the structure of galactic clusters and superclusters along filaments and gigantic bubble structures by a self-organization process. In this process, galactic clusters and superclusters probably developed together with the formation of gravity as a partially shielded residual interaction.

The observations support an evolutionary theory in which large spatial structures, such as galaxies and galactic clusters have by no means developed only upwards from the agglomeration of building blocks of the microcosm. Partial shielding of huge vortices in the quantum-fluid component of the æther via formation of vortex filaments could cause the formation of galactic nuclei as topological defects in the quantum æther. At their electroweak phase boundary, above a critical magnitude instabilities occur, which leads to fermiogenesis. This is the other essential side of evolution "downwards." The systemic interaction of these evolutionary processes "top down" and " bottom up" may form a complex process of formation of the visible forms of matter in the cosmos over trillions of years.

The process proposed here is the existence and partial shielding of macroscopic superfluidic motions that are self-consistent, along with the emerging mass structures. However, contradictions between the spatial structure of the gravitational field and the distributions of mass are constantly emerging, which at times are discharged as a process of mass generation in fermiogenesis in active galaxy nuclei.

This picture of the systemic self-organization process of all structural levels of matter in the interaction of processes in the macrocosm and microcosm is contrary to the still widely held view of a one-way street of development from microcosm to macrocosm. The opposite is also true: structure and development of matter in the microcosm is also determined by the structure formation in the macrocosm. The question of whether the chicken or the egg, whether the structures of the microcosm or the macrocosm, came first dissolves in the interpretation of their evolution by a mutually connected dialectical process.

8.11 How does the cosmos develop to a vast size?

The larger the structure, the larger the time-scale for its development. While the conversion of elementary particles takes place in the femtosecond to nanosecond range, the evolutionary process of stars already takes several billion years. The evolutionary cycles of galaxies correspond to time-scales of hundreds of billions of years. The thermalization time of microwave background radiation as the thermal radiation of the known part of the cosmos indicates a time scale of 500 billion years. It is an indication that the part of the universe observable for us today with the different interlaced evolutionary cycles of the structural levels has developed over time scales up to trillions of years.

But does the universe everywhere look like the cosmos known in our region? These are speculative questions to which one can give only very preliminary answers. The length scale for an infinite red-shift of photons, extrapolated according to the Hubble Law, is about 13 billion light-years (4,000 megaparsecs). The red-shift of the photons represents, as far as we know today, a limit for the observation of very distant regions of the cosmos.

Do even larger structures occur than the "Great Walls" of the galactic superclusters? It can be assumed that the structure formation of matter continues in systems of even larger sizes. Even larger dimensions of the structure formation may be connected with larger-scale motion processes of the quantum æther, which are only speculative today. However, it is quite justified to assume that an even larger structural level of matter could be associated with macroscopic phase changes of the quantum æther, so that different states of the quantum æther form in other parts of the universe than in the part visible to us. All previous ideas in the history of astronomy, which assumed that nothing new would occur on a certain structural level of matter, have so far proved to be wrong.

The question of whether the cosmos looks the same everywhere is closely related to the question of the origin of the so-called "fundamental natural constants." For example, the speed of light is a material parameter of the quantum æther. Such parameters are not really constant, but change as a function of the size, the gravitational field, or other state changes of the quantum æther. Therefore, it is justified to assume that with a change of the quantum æther, for example at very large dimensions in the cosmos, the value of such "constants" also changes. This would lead to a picture of a very rich and diverse cosmos, which produces matter structures that are inconceivable for us today. Materialistically, these questions aim at an improved understanding of the nature of the quantum æther, as well as deeper levels of matter, and how they have evolved in evolution with the macrocosm with just those properties we find today. Such an improved understanding will be an important step toward a comprehensive picture of the infinite evolution of matter in space and time.

8.12 Fine tuning of the natural constants: Big Bang with "intelligent design"

In opposition to this is the idealistic conception of the "basic laws of nature and natural constants," which are somehow given "externally." It is the idealistic antithesis to the evolutionary theory of matter, in which the strength of interactions, sizes of particle masses, charges, and other "fundamental parameters" have evolved self-consistently with the structure formation of various stages of matter in the microcosm and macrocosm in a process of self-organization.

The representatives of the fine-tuning thesis claim that only tiny changes of the "fundamental natural constants" would lead to a fundamentally different universe, in which perhaps life would not be possible. At first, they took up materialistic insights. In 1952, for example, the astrophysicist Hoyle correctly predicted that the carbon nucleus must have an excited state at about 7.7 MeV, in order for carbon to be produced at all in the nuclear-fusion reactions of stars. In fact, such a state was later found experimentally with the predicted energy. This excitation state depends on the exact strength of the strong nuclear force. Since the existence of carbon nuclei is essential for the formation of organic molecules and life, it would follow that the strong nuclear force in the cosmos may change little without undermining the formation of carbon in the nuclear fusion of stars.

Such a materialistic view, however, is turned upside down by the fine-tuning theorists when they arbitrarily select and vary individual quantities from the manifold properties of matter [Vaas 2006]. Tegmark [Tegmark 1998] investigated the influence of a hypothetical change in the ratio of electron mass to proton mass. He came to the conclusion that life based on complex molecules can only develop in a narrow window. A very large mass ratio would lead to the fact that no stable molecular systems could exist, due to the large oscillation energy of the atomic nuclei. In contrast to Tegmark, Stenger [Stenger 2000] comes by variation of four "fundamental constants" to the result that much larger fluctuations of the constants are allowed. In fact, not only four, but also all structural variables of the universe would have to be self-consistent with each other, because they are interdependent and interact with each other. The idea of "fundamental natural constants" that require "fine tuning" is an open expression of the ideology of the world formula that places "fundamental theories and parameters" as an intellectual principle above material nature.

The idealistic view of the laws of nature and constants of nature as something externally opposed to matter explains the evolution of the cosmos either as a rather improbable accidental event, or is subject to theological ideologies, even religious ones, of "intelligent design." The ideological discussion about this is summarized in the thesis of the anthropic principle. It was formulated in 1974 by Carter: *"… We must be prepared to consider the fact that our place in the universe is necessarily privileged in the sense that it is compatible with our existence as observers"* [Carter 1974]. We had already made it clear in Chapter 1 that the connection between characteristics of the observable universe, and the necessity of the existence of a conscious observer who can recognize this universe, is unscientific and represents a form of idealistic world view, creationism in cosmology. The physicist Tipler even argues that "the universe must be exactly such that it must produce life at a certain stage of development" [Barrow & Tipler 1986]. In the book *The Physics of Immortality: Modern Cosmology, God and the Resurrection of the Dead*, the absurd thesis is pushed to the extreme that the universe as a whole is steered to a "god-like" state by the development of intelligent life [Tipler 1994]. The thesis of predestination not only of the cosmos but also of man is today the heart of various religious and esoteric ideologies. According to it, neither the development of matter nor that of man is open. That the whole cosmos only exists in order to produce man, like a Mr. Tipler with his imaginary gods, at some point can hardly be surpassed in anthropocentrism and self-exaltation.

8.13 Self-organization versus heat death

The infinite development of matter, however, is neither theologically oriented toward a goal, nor is it subject to a general degeneration process through an increase in entropy in a "heat death." In fact, in many systems, an increase in the complexity and diversity of matter structures occurs over long periods of time. This is true for the formation of our solar system with a variety of planets and a rich chemical complexity formed from originally less complex interstellar hydrogen. The chemical complexity present in our and other solar systems today has taken a number of generations of stars and tens of billions of years. The great chemical complexity of the chemical elements is a necessary prerequisite for the development of complex molecules in space and on the Earth's surface, as a prerequisite for the emergence of life. The evolution toward higher complexity can be particularly observed in the evolution of the biosphere, life and species in a process that has now lasted 3.6 billion years. As the Belgian Nobel Prize winner Christian de Duwe points out in his remarkable book *Vital Dust: Life as a Cosmic Imperative*, the development of life under such conditions is likely [Duwe 1995].

A further counter-thesis to the dialectical-materialist theory of the evolution of matter in self-organization processes is the idealist imposition of the second law of thermodynamics onto the total system of matter in the cosmos. This states that entropy can never decrease in processes in closed systems close to equilibrium. As a matter of course, a coffee cup never spontaneously becomes hot by extracting heat from its surroundings. A hot coffee cup cools down by releasing heat to the environment until equilibrium is reached. The entropy law of thermodynamics is a probability statement of the transformation of ordered into disordered movements. In a completely closed system, what is probable is the equal distribution of energy over all microscopic forms of motion, and therefore an approximation to thermodynamic equilibrium. This law of probability is incorporated into the concept of "free energy," which determines what portions of energy forms are available for transformation into another form.

Clausius was one of the first to apply this law of equilibrium to the universe as a whole as an "absolutely closed system": *"The more the universe reaches the limit state in which entropy is at a maximum, the more the opportunities for further change decrease."* Then *"... the universe is in a state of unchangeable death."* [Clausius 1865]

Herman Haken, one of the leading theorists of the self-organization of matter, criticizes this notion: *"In the opinion of almost all physicists, the world was created about ten billion years ago in a Big Bang as a tremendously hot fireball in which there was no order. So at the beginning of the world, chaos, thus no order. And afterwards, the disorder is supposed to increase even more, up to its maximum. Where is there still room for ordered meaningful structures, thus especially for life?"* [Haken 1984]

8.14 Controversies about the entropy of gravitating systems

The laws of probability do not contradict an increase in complexity through self-organization at all – on the contrary. The disordered random structure of an interstellar plasma probably changes under gravitational collapse into a complex state of a solar system with all the manifold structures up to the development of life. The gravitational collapse that occurs in the further evolution of stars and galaxies in "black holes" forms systems that "swallow" all possible forms of matter from photons to stellar corpses like a large "trash can" of the cosmos. Theoretical studies show that in systems under gravitational collapse, entropy can even decrease [Landsberg 1984, Thirring 1991, 2011].

However, this is strenuously denied by the advocates of the second law of thermodynamics. They argue that entropy must continue to increase, even after a collapse into a "black hole" because thermal photons are emitted [Lineweaver 2008]. But instead of using statistical mechanics to make a concrete analysis of the probable processes of gravitational collapse, entropy terms are invented for "black holes" in order to save the second law [Bekenstein 1973, Hawking 1974]. They are not based on any experimental observation or evidence. Under the dogma of "endless increasing entropy", even arbitrary entropy parameters are defined for the gravitational field [Wallace 2009, Acquaviva 2015]. These theories also detach themselves from all experimental observation.

In fact, the big bang cosmology can only be maintained by making the second law of thermodynamics a central dogma of evolutionary theory. To this end, the "arrow of time" is derived from the alleged growth of entropy in absolutely every process in the cosmos. A modern textbook of cosmology thus asserts, without any concrete investigation: *"The behavior of gravity is different from that of the dissolution of our sugar cube in tea, which is caused by electromagnetic forces. In the case of tea, sugar tends to dissolve as uniformly as possible. In the case of a gravitating system, the tendency is exactly the opposite: it condenses. Therefore, the agglomeration of matter in galaxies and stars in the expanding universe is accompanied by an increase in the entropy of the universe. Similarly, the generation of black holes suggests an increase in entropy. Thus, the entropy of the universe collapsed at the end is very large, while its initial entropy was relatively small."* [Contopoulos 1987, p. 184] The dogmatism and prejudice of such claims can hardly be surpassed: the application of the concept of entropy, which derives from the study of the achievement of equilibrium in closed systems, to an open system of structural levels of matter, in which imbalance and evolution occurs again and again, is by no way sientifically justified.

The asymmetry of time is also used to justify why there must be an asymmetry of singularities in space-time: *"The universe produces 'black holes' but not 'white holes'"* [ibid., p.185]. The dogma of increasing entropy forbids the existence of *"white holes"* [Penrose 2010]. Physicists such as Roger Penrose devoted much of their career to creating an actual barrier against the study of particle formation in active galactic nuclei.

8.15 Nonequilibrium and evolution are the determining factor

The evolution of matter in the cosmos is determined by the unfolding of contradictions. Just as the motion of matter is indestructible, in an open system of structural levels of matter, the occurrence of contradictions of system dynamics, and thus of nonequilibrium, is the determining factor. The existence of infinitely many different structural levels of matter leads to the fact that the nonequilibrium on a certain system level (for example the sun) evokes the formation of nonequilibrium structures on another system level (for example the molecules and the biosphere on earth). In general, the tendency toward equilibrium on one structural level of matter is canceled out by developments on the other. The basic thesis of the universe as a closed system is untenable.

But what about the thermalization of photons on their way through the vastness of space via various transformation processes and the resulting thermal radiation as the microwave background radiation? Isn't this process irreversible? Would this not increase the thermal radiation of the intergalactic medium over time?

If the microwave background actually occurs homogeneously in the cosmos, the number of photons associated with it is huge. It far exceeds the number of photons that could be produced by nuclear fusion of all hydrogen atoms present today. This reflects the fact that this thermal bath of photons must indeed have developed over periods of time much longer than those of the evolutionary cycles of stars and galaxies. But even the accumulation of microwave photons will not be a one-way street. Experience shows that there are no one-way street developments with an absolute end in nature. Whatever process converts disorderly motion of low-energy photons into other forms of motion, is up to concrete research to show. Perhaps very low-energy photons below a frequency threshold release energy over a long period of time to the zero-point excitations of the quantum æther. Such a process is plausible, because at very small photon energies a rigid threshold of the quantum of action between unstable and stable photons would lead to contradictions.

The universe is by no means doomed to heat death, nor does it need the construct of the Big Bang to reset the clock of aging to zero. Rather, it is characterized by cyclic processes, nonequilibrium and the interaction of different parts.

It behaves like an open system at all structural levels. Thus, it produces an infinite evolution, which at least sometimes leads to increasing complexity and diversity of structures and processes, with an open outcome. It is up to human beings to bring their scientific world view, research method and way of thinking into concordance with reality. Understanding nature as a system of interlocking cycles and cycles of development will also help humanity to shape its own metamorphosis with nature sustainably, built up a closed loop economy and in conscious long term harmony with the laws of nature and society. It is urgently necessary to fight for the social changes in modes of production and way of life that are necessary to achieve this.

9 Mathematical appendix

9.1 General properties of topological structures

Any kind spatially inhomogeneous disturbance or excitation of a matter system can be described as a continuous field. The fields of topological defects or structures differ qualitatively from the fields of arbitrary excitations in a way determined by their stability, which is illustrated in Figure 101.

Figure 101a illustrates an ideal crystal with strictly periodically ordered atoms. In such a state, the field of displacements disappears identically for all space positions, $u(r) = 0$ (Figure 101b). If, on the other hand, a crystal is elastically deformed, an inhomogeneously distorted state is created, with a displacement field such as, for example, in Figure 101c. Both states can be described mathematically by projecting the angle φ of the displacement field along a closed curve onto a torus (Figure 101d). The undisturbed crystal (a, b) corresponds to a circle line (yellow line, $\varphi = 0$ for all θ) on the torus, while the disturbance (c) is described by moving at different angles θ on the surface of the torus (orange line $\varphi(\theta)$).

When the disturbance/excitation qualitatively changes to a topologically stable structure, such as a vacancy in a crystal (e) with displacement field shown in (f), a qualitatively new property occurs: the displacement field is then characterized by a complete revolution on the torus surface (yellow line in h)), for a revolution enclosing the core of the defect/structure (yellow circle in (f)). The same applies to a vortex as a topologically stable structure in a superfluid, whose velocity field $v(r)$ can be seen in (g), and on which the torus performs the same revolution. The stability of the topological structure is mathematically reflected in the following way: When revolving in an area that does not contain the defect nucleus (black circles in (f) and (g)), the field disappears in the limit of the radius circle/torus $r \rightarrow 0$. If the topological structure is contained in the rotation (yellow circle), the field for $r \rightarrow 0$ does not disappear, because the rotation around the entire torus surface remains. Topological defects/structures therefore exhibit a connectivity of distortions of their ground state which cannot be made to disappear by local changes or relaxation.

9.2 Topological structures in quantum fluids

Anderson–Toulouse Vortex in superfluid helium-3A

This topological defect has the structure of a vortex tube, but with a soft core. Therefore, this structure is called a Skyrmion. It has two quanta of circulation.

The vector field of the orbital angular momentum of the quantum fluid has the following structure at the vicinity of the defect

https://doi.org/10.1515/9783110644203-009

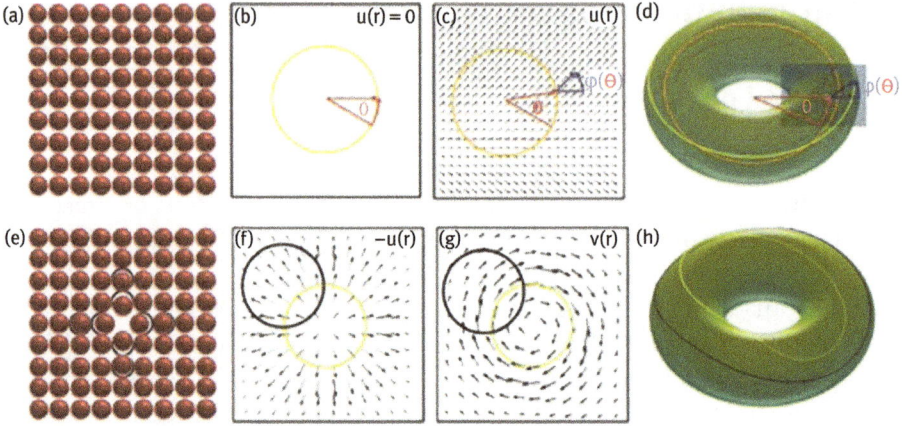

Figure 101: Disturbances/excitations of a medium (crystal (a–f), superfluid (g)) with change into topologically stable structures. The latter can be distinguished from simple perturbations/fields by a revolution on a torus. See text for an explanation.

$$\vec{l}(r, \varphi) = \hat{z}\cos(\eta(r)) + \hat{r}\sin(\eta(r)).$$

The following applies to the function $\eta(r)$: $\eta(0) = 0$ and $\eta(\infty) = \pi$.

For Figure 32a in Chapter 3, the following function was selected

$$\eta(r) = \pi + 2\operatorname{atan}(r).$$

The vortex field of the velocity of the superfluid is given as

$$\vec{v}_s(r, \varphi) = \frac{\hbar}{2mr}(1 - \cos\eta(r))\hat{\varphi}.$$

Despite the soft vortex core, the circulation is quantized:

$$k = \oint d\vec{r}\,\vec{v}_s = 2k_0,$$

with the quantum of circulation $k_0 = \frac{h}{2m}$.

Mermin–Ho vortex in superfluid helium-3A

The structure of the orbital angular momentum is now described by the function

$$\eta(r) = \frac{1}{2}\pi + 2\operatorname{atan}(r).$$

The vector field of the orbital angular momentum and the vortex field of the velocity of the superfluid can be calculated by using the same expressions as in the Anderson–Toulouse vortex.

The defect has a quantum of circulation κ_O and is visualized in Figure 32b. This type of topological defect can also occur as a pair of vortices with circular resp. hyperbolic current distribution. See [Volovik 2003] Figure 16.7, p. 204.

Spherical hedgehog defect (t'Hooft–Polyakov monopoly) in superfluid helium-3B

This spherical defect has a quantum of circulation κ_O and represents a magnetic monopoly in helium-3B (see Figure 32c). The vector field of the orbital angular momentum of the quantum fluid is induced only in the immediate vicinity of the monopoly and disappears with increasing distance. It can be described as

$$l_x = \cos(k\,y) \cdot \sin(k\,x) \cdot \cos(k\,z) \cdot exp(-h\,r);$$
$$l_y = \cos(k\,x) \cdot \sin(k\,y) \cdot \cos(k\,z) \cdot exp(-h\,r);$$
$$l_z = \sin(k\,z) \cdot exp(-h\,r);$$

With

$$r = \sqrt{x^2 + y^2 + z^2}$$

The vortex field of the velocity of the superfluid is

$$\vec{v}_s = \beta\,\vec{\nabla}F,$$

with

$$F = atan\left(\frac{l_y}{l_x}\right), \quad \beta = \frac{l_z}{l} \text{ and } 1 = \sqrt{l_x^2 + l_y^2 + l_z^2}$$

The magnetic field of the monopole is given as

$$\vec{B} = \vec{\nabla} \times \vec{l}.$$

Planar textures in superfluid helium-3A

This type of defect represents a soliton in the superfluid helium-3A phase (Figure 32d). A similar structure can also occur as a phase boundary between different phases, such as helium-A and helium-B.

The vector field of the orbital angular momentum of the quantum fluid is

$$l_x = 0;$$
$$l_y = 0;$$
$$l_z = \operatorname{atan}(2x).$$

The magnetic field of the texture is given as:

$$\vec{B} = \vec{\nabla} \times \vec{l}.$$

And the vortex field of the current density of the superfluid is

$$\vec{j} = \vec{\nabla} \times \vec{B}.$$

9.3 Causes of matter waves and inertial mass with uniform and accelerated motion

The spectral density of the electromagnetic zero-point field as a function of frequency ω is given by

$$\rho_{ZPF}(\omega) = \frac{\hbar \omega^3}{2 \pi^2 c^3}.$$

De Broglie suggested that a particle having a mass m_0 and a rest energy $m_0 c^2$ is associated with a frequency known as the Compton frequency, which is given according to the Einstein–de Broglie equation as

$$\hbar \omega_c = m_0 c^2.$$

This corresponds to a locally oscillating field which is linked to the particle, for example, the revolution frequency of a self-contained eddy current density of a condensate. Through constant emission and absorption of unstable photons, the particle is in resonant interaction with the surrounding electromagnetic zero-field. It is assumed in this simple model that a resonance of the interaction occurs at the Compton frequency.

If the particle moves in the +x-direction, this resonance frequency of the interactions between the zero-point field arriving at the particle from the front is shifted to a lower frequency in the resting laboratory system

$$\omega_- = \gamma \omega_c \left(1 - \frac{v}{c}\right).$$

This zero-point oscillation with a lower frequency in the laboratory system is blue shifted in the system of the moving particle due to the Doppler effect. Consequently, the resonance occurs at ω_-. Conversely, a mode hitting the moving particle from behind is redshifted. This is why resonance occurs in the laboratory system for modes with a higher frequency.

$$\omega_+ = \gamma \omega_c \left(1 + \frac{v}{c}\right).$$

In both equations, the Lorentz factor

$$\gamma = \frac{1}{\sqrt{1 - \frac{v^2}{c^2}}}$$

occurs. The superposition of the two Doppler-shifted zero-point frequencies leads to a speed-dependent modulation of the Compton frequency given by

$$\omega_B = \gamma \omega_c \frac{v}{c}$$

and is the de Broglie frequency of a matter wave. The wavelength of this modulation is

$$\lambda_B = \frac{\lambda_C}{\gamma} \frac{c}{v},$$

and can be interpreted as the de Broglie wavelength $\lambda_B = h/p$ of a particle with momentum.

$$p = m_0 \gamma v$$

It corresponds to the superposition of two waves propagating in opposite directions with the same amplitude in the laboratory system. In the resting system of the particle, it appears as a standing wave. More details on this model can be found in [de la Pena 1996] and [Haisch 2000b], respectively.

In the latter article, also, the connection to the inertial mass m_i of the particle as reaction force to an accelerated motion in the zero-point field is made, giving

$$m_i = \frac{V_0}{c^2} \int \eta(\omega) \rho_{ZPF}(\omega) d\omega$$

where $\eta(\omega)$ a frequency-dependent scattering probability and V_0 a scattering volume represents. A change in inertial mass at high speeds near the speed of light can result from changes in scattering volume, scattering probability, or a change in the frequency spectrum at high frequencies.

9.4 Shielding lengths and masses of the exchange quanta

A superconductor is an insighfull system for understanding the connection between particle like topological structures, their interaction by exchange quanta and partial shielding. Here, two length scales are relevant and determine the formation of magnetic vortex tubes as topological defects. The London magnetic penetration depth λ_L describes the length scale of the shielding of the magnetic field and the coherence length ξ the length scale of the variation of the density of the superconducting condensate close to the core of the vortex tube (see Figure 102). In London theory, the magnetic penetration depth is given as

$$\lambda_L(T) = \sqrt{\frac{mc^2}{4\pi n_s(T)e^2}}$$

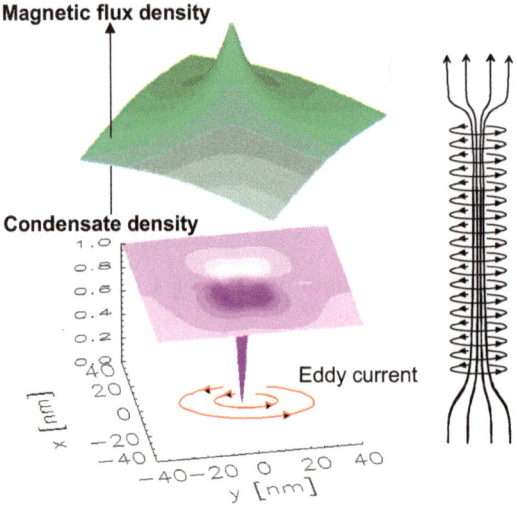

Figure 102: Structure of magnetic vortex tubes in superconductors. Above: The magnetic flux density distribution in the interior is shielded on the length scale of the magnetic penetration depth. Below: In the center of the vortex tube a normal conducting core is formed with the radius of the coherence length. Right: Schematic representation of the vortex tube with the enclosed magnetic field and superconducting eddy currents.

where m is the mass of the electron, $n_S(T)$ is the temperature-dependent density of the superconducting condensate, and e is the electrical charge of the electron. This magnetic shielding length is, therefore, smaller the higher the density of electrons in the superconducting condensate.

The partial shielding of the magnetic field corresponds to the generation of an effective photon mass in the superconducting state of an electron liquid of

$$m_{ph} = \frac{\hbar}{c\lambda_L},$$

The mass is noticeable in an attenuation of electromagnetic waves when they penetrate the superconductor. The reason is the generation of plasma oscillations of the superconducting condensate by the electromagnetic wave of the photon.

The coherence length describes the average distance of the electrons condensed to Cooper pairs in the superconductor. It is expressed in the Ginzburg–Landau theory of superconductivity by

$$\xi(T) = \sqrt{\frac{\hbar^2}{2(2m)\alpha(T)}}$$

where the temperature-dependent constant $\alpha(T)$ is proportional to the density of the condensed electrons, $\alpha(T) \sim n_S(T)$. The more electrons are condensed, and the larger the mass of the Cooper pair is $2m$, the greater is the phase stiffness of the condensate and the smaller is the coherence length.

The coherence length is the smallest length scale on which an inhomogeneity can develop in the superconducting electron density. It describes, for example, the diameter of the normally conducting core of a vortex tube.

With the formation of the normally conducting core by partial shielding of the electromagnetic interaction, an inertial mass of the vortex tube is created. Its connection with the coherence length is given by

$$m_\mu = \frac{\hbar}{\xi c},$$

When applying this well-tested theory of superconductivity to the still little understood processes in the quantum æther, m_μ corresponds to the mass of the Higgs boson. In the superconductor, the correlation with the size of the superconducting energy gap 2Δ is given by

$$m_\mu c^2 = \sqrt{6}\Delta \frac{c}{v_F}$$

where v_F is the Fermi velocity of the electrons. Energy gaps also occur in the quantum æther and are linked to the mass of the respective fermions. However, the relationship between the mass of the Higgs boson and the different masses of the fermions is complex. In contrast to the plasmon (=photon with mass), which is an excitation of the phase of the superconducting condensate, the Higgs boson is an excitation of the amplitude of the condensate.

The shielding length of the gravitational field can be determined from the redshift of the photon according to the Hubble law using two alternative methods both leading to very similar results.

The photon mass can be expressed directly via the Hubble constant H_0 by

$$m_{ph} c^2 = \hbar H_0$$

With $H_0 = 74.3$ km/s MPc^{-1} (=2.4 × 10^{-18} 1/s) this results in a mass of 8.9 × 10^{-69} ·kg or an unimaginably small energy of 5 × 10^{-33} eV. The Hubble constant thus corresponds to a shielding length of 3.9 × 10^{25} m of photons. Alternatively, using the theory of partially shielded gravitation of [Preston & Potter 2006] (see Sections 9.4 and 9.5), gives a shielding length of

$$L = k^{-\frac{1}{2}}.$$

From the course of the redshift with the distance in Figure 99 follows $k = 6.5 \times 10^{-53}$ m^{-2} and thus a value for the shielding length of 1.2 × 10^{26} m (approx. 4000 Mpc). All particle masses related to partial screening of the corresponding fields can be found in Table 2.

Table 2: Examples of particle masses and shielding lengths for the structure formation of matter on different length scales by partial shielding of excitations of the underlying condensates.

Shielding effect	Involved excitation	Energy/mass of the boson	Shielding length
Weak nuclear power	W-Boson	80.4 GeV	2.5 × 10^{-18} m Electroweak phase transition
Weak nuclear power	Higgs-Boson	125 GeV	1.6 × 10^{-18} m Core of the electron
Electromagnetic field in a superconductor	Plasmon as a photon with mass	2 eV (=3.5 × 10^{-36} kg)	10^{-7} m Magnetic vortex tube
Electromagnetic field in a superconductor	Amplitude of the condensate	130 eV (=2.3 × 10^{-34} kg)	10^{-9} m Normal conducting vortex core
Gravitational field	Redshift Photon	1.7 × 10^{-33} − 5 × 10^{-33} eV (=(3–9) × 10^{-69} kg)	3.9 × 10^{25}–1.2 × 10^{26} m (1300–4000 Mpc) Galaxy supercluster

9.5 Theory of partially shielded gravity

From the general-relativistic Hamilton–Jacobi equation for the action of a gravitational field [Landau & Lifshitz 1965] under consideration of energy-momentum preservation, a wave equation for a gravitational wave field can be derived [Preston & Potter 2006]. Within the Schwarzschild approximation, the outer field of a point source is given as

$$V(r) = -\frac{GM}{r} + \frac{l(l+1)H^2c^2}{2r^2}$$

whereby $\sqrt{l(l+1)}Hc$ replaces the classical angular momentum L, c = speed of light, G gravitational constant and Hc is a measure for the angular momentum per mass. The result expresses a quantization of the orbital angular momentum of the underlying superfluid medium.

For large length scales, the gravitational field merges into that of an inner observer. Therefore, a metric is used which is the same for all observers. For the simplest model of a homogeneous sphere filled with a gaseous matter distribution, the potential is

$$V(r) = -\frac{k c^2 r^2}{2(1-k r^2)} + \frac{l(l+1)H^2 c^2}{2 r^2 (1-k r^2)},$$

with

$$k = 8 \pi G \frac{\rho}{3 c^2}$$

and mass density ρ. The first term is a portion that occurs when the speed of light (the so-called metric) is independent of the location. Even without a macroscopic angular momentum ($l = 0$), the result is a repulsive potential

$$V(r) = -\frac{k c^2 r^2}{2(1-k r^2)}.$$

This reflects a partial shielding of gravity on large length scales. In order to give an approximation of the potential curve of a massive object (see Figure 97), the potentials for a Schwarzschild solution and for the inner solution were then put together piece by piece. Mass and size of the active galaxy M87 were used as a specific example.

This results in:

Portion of the galaxy nucleus:

$$V(r) = -G M_c m \left(\frac{3 R_s^2 - r^2}{2 R_s^3}\right) \text{ for } r \le R_s$$

$$V(r) = -G M_c m \frac{1}{r} \text{ for } r > R_s$$

Portion of the total galaxy:

$$\text{for } r \le R_g \quad V(r) = -G M_c m \frac{1}{r} - G M_g m \frac{3 R_g^2 - r^2}{2 R_g^3}$$

$$\text{for } r > R_g \quad V(r) = -G M_c m \frac{1}{r} - G M_g m \frac{1}{r} + H^2 c^2 \frac{1}{r^2 (1-k r^2)} - k r^2$$

Determination of the quantities for M87:

Mass of the galaxy nucleus $M_c = 3 \times 10^{11} \, M_s$ with mass of the sun $M_s = 1.9 \times 10^{30}$ kg. Mass of the total galaxy: $M_g = 5 \times 10^{12} \, M_s$

Schwarzschild radius: $R_s = \frac{GM_c}{c^2}$

Radius of the galaxy: $R_g = 10^{20}$ m

Gravitational constant: $G = 6.67 \, 10^{-11}$ N m^2 / kg^2

Hubble constant: $H = 2.4 \times 10^{-18}$ 1/s und $k = 5 \times 10^{-53}$ m^{-2}.

In order to express the energy of the gravitational potential $V(r)$ by its effect on a mass probe m in Figure 97, the electron mass $m = 9.1 \times 10^{-31}$ is assumed and $V(r)$ is divided by the elementary charge = $1.602 \, 10^{-19}$ C in order to yield electron volt as the energy unit.

9.6 Hubble law and interpretation using the theory of partially shielded gravity

In order to establish a relationship between the redshift z of the light and the distance of the light source, the distance must be measured independently. To do this, the apparent brightness of a star is compared with its assumed absolute brightness. Supernova explosions of a certain type have a known brightness curve and are therefore used as a "standard light source" with a known absolute brightness. An object with luminous intensity L and spherical emission of light generates a photon flux density

$$F = \frac{L}{4\pi r^2}$$

at a distance r. The relationship between the apparent brightness of an object m and its absolute brightness M is:

$$m - M = 5 \log_{10}\left(\frac{r}{10\,pc}\right) = 5 \log_{10}(r) + 5,$$

where the distance r is given in units of megaparsec (=3.086 × 10^{22} m or 3.261 × 10^6 light years). The relationship is normalized so that apparent and absolute brightness become identical ($m - M = 0$) for $r = 10$ pc. $m - M$ is also called the distance modulus.

The apparent brightness of distant objects such as galaxies or stars depends not only on the distance but also on other processes that absorb light or convert photon energy into other forms. Therefore, the determination of distances is associated with great uncertainties.

For distances at which the redshift of photon energy begins according to Hubble's law, the relationship between photon flux density and distance must be modified. As the energy of the photons decreases with distance, the flux density is reduced by a factor $(1 + z)$. If the speed of light of the photons is reduced by running against a gravitational potential, an additional factor $(1 + z)$ is required. This modifies the relationship between the distance modulus m-M and the distance [Peebles 1993]:

$$m - M = 5 log_{10}(r) + 2,5\, log_{10}(1 + z)^n + 25,$$

where $n = 1$ in the first case and $n = 2$ in the second. Within the framework of the Big Bang model, $n = 2$, however with the wrong argument of an expansion of the space that the light passes through.

Resolved according to the distance r the following results:

$$r = \frac{10^{-5}}{(1+z)^{\frac{n}{2}}} 10^{\frac{m - M}{5}}.$$

The relationship between redshift and distance in Figure 99 thus depends on the assumed processes of light propagation through the cosmos. For $n = 1$ there is an almost linear correlation between redshift and distance, which is described in good approximation by the Hubble law:

$$z = \frac{v}{c} = \frac{H_0 r}{c}.$$

However, there are also numerous outliers that indicate that there are other causes of redshift.

In Figure 99, $n = 2$ was set to take in account of the change in the speed of light in the gravitational field (see Section 5.3) and a Hubble constant of $H_0 = 74,3$ km/s Mpc^{-1} was chosen for the Hubble law, which is currently the best value. For the description of the redshift of the photons by the partially shielded gravitational potential introduced in Section 9.4 according to [Preston & Potter 2006], we yield

$$z = \frac{r H_0}{c} \frac{1}{1 - k_1 r^2},$$

with $H_0 = ck^{0.5}$ and the constant $k = \frac{8\pi G\rho}{3c^2}$ already introduced in Section 9.4.

For the theoretical modeling of the relationship between redshift and distance in Figure 99, $k = 6.5 \times 10^{-53}$ m^{-2} was chosen (this results in $H_0 = 74.4$ km/s Mpc) and $k_1 = 1.8 \times 10^{-53}$ m^{-2}. Compared to the work of H. Preston and F. Potter [Preston & Potter 2006], k and k_1 are slightly different, which provides a better description of the experimental data. It may originate from a deviation of the mass density distribution from a homogeneous sphere. The experimental data in Figure 99 are from [Perlmutter et al. 1999, Riess et al. 2004, Super Nova Project 2010].

9.7 Model of the electron as a topological structure

The model describes the structure of the angular momentum field in the electroweak phase of the quantum æther in the core of the electron. This is given as an ellipsoidal hedgehog structure.

$$l_x = cos(k\ x) \cdot sin(k\ y) \cdot exp(-r/l_{ew});$$
$$l_y = -cos(k\ y) \cdot sin(k\ x) \cdot cos(k\ z) \cdot exp(-r/l_{ew});$$
$$l_z = 0,$$

with $r = \sqrt{x^2 + y^2 + z^2}$ and the electroweak shielding length l_{ew}, that is determined by the mass of the Z- and W-bosons.

The vortex field of the velocity of the superfluid is

$$\vec{v}_s \sim \vec{\nabla}F \cdot exp(-z/l_h),$$

with $F = -atan\left(\frac{l_y}{l_x}\right)$ and the Higgs shielding length l_h, which is determined by the mass of the Higgs boson and thus by the energy gap of the Higgs condensate.

The distribution of the magnetic moment of the electron is proportional to

$$\vec{m} \sim \vec{\nabla} \times \vec{l}.$$

The electrical field distribution is obtained from

$$\vec{E} = \vec{\nabla}\phi,$$

with

$$\phi = \vec{l} \cdot \vec{v}_s.$$

Bibliography

[Acquaviva 2015] G. Acquaviva, G. F. R. Ellis, R. Goswami, and A. I. M. Hamid, Constructing black hole entropy from gravitational collapse, Phys. Rev. D 91 (2015) 064017.

[A. Assis & M. Neves 1995] A. Assis und M. Neves, History of 2.7 K temperature prior to Penzias and Wilson, Apeiron, 2 (1995) 79–84.

[Abe 2014] K. Abe et al. (Super-kamiokande collaboration), Search for proton decay via $p \rightarrow v + K+$ using 260 kiloton·year data of super-kamiokande, Phys. Rev. D 90 (2014) 072005.

[Achard 2005] P. Achard et al., Measurement of the running of the electromagnetic coupling at large momentum-transfer at LEP, Phys. Lett. B 623 (2005) 26–36.

[Adler 1969] S. Adler Axial-vector vertex in spinor electrodynamics, Phys. Rev. **177** (1969) 2426.

[Adler 1970] S. L. Adler et al., Photon splitting in a strong magnetic field, Phys. Rev. Lett. 25 (1970) 1061–1065.

[Adler 1987] C. G. Adler, Does mass really depend on velocity, dad?, Am. J. Phys. 55 (1987) 739.

[Aharmim 2005] B. Aharmim et al., Phys. Rev. C 72 (2005) 055502.

[Aharonian 2006] F. Aharonian et al., Science 314 (2006) 1424.

[Akiyama 2015] K. Akiyama et al., 230 GHz VLBI observations of M87: Event-horizon-scale structure at the enhanced very-high-energy-ray state in 2012, arXiv:1505.03545v3.

[Akiyama 2019] K. Akiyama et al, First M87 Event Horizon Telescope Results. I. The Shadow of the Supermassive Black Hole, The Astrophysical Journal Letters 875 (2019) L1.

[Alfvén & Arrhenius 1976] H. Alfvén and G. Arrhenius, Evolution of the solar system, NASA Publication SP-345 (1976).

[Alfven 1981] H. Alfvén, Cosmic Plasma, Reidel, Holland, 1981.

[Alfvén 1984] Kosmologie und Antimaterie. Über die Entstehung des Weltalls, Umschau-Verlag, Frankfurt am Main 1984.

[Alfven 1990] O. G. H. Alfven, Cosmology in the plasma universe: An introductory exposition, IEEE Trans. Plasma Sci. 18 (1990) 5.

[Allen & Jones 1938] J. F. Allen and H. Jones, Nature 141 (1938) 243–244.

[Ambarzumjan 1976] V. A. Ambarzumjan, Probleme der modernen Kosmologie, Basel 1976.

[Amsler 2008] C. Amsler et al, Review of Particle Physics, Physics Letters B 667 (2008) 1–6.

[Anderson 1992] Anderson J.D. et al., Recent developments in solar-system tests of general relativity. In: Sato H., Nakamura T. (ed.), Proc. Sixth Marcel Grossmann Meeting. World Scientific, Singapore (1992).

[Anderson 1997] P. W. Anderson, When the electron falls apart, Physics Today, October 1997, 42.

[Anderson et al. 1995] M. H. Anderson, J.R. Ensher, M.R. Mathews, C. E. Wieman, and E. A. Cornell, Science 269 (1995) 198.

[Arndt 1999] M. Arndt, O. Nairz, J. Vos-Andreae, C. Keller, G. van der Zouw et al. Wave-particle duality of C60 molecules, Nature 401 (1999) 680.

[Arp 1997] H. Arp, Quasar creation and evolution into galaxies, J. Astrophys. Astr. 18 (1997) 393–406.

[Arp 1998] H. Arp, Seeing red, redshifts, cosmology and academic science, Apeiron, Montreal 1998.

[Arp 1998a] H. Arp, Redshifts of new galaxies, arXiv: astro-ph/9812144v1(1998).

[Aryal & Saurer 2005] B. Aryal and W. Saurer, Spin vector orientations of galaxies in seven Abell clusters of BM type III, Astron. Astrophys., 432 (2005) 841.

[Ashby 2010] N. Ashby, The Sagnac effect in the global positioning system. In G. Rizzi and M. L. Ruggiero (Eds.), Relativity in Rotating Frames. London: Kluwer Academic Publishers, 2010.

[Ashman & Zepf, 1992] K. M. Ashman and S. E. Zepf, Astrophys. J. 384 (1992) 50.

https://doi.org/10.1515/9783110644203-010

[Aspect et al. 1981] A. Aspect et al., Experimental tests of realistic local theories via Bell's theorem, Phys. Rev. Lett. 47 (1981) 460.

[Aspect et al. 1982] A. Aspect, J. Dalibard, and G. Roger, Experimental test of bells inequalities using time-varying analyzers. Phys. Rev. Lett. 49 (1982) 1804.

[Bahcall 1997] J. N. Bahcall et al., Hubble space telescope images of a sample of 20 nearby luminous quasars, Astrophys. J. 479 (1997) 642.

[Bailey 1977] J. Bailey, K. Borer, F. Combley, H. Drumm, F. Krienen, F. Lange, E. Picasso, W. von Ruden, F. J. M. Farley, J. H. Field, W. Flegel & P. M. Hattersley, Measurements of relativistic time dilatation for positive and negative muons in a circular orbit, Nature 268 (1977) p. 301–305; doi:10.1038/268301a0.

[Barrow & Tipler 1986] John D. Barrow and Frank J. Tipler, The anthropic cosmological principle, Oxford University Press, 1986, p. 21.

[Baryshev 2006] Y. Baryshev, Physics of gravitational Interaction: Geometry of space or quantum field in space, Proceedings of the 1. Crisis in Cosmology Conference, AIP conference proc. vol. 822 (2006).

[Bayer 1993] H. C. von Bayer, Das Atom in der Falle, RoRoRo, 1993.

[Baynes 2012] F. N. Baynes, M. E. Tobar, and A. N. Luiten, Oscillating test of the isotropic shift of the speed of light, Phys. Rev. Lett. 108 (2012) 260801.

[Bekenstein 1973] Bekenstein J. D., Black holes and entropy, Phys. Rev. D 7 (1973) 2333.

[Bell & Jackiw 1969] J. S. Bell and R. Jackiw, A PCAC puzzle: $\Psi_0 \to \gamma\gamma$ in the σ model, Nuovo Cim. **A 60** (1969) 47.

[Belusov 1959] Belousov B. P., A periodic reaction and its mechanism in collection of short papers on radiation medicine for 1958, Med. Publ., Moscow, 1959.

[Berger 1992] Berger, Teilchenphysik, Springer 1992.

[Berger 2015] Ch. Berger, Photon structure function revisited, J. Mod. Phys., 6 (2015) 1023–1043.

[Bertotti 2003] B. Bertotti, L. Less, P. Tortora, A test of general relativity using radio links with the Cassini spacecraft, Nature 425 (2003) 374.

[Bertozzi 1994] W. Bertozzi, Speed and Kinetic Energy of Relativistic Electrons, American Journal of Physics 32 (1964) 551.

[Bethke 2007] S. Bethke, Experimental tests of asymptotic freedom, Prog. Part. Nucl. Phys. 58 (2007) 351–386.

[Binggeli 1999] H. Binggeli, The Virgo Cluster Home of M87, Lect. Notes Phys. 530 (1999) 9.

[Biretta 1999] J. A. Biretta et al., Astrophys. J. 520 (1999) 621.

[Bjorken 1987] J. D. Bjorken und S. D. Drell, Relativistische Quantenmechanik, BI Wissenschaftsverlag 1987.

[Blanchard 2006] A. Blanchard, The Big Bang picture: A remarkable success of modern science, Proceedings of the 1. Crisis in Cosmology Conference, AIP Conference Proc. 822 (2006) 148.

[Bloch 2000] I. Bloch et al., Nature 403 (2000) 166.

[Blochinzew 1953] D. T. Blochinzew, Kritik der philosophischen Anschauungen der sogenannten „Kopenhagener Schule" in der Physik, Sowjetwissenschaft, Berlin 1953, p. 545–574.

[Bohm & Hiley 1993] D. Bohm and B. J. Hiley, The Undivided Universe, An Ontological In- terpretation of Quantum Theory, Routledge & Kegan & Paul, London, 1993.

[Bohm 1952] D. Bohm, Phys. Rev. 85 (1952) 180.

[Bohm 1957] D. Bohm, Causality and chance in modern physics, 1957.

[Bohm 1980] D. Bohm, Wholeness and the implicate order, 1980, Routledge.

[Böhm 2003] M. Böhm, A. Denner und H. Joos, Gauge theories of the strong and electroweak interaction, Teubner, 2003.

[Böhringer 1999] H. Böhringer, The intracluster medium in the X-ray halo of M 87, Lect. Notes Phys. 530 (1999) 83.

[Bond et al. 1996] J. R. Bond et al., How filaments are woven into the cosmic web, Nature 380 (1996) 603.

[Bondi, Gold und Hoyle 1955] Bondi, Gold, und Hoyle, Observatory 75 (1955) 80.

[Bourilkov 2001] D. Bourilkov, Phys. Rev. D 64 (2001) 071701R.

[Boyanovski 2006] D. Boyanovsky, H. J. de Vega, and D. J. Schwarz, Phase transitions in the early and present universe, Ann. Rev. Nucl. Part. Sci. 56 (2006) 441–500.

[Boyer 1969] T. H. Boyer, Cutoff-Independent Character of Electromagnetic Zero-Point Forces, Phys. Rev. 185 (1969) 2039.

[Breitenbach 1997] G. Breitenbach, S. Schiller, and J. Mlynek, Measurement of the quantum states of squeezed light, Nature 387 (1997) 471.

[Breitenbach Web Site] http://www.gerd-breitenbach.de

[Brodie & Strader 2006] J. P. Brodie and J. Strader, Extragalactic globular clusters and galaxy formation, Annu. Rev. Astron. Astro- phys., 44 (2006) 193.

[Burbidge 1957] M. Burbidge et al., Rev. Mod. Phys. 29 (1957) 547.

[Burkhardt 2011] H. Burkhardt and B. Pietrzyk, Recent BES measurements and the hadronic contribution to the QED vacuum polarization, Phys. Rev. D 84 (2011) 037502.

[Carilli 1996] C. L. Carilli und P. D. Barthel, The Astron. Astrophys. Rev. 7 (1996) 1.

[Carr 2007] B. Carr, (Editor): Universe or Multiverse? Cambridge University Press, 2007.

[Carter 1974] B. Carter, Large number coincidences and the anthropic principle in cosmology, IAUS 63 (1974) 291.

[Lerner CCC 2006] Proceedings of the 1. Crisis in Cosmology Conference, ed. by E. Lerner and J. B. Almeida, AIP 822 (2006).

[Campana 2007] S. Campana et al., A metal-rich molecular cloud surrounds GRB 050904 at redshift 6.3, The Astrophysical Journal 654 (2007) L17–L20.

[CERN 2004] Ch. Sutton and P. Zerwas, The W and Z at LEP, CERN Courier, 3rd May 2004.

[CERN 2005] M. Grünewald, Experiments finally unveil a precise portrait of the Z, CERN Courier, 2nd November 2005.

[CERN 2013] ATLAS Collaboration, Measurements of Higgs boson production and couplings in diboson final states with the ATLAS detector at the LHC, Phys. Lett. B 726 (2013) 88.

[Chandrasekhar 1939] S. Chandrasekhar, Introduction into the study of stellar structure, Dover, New York, 1939.

[Chaplin 2008] Martin Chaplin, http://www.lsbu.ac.uk/water/phase.html, Dezember 2008.

[Chemla & Shah, 2001] D. S. Chemla and J. Shah, Nature, 411 (2001) 549.

[Cheung 2013] E. Cheung, E. Athanassoula, L. Masters, R. C. Nichol, A. Bosma, E. F. Bell, S. M. Faber, D. C. Koo, C. Lintott, Th. Melvin, K. Schawinski, R. A. Skibba, and K. W. Willett, Astrophys. J. 779 (2013) 162.

[Chou 2010] C. W. Chou, D. B. Hume, T. Rosenband, and D. J. Wineland, Optical clocks and relativity, Science 329 (2010) 1630.

[Clausius 1865] R. Clausius, Über Verschiedene für die Anwendung bequeme Formen der Hauptgleichungen der mechanischen Waer- metheorie. Annalen der Physik und Chemie, 7 (1865) 389–400.

[Cole 2003] D. C. Cole and Y. Zou, Quantum mechanical ground state of hydrogen obtained from classical electrodynamics, Phys. Lett. A317 (2003) 14.

[Combes 2005] Combes F., Secular evolution of galaxies. ArXiv Astrophys. e-prints, 30 (2005) 31.

[Comte 1830] August Comte, Die Soziologie, Die positivistische Philosophie im Auszug. Edited by Friedrich Blaschke, Kröner, 2nd ed. 1974, p. 8.

[Consoli & Field 1997] M. Consoli and J. H. Field, J. Phys. G: Nucl. Part. Phys. 23 (1997) 41–67

[Contopoulos 1987] G. Contopoulos and D. Kotsakis, Cosmology: The structure and evolution of the universe, Springer 1987.

[Davies 1975] P. C. W. Davies, J. Phys. A 8 (1975) 609.

[de la Pena 1996] L. de la Pena and A. M. Cetto, The quantum dice: An introduction to stochastic electrodynamics, Kluwer Academic Publishers, 1996.

[deBroglie 1960] Louis de Broglie, Non-linear wave mechanics, A Causal Interpretation, Elsevier, Amsterdam, 1960.

[Delgado-Serrano 2010] R. Delgado-Serrano, The evolution of the Hubble sequence, morpho-kinematics of distant galaxies, Paris, 2010.

[DeMeo 2000] J. DeMeo, Dayton Miller's ether-drift experiments: A fresh look, 2000, http://www.orgonelab.org/miller.htm

[Dickhut 1987] W. Dickhut, Materialistische Dialektik und bürgerliche Naturwissenschaft, Neuer Weg, Essen 1987.

[Dickhut 1988] W. Dickhut, Die dialektische Methode in der Arbeiterbewegung, Reihe Revolutionärer Weg Bd. 6, Verlag Neuer Weg, 1988.

[Dietrich 2003] M. Dietrich et al., APJ 589 (2003) 722.

[Dirac 1930] P. M. Dirac, A theory of electrons and protons, Proc. Roy. Soc. A 126 (1930) 360.

[Dirac 1951] P. A. M. Dirac, Is there an ether?, Nature, 168 (1951) 906–907.

[Dirac 1987] P. A. M. Dirac, Reminiscences about a great physicist, by B. N. Kursunoglu and E. P. Wigner, Cambridge University Press 1987.

[Dittrich 2000] W. Dittrich and H. Gies, Probing the quantum vacuum: Perturbative effective action approach in quantum electrodynamics and its applications, Springer 2000.

[Dürr 2001] D. Dürr, Bohmsche Mechanik als Grundlage der Quantenmechanik, Springer 2001.

[Duwe 1995] Ch. de Duwe, Vital Dust: Life as a Cosmic Imperative, Basic Books 1995.

[Eddington 1919] F. W. Dyson, A. S. Eddington, and C. Davidson, A Determination of the deflection of light by the Sun's gravitational field, from observations aade at the total eclipse of May 29, 1919, Philosophical Transactions of the Royal Society of London. Series A, (1920) 291–333.

[Edwards 2002] M. R. Edwards (Herausgeber), Pushing gravity: New perspectives on the Sage's theory of gravitation, Apeiron, Montreal 2002.

[Einstein 1911] A. Einstein, Relativitätstheorie, Naturforschende Gesellschaft, Zurich, quarterly, 56 (1911) 1–14.

[Einstein 1915] A. Einstein, Relativitätstheorie, in: Die Physik, Teubner, Leipzig, 1915.

[Einstein 1922] A. Einstein, Grundzüge der Relativitätstheorie, Vieweg, 1922.

[Einstein 1922] A. Einstein, Sidelights on relativity, Dutton New York, 1922, repr. 1983.

[Einstein 1955] A. Einstein, Einleitende Bemerkung über Grundbegriffe in: Louis de Broglie and the Physicists, Claasen, 1955, p. 14.

[Einstein 1960] A. Einstein, Ideas and opinions, New York, 1960, p. 375.

[Einstein und Infeld 1938] A. Einstein and L. Infeld, Die Evolution der Physik, New York 1938, quoted from the edition 2004, p. 156.

[Eisele 2009] Ch. Eisele, A. Yu. Nevsky, and S. Schiller, Laboratory test of the isotropy of light propagation at the 10–17 level, Phys. Rev. Lett. 103 (2009) 090401.

[Engelhardt 2013] W. Engelhardt, Classical and relativistic derivation of the Sagnac effect, arXiv:1404.4075 (2014)

[Engels 1885] F. Engels, Dialektik der Natur, Marx/ Engels, Werke, Bd. 20. First published in the Soviet Union in 1925.

[Eötvös 1890] R. v. Eötvös, Mathematische und Naturwissenschaftliche Berichte aus Ungarn, 8 (1890) 65.

[Fahr 1995] H. J. Fahr, Universum ohne Urknall, Springer 1995.

[Fahr 2015] H. J. Fahr and M. Sokaliwska, Remaining problems in interpretation of the cosmic microwave background, Phys. Res. Int. 2015, Article ID 503106, (2015).

[Farago 1957] P. S. Farago and L. Janossy, Review of the experimental evidence for the law of variation of the electron mass with velocity, Il Nuovo Cimento V (1957) 1411.

[Ferrarese und Ford 2005] L. Ferrarese and H. Ford, Supermassive black holes in galactic nuclei: Past, present and future research, Space Sci. Rev. 116 (2005) 523.

[Feynman 1985] R. Feynman, QED: The strange theory of light and matter. Princeton University Press, 1985.

[Foulkes 2001] W. M. C. Foulkes, L. Mitas, R. J. Needs, and G. Rajagopal, Rev. Mod. Phys. 73 (2001) 33–83.

[Gabrielse 2006] G. Gabrielse, D. Hanneke, T. Kinoshita, M. Nio, and B. Odom, New determination of the fine structure constant from the electron g value and QED, Phys. Rev. Lett. 97 (2006) 030802, Erratum, Phys. Rev. Lett. 99 (2007) 039902.

[Gamov 1961] Gamov, G., The creation of the universe, Viking, revised edition, 1961.

[Geller 1972] K. N. Geller und R. Kollarits, Experiment to measure the increase in electron mass with velocity, Am. J. Phys. 40 (1972) 1125.

[Genz 1994] H. Genz, Die Entdeckung des Nichts, Hanser, 1994.

[Georgi 1986] H. Georgi, in "Teilchen Felder Symmetrien", *Spektrum der Wissenschaft*, 1986;

[Gift 2014] S. J. G. Gift, Time transfer and the Sagnac correction in the GPS, Appl. Phys. Res. 6(6) 2014. ISSN 1916–9639 E-ISSN 1916–9647 Published by Canadian Center of Science and Education.

[Glansdorff & Prigogine 1971] Glansdorff, P., Prigogine, I. Thermodynamic theory of structure, stability, and fluctuations, Wiley-Interscience, London, 1971.

[Gottstein 2007] G. Gottstein, Physikalische Grundlagen der Materialkunde, Springer 2007.

[Greene 2004] B. Greene, The fabric of the cosmos, space, time, and the texture of reality, New York: A. A. Knopf (2004) Quote is from German edition, Siedler, 2004, p. 33.

[Greiner 1985] W. Greiner, B. Müller, und J. Rafelski, Quantum electrodynamics of strong fields, Springer Verlag 1985.

[Guillaume 1896] C.-E. Guillaume, La Nature 24 (1896) 234.

[Guth 1981] A. Guth, Phys. Rev. D 23 (1981) 347.

[Guye 1921] C. E. Guye, S. Ratnowsky und B. Lavanchy, Mem. SOC. Phys. Geneve 39 (1921) 273.

[Hackermüller 2004] L. Hackermüller, K. Hornberger, B. Brezger, A. Zeilinger, and M. Arndt, Decoherence of matter waves by thermal emission of radiation, Nature 427 (2004) 711–714.

[Hafele und Keating 1971] J. C. Hafele, R. E. Keating, Science 177 (1972) 166.

[Haisch 1999] B. Haisch, A. Rueda, Causality and locality in modern physics, 171–178 (1998), Kluwer Academic Publishers, editors G. Hunter, S. Jeffers, and J. P. Vigier.

[Haisch 2000] B. Haisch, A. Rueda, and Y. Dobyns, arXiv: gr-qc/0009036 (2000).

[Haisch 2000b] B. Haisch und A. Rueda, On the relation between a zero-point-field-induced inertial effect and the Einstein-de Broglie formula, arXiv: gr-qc/9906084v3 (2000).

[Haken 1981] H. Haken, Erfolgsgeheimnisse der Natur: Synergetik, die Lehre vom Zusammenwirken, DVA 1981.

[Haken 1984] H. Haken, Erfolgsgeheimnisse der Natur Ullstein Taschenbuch (1984).

[Haken 1987] H. Haken und Wolf, Atom und Quantenphysik, Springer 1987.

[Haken 1990] H. Haken, Synergetik, Springer 1990.

[Haken 2004] H. Haken, Synergetics: Introduction and advanced topics, Springer 2004.

[Halpern 1933] O. Halpern, Scattering processes produced by electrons in negative energy states, Phys. Rev. 44 (1933) 855–856.

[Hamann 2007] F. Hamann et al., ASP Conference series, arXiv: astro-ph/0701503v1 (2007).

[Hammer 2009] F. Hammer, H. Flores, M. Puech, Y. B. Yang, E. Athanassoula, M. Rodrigues and R. Delgado-Serrano, The Hubble sequence: just a vestige of merger events? A & A 507 (2009) 1313–1326.

[Hanneke 2008] D. Hanneke, S. Fogwell, and G. Gabrielse, New measurement of the electron magnetic moment and the fine structure constant, Phys. Rev. Lett. 100 (2008) 120801.

[Hansen & Kawaler 1994] C. J. Hansen und S. B. Kawaler, Stellar interiors, physical principles, structure and evolution, Springer, 1994.

[Harris 2006] D. E. Harris und H. Krawczynski, Annu. Rev. Astro. Astrophys. 44 (2006) 463.

[Hawking 1974] S. W. Hawking, Black hole explosions. Nature 248 (1974) 30.

[Hawking 1997, S. 93] S. W. Hawking, Eine kurze Geschichte der Zeit. RoRoRo Sachbuch Science 1997.

[Hawking 2001] S. W. Hawking, The universe in a nutshell, Bantam Press, 2001.

[Hazelett 1979] R. Hazelett, D. Turner, The Einstein myth and the Ives papers: A counter revolution in physics, Greenwich, 1979.

[Hecht 2009] E. Hecht, Einstein never approved of relativistic mass, Phys. Teach. 47, (2009) 336.

[Hegel 1813] G. W. Fr. Hegel, Science of logic, Berlin 1813,. English edition: Cambridge University Press (2010).

[Heisenberg 1927] W. Heisenberg, Über die Grundprinzipien der Quantenmechanik, Forschungen und Fortschritte 3 (1927) 83.

[Higgs 1964] P. Higgs, Phys. Rev. Lett. 13 (1964) 508–509.

[Hoddeson et al. 1992] L. Hoddeson (eds.) et al., Out of the crystal maze: Chapter from the history of solid-state physics, Oxford University Press 1992.

[Holmlid 2005] L. Holmlid, Redshifts in space caused by stimulated Raman scattering in cold intergalactic Rydberg matter with experimental verification, J. Exp. Theoret. Phys. 100 (2005) 637.

[Hossenfelder 2018] S. Hossenfelder, Lost in math: How beauty leads physics astray, Basic Books (2018).

[Hoyle and Tayler 1964] F. Hoyle and R. Tayler, Nature 203 (1964) 1108.

[Hoyle, Burbidge & Narlikar 2000] A different approach to cosmology, Cambridge University Press, 2000.

[Hu et al. 2006] F. X. Hu, G. X. Wu, G. X. Song, Q. R. Yuan, and S. Okamura, Orientation of galaxies in the local supercluster: A review, Astrophys. Space Sci. 302 (2006) 43.

[Hu 2011] B. L. Hu, Gravitation and nonequilibrium thermodynamics of classical matter, Int. J. Mod. Phys. D. 20 (2011) 697–716.

[Hubble 1936] E. P. Hubble, The realm of the nebulae, Yale University Press, New Haven, 1936.

[Hubert & Schäfer 1998] A. Hubert and R. Schäfer, Magnetic domains: The analysis of magnetic microstructures, Springer (1998).

[Hund 1975] F. Hund, Geschichte der Quantentheorie, BI Wissenschaftsverlag 1975.

[Ibison 2006] M. Ibison, Thermalisation of starlight in the steady state cosmology, AIP Conference Proc. 822 (2006) 171.

[Ives 1937–1952] H. E. Ives, Collected papers, republished in D. Turner and R. Hazelett, The Einstein Myth and the Ives Papers: A Counter-Revolution in Physics (Old Greenwich, Conn.; Devin-Adair, 1979).

[Jacobson 2006] T. A. Jacobson und R. Parentani, Spektrum der Wissenschaft April/2006, p. 40.

[Jakubith 1990] S. Jakubith, H. H. Rotermund, W. Engel, A. von Oertzen, and G. Ertl, Spatiotemporal concentration patterns in a surface reaction: Propagating and standing waves, rotating spirals, and turbulence, Phys. Rev. Lett. 65 (1990) 3013–3016.

[Jensen 2004] W. Jensen, 2004, arXiv:astro-ph/0404207v1

[Johansen 2002] T. H. Johansen, M. Baziljevich, D. V. Shantsev, P. E. Goa, Y. M. Galperin, W. N. Kang, H. J. Kim, E. M. Choi, and S. I. Lee, Europhys. Lett. 59 (2002) 599.

[Jooss & Lutz 2006] Ch. Jooss and J. Lutz, The evolution of the universe in the light of modern microscopic and high-energy physics, AIP Proceedings, 822 (2006) 200.

[Jooss 2004] Ch. Jooss, K. Guth, M. A. Schofield, M. Beleggia, and Y. Zhu, Direct measurements of electric potentials at grain boundaries: Mechanism for current improvement in high Tc superconductors, Phys. C. 408–410 (2004) 443.

[Jooss 2005] Ch. Jooss, unveröffentlichtes Manuskript, 2005.

[Junor 1999] W. Junor et al., Nature 401 (1999) 891.

[Kacher 2013] von H. Kacher, H. Meyer, Skriptum Atomphysik: Eine Einführung in Grundlagen und Anwendungen, Springer 2013.

[Kafka 1989] Peter Kafka, Das Grundgesetz vom Aufstieg, Hanser 1989.

[Kalies 2019] G. Kalies, Vom Energieinhalt ruhender Körper, De Gruyter (2019).

[Kataoka 2007] J. Kataoka et al., Astrophys. J. 672 (2008) 787–799.

[Keel 2003] Keel, Alternate approaches and the redshift controversy, 2003, http://www.astr.ua. edu/keel/galaxies/arp.html

[Kelvin 1900] Nineteenth-century clouds over the dynamical theory of heat and light, The London, Edinburgh and Dublin Philosophical Magazine and Journal of Science, 2 (1901) 1.

[Ketterle et al. 1999] W. Ketterle, D. S. Durfee und D. M. Stamper-Kurn, Making, probing and understanding Bose–Einstein condensates, in Proceedings of the International School of Physics Enrico Fermi, IOS Press, 1999.

[Kibble 1964] G. S. Guralnik, C. R. Hagen, and T. W. B. Kibble, Phys. Rev. Lett. 13 (1964) 585–587.

[Kim 2020] J. -Y Kim et al, Event Horizon Telescope imaging of the archetypal blazar 3C 279 at an extreme 20 microarcsecond resolution, Astronomy & Astrophysics (2020) DOI: 10.1051/0004-6361/202037493

[King 2015] S. F. King, Neutrino Mass Models: A road map, J. Phys. Conf. Ser.136 (2008) 022038.

[Kirhakos 1999] S. Kirhakos et al., The host galaxies of three radio-loud quasars: 3C 48, 3C 345, and B2 1425+2671, Astrophys. J. 520 (1999) 67.

[Kleemann 2006] M. Kleman and O. D. Lavrentovich, Topological point defects in nematic liquid crystals, Philos. Mag. 86 (2006) 4117.

[Kleinert 2013] H. Kleinert, Superfluid 3He, In: Collective Fields, 2013, http://users.physik.fu-berlin. de/~kleinert/63/63.pdf

[Klinkhamer 2005] F. R. Klinkhamer und G. E. Volovik, Merging gauge coupling constants without grand unification, J. Exp. Theor. Phys. Lett. 81 (2005) 551–555.

[Kniazev et al. 2003] A. Y. Kniazev et al., Astrophys. J. 593 (2003) L73–L76.

[Koide 2006] S. Koide et al., Phys. Rev. D 74 (2006) 044005.

[Kopnin 1991] N. B. Kopnin and M. M. Salomaa, Mutual friction in superfluid He3: Effects of bound states in the vortex core, Phys. Rev. B 44 (1991) 9667.

[Kostro 1998] L. Kostro, Open questions in relativistic physics, edited by F. Selleri, Apeiron 1998.

[Kostro 2000] L. Kostro, Einstein and the Ether, Apeiron Montreal, 2000.

[Kotanyi 1981] C. Kotanyi C., PhD Thesis, Rijksuniversiteit, Groningen, 1981.

[Kracklauer 1992] A. F. Kracklauer, An Intuitive Paradigm For Quantum Mechanics, Physics Essays 5 (1992) 226.

[Krichbaum 2007] T. P. Krichbaum, How compact are the cores of AGN? Sub-parsec scale imaging with VLBI at millimeter wavelength, ASP Conf. Series, Vol. 386 (2008) 186–194, arXiv:0708.3915v1 [astro-ph].

[Krusius 1998] M. Krusius, T. Vachaspati, and G. E. Volovik, Flow instability in 3He-A as analog of generation of hypermagnetic field in early Universe, arXiv: cond-mat/9802005v1 (1998).

[Kuhn 1962] T. S. Kuhn, The structure of scientific revolutions, The University of Chicago Press, 1962, 3rd edition 1996.

[Lamb und Retherford, 1947] W. E. Lamb Jr. and R. C. Retherford, Fine structure of the hydrogen atom by a microwave method, Phys. Rev. 72 (1947) 241.

[Lamoreaux 1997] S. K. Lamoreaux, Phys. Rev. Lett. 78 (1997) 5.

[Landau 1958] L. Landau, Course of theoretical Physics, Vol. 5, Statistical Physics, Part 1, 1st German edition 1958, 8th edition Harry Deutsch Verlag 1991.

[Landau & Lifshitz 1939] L. D. Landau and E. M. Lifshitz, Course of theoretical physics, Vol. 2, The Classical Theory of Fields, 1. Russian edition 1939, 1st English edition 1951.

[Landau & Lifshitz 1965] L. D. Landau and E. M. Lifshitz, Classical theory of fields, Pergamon Press Oxford, 1965, p. 246 & 307.

[Landau & Lifshitz 1992] L. D. Landau und E. M. Lifshitz, Course of theoretical physics, Vol. 9, Statistical Physics, Part II, from the German edition Harry Deutsch 1992.

[Landsberg 1984] P. Landsberg, Is equilibrium always an entropy maximum? J. Stat. Phys. 35 (1984) 159.

[Langmuir 1928] I. Langmuir, Oscillations in ionized gases. In: Proceedings of the National Academy of Science. 14 (1928) 627–637.

[Laughlin 2005] R. B. Laughlin, A different universe: Reinventing physics from the bottom down, Basic Books, New York, 2005, S. ix–xi.

[Laughlin 2008] Robert Laughlin, Der Urknall ist nur Marketing, Spiegel 1, 2008, p. 120.

[Laughlin und Pines 2000] R. B. Laughlin and David Pines, The theory of everything, PNAS 97 (2000) 28.

[Lee & Erdogdu 2007] J. Lee and P. Erdogdu, The alignments of the galaxy spins with the real-space tidal field reconstruction from the 2MASS redshift survey, Astrophys. J. 671 (2007) 1248.

[Leggett 2006] A. J. Leggett, Quantum liquids, Oxford Graduate Texts, 2007.

[Lenin 1908] V. I. Lenin, Materialismus und Empiriokritizismus, Lenin, Werke, Volume 14.

[Lenin 1914] V. I. Lenin, Conspecturs of Hegel's book The Science of Logic, Collected Works, Volume 38, created in the years 1914/1915.

[Lerner 1992] E. J. Lerner, The Big Bang never happened, Vintage Books, New York, 1992.

[Lerner 2006] E. J. Lerner, Evidence for a Non- Expanding Universe: Surface Brightness data from HUDF, AIP Conference Proc. 822 (2006) 60.

[Levine 1997] I. Levine; TOPAZ Collaboration (1997), Measurement of the electromagnetic coupling at large momentum transfer. Phys. Rev. Lett. 78 (1997) 424.

[Li 2006] Y. Li et al., Formation of $z \sim 6$ quasars from hierarchical galaxy mergers, Astrophys. J. 665 (2007) 187–208.

[Linde 1990] A. Linde, Inflation and quantum cosmology, Academic Press, Boston, 1990.

[Lineweaver 2008] Ch. H. Lineweaver and Ch. A. Egana, Life, gravity and the second law of thermodynamics, Phys. Life Rev. 5 (2008) 225–242.

[Longair 1998]. M. S. Longair, Galaxy formation, Springer 1998.

[Lopez-Corrredoira & Gutierrez 2002] M. Lopez-Corrredoira und C. Gutierrez, A & A, 390 (2002) L15.

[Lutz 1991] J. Lutz, W.-D. Rochlitz, G. Balzer-Jöllenbeck, Ratlos vor der großen Mauer, Das Scheitern der Urknall Theorie, Verlag Neuer Weg, 1991.

[Mach 1883] Ernst Mach: Die Mechanik in ihrer Entwicklung historisch und kritisch dargestellt. F.A. Brookhaus, Leipzig, 9th Auflage - 1933. – Kapitel 4, Sektion 4, p. 457–458.

[Mahdavi 2007] A. Mahdavi et al., 2007, http://chandra.harvard.edu/photo/2007/a520/

[Mahulikar & Herwig 2004] S. P. Mahulikar and H. Herwig, Conceptual investigation of the entropy principle for identification of directives for creation, existence and total destruction of order, Phys. Scrip. 70 (2004) 212–221.

[Mandel 2003] O. Mandel, M. Greiner, A. Widera, T. Rom, Th. W. Hänsch, and I. Bloch, Controlled collisions for multiparticle entanglement of optically trapped atoms, Nature 425 (2003) 937.

[Mannheim 2006] P. D. Mannheim, Alternatives to dark matter and dark energy, Prog. Particle Nucl. Phys. 56 (2006) 340.

[Markevitch 2007] M. Markevitch et al., 2007, http://chandra.harvard.edu/photo/2006/1e0657

[Marmet 1996] P. Marmet, Stellar aberration and Einstein's relativity, Phys. Essays 9 (1996) 96–99.

[Martel 1999] A. R. Martel, S. A. Baum, W. B. Sparks, E. Wyckoff, J. A. Biretta, D. Golombek, F. D. Macchetto, S. de Koff, P. J. McCarthy, and George K. Miley, Astrophys. J. Suppl. Ser. 122, (1999) 81.

[Martyushev & Seleznev 2006] L. M. Martyushev, V. D. Seleznev, Maximum entropy production principle in physics, chemistry and biology, Phys. Rep. 426 (2006) 1–45.

[Mathiesen 2006] B. Mathiesen, A cosmic coincidence resurrects the cyclical Universe, 5 June, 2006, www.physorg.com/news68731082.html

[Mathur 2000] S. Mathur, New Astronomy Rev. 44 (2000) 469.

[McHardy 2006] I. McHardy et al., Simultaneous X-ray and infrared variability in the quasar 3C273 II: Confirmation of the correlation and X-ray lag, Mon. Not. Roy. Astron. Soc. 375 (2007) 1521–1527.

[McWilliam 1997] A. Mc William, Abundance ratios and galactical chemical evolution, Annu. Rev. Astron. Astrophys. 35 (1997) 503.

[Meißen 2001] F. Meißen, Untersuchung der Musterbildung bei der katalytischen CO-Oxidation auf Pt 110 mittels Niederenergie- Elektronenmikroskopie, Dissertation TU Berlin, 2001.

[Meissner 1992] W. Meissner, Wie tot ist Schrödingers Katze?, B I Wissenschaftsverlag, 1992.

[Meyenn 1994] K. von Meyenn, Quantenmechanik und Weimarer Republik, Vieweg, 1994.

[Michelson and Morley 1883] A. A. Michelson and E. W. Morley, J. Phys. 7 (1883) 444.

[Michelson and Morley 1887] A. A. Michelson and E. W. Morley, Am. J. Sci. 203 (1887) 333.

[Miller 1933] D. C. Miller, The ether-drift experiment and the determination of the absolute motion of the earth, Rev. Mod. Phys. 5 (1933) 203.

[Murayama 2002] H. Murayama, Origin of neutrino mass, Phys. World 35, May 2002.

[Nasa 1999] Andrew Fruchter and the ERO Team, Sylvia Baggett, Richard Hook, Zoltan Levay, 1999, http://spacetelescope.org/images/heic9910a/

[Nernst 1916] W. Nernst, Über einen Versuch, von quantentheoretischen Betrachtungen zur Annahme stetiger Energieänderungen zurückzukehren. In: Verhandlungen der Deutschen Physikalischen Gesellschaft. Jg. 18, Heft 4. S. 83–116, Braunschweig, Vieweg, 1916

[Nernst 1937] W. Nernst, Weitere Prüfung der Annahme eines stationären Zustandes im Weltall, Zeitschrift für Physik 106 (1937) 633–661.

[Nernst 1938] W. Nernst, Annalen der Physik 32 (1938) 44–48.

[Nicolis 1999] Nicolis, C. Entropy production and dynamical complexity in a low-order atmospheric model, Quart. J. Royal Meteorol. Soc. 125 (1999) 1859.

[Nishino 2009] H. Nishino et al. (Super-kamiokande collaboration), Search for proton decay via $p \rightarrow e + \pi^0$ and $p \rightarrow \mu + \pi^0$ in a large water cherenkov detector, Phys. Rev. Lett. 102 (2009) 141801.

[Nordstroem 2004] B. Nordstroem, M. Mayor, J. Andersen, J. Holmberg, F. Pont, B. R. Jør-gensen, E. H. Olsen, S. Udry, and N. Mowlavi, Astron. Astrophys. 418 (2004) 989.

[Ohanian 1986] H. C. Ohanian, What is spin? Am. J. Phys. 54 (1986) 501.

[Orowan 1934] E. Orowan, Zur Kristallplasti- zität, Z. Phys. 89 (1934) 605.

[Pagel 1997] B. E. J. Pagel, Nucleosynthesis and chemical evolution of galaxies, Cambridge University Press, 1997.

[Passon 2005] O. Passon, Bohm'sche Mechanik: Eine elementare Einführung in die deterministische Interpretation der Quantenmechanik, Harri Deutsch, 2005.

[Peebles 1993] P. J. E. Peebles, Principles of physical cosmology, Princeton University Press, 1993.

[Penrose 2010] R. Penrose, Cycles of time: An extraordinary new view of the universe, Bodley Head, London, 2010.

[Perlmutter 1999] S. Perlmutter et al., Measurement of Ω and Λ from 42 high-redshift supernovae, Astrophys. J. 517 (1999) 565.

[Philippidis 1979] C. Philippidis, C. Dewdney and B. Hiley, Nuovo Cimento B 52 (1979) 15.

[Piccinelli 2004] G. Piccinelli und A. Ayala, Electroweak baryogenesis and primordial hypermagnetic fields, Lect. Notes Phys. 646 (2004) 293–308.

[Pines & Nozieres 1966] D. Pines and P. Nozieres, Theory of quantum liquids, Benjamin, New York, 1966.

[Planck 1911] M. Planck, Verhandl. Deutsch. Phys. Ges. 13 (1911) 138.

[Planck 1912] M. Planck, Ann. D. Phys. 37 (1912) 642.

[Planck 1949] M. Planck, Vorträge und Erinnerungen, 5th edition, Stuttgart 1949, p. 228.

[Poincare 1908] H. Poincaré, La dynamique de l'électron. In: Revue générale des sciences pures et appliquées. 19 (1908) 386–402.

[Polanyi 1934] M. Polanyi, Über eine Art Gitterstörung, die einen Kristall plastisch machen können, Z. Phys. 89 (1934) 660.

[Popper 2005] K. R. Popper, The logic of scientific discovery, Taylor & Francis, 2005, first published in German in 1935.

[Popper 1965] K. R. Popper, Das Elend des Historizismus, Tübingen, 1965.

[Pound 1960] R. V. Pound and G. A. Rebka, Jr., Apparent weight of photons, Phys. Rev. Lett. 4 (1960) 337.

[Prantzos 2007] N. Prantzos, Origin and evolution of the light nuclides, Space Sci. Rev. 130 (2007) 27.

[Preston & Potter 2006] H. Preston und F. Potter, Quantum celestial mechanics: Large-scale gravitational quantization states in galaxies and the universe, Proceedings of the 1. Crisis in Cosmology Conference, AIP Conference Proc. Vol. 822 (2006).

[Prigogine & Stengers 1993] N. Prigogine and I. Stengers Das Paradox der Zeit, Zeit, Chaos und Quanten, Piper, 1993.

[Prigogine 1977] G. Nicolis, und I. Prigogine, Self-organization in nonequilibrium systems, Wiley-Interscience, New York, 1977.

[Radecke 1997] H. D. Radecke, Evidence for an extended extragalactic gamma ray source, Astrophys. Space Sci. 249 (1997) 303.

[Rafelski 1985] J. Rafelski, B. Müller, Die Struktur des Vakuums, Verlag Harri Deutsch, 1985.

[Rauch 2000] H. Rauch und S. A Werner, Neutron interferometry: Lessons in experimental quantum mechanics, Oxford University Press, 2000.

[Rauch 2008] H. Rauch, Eur. Phys. J. Special Topics 159 (2008) 27–36.

[Regener 1933] E. Regener, Zeitschrift für Physik 80 (1933) 666–669.

[Renzini 1997] A. Renzini, Iron as a tracer in galaxy clusters and group, Astrophys. J. 488 (1997) 35.

[Richter 2006] B. Richter, Physics Today, October 2006, p. 8.

[Riess 2004] Riess, A. G. et al., Type Ia Supernova discoveries at z > 1 from the hubble space telescope: Evidence for past deceleration and constraints on dark energy evolution, Astrophys. J. 607 (2004) 665.

[Roepstorff 1994] G. Roepstorff, Path integral approach to quantum physics : An introduction, Springer, 1994.

[Rolfs & Rodney 1988] C. E. Rolfs and W. S. Rodney, Cauldrons in the cosmos, University of Chicago Press, 1988.

[Ruder 1990] H. Ruder, Die spezielle Relativitätstheorie, Lecture manuscript, Tübingen, 1990.

[Ryckman 2005] Th. Ryckman, The reign of relativity: Philosophy in physics 1915–1925, Oxford University Press, 2005.

[Sakharov 1967] A. D. Sakharov, Vacuum quantum fluctuations in curved space and the theory of gravitation, Dokl. Akad. Nauk 177 (1967) 70.

[Savaglio 2006] S. Savaglio, GRBs as cosmological probes—cosmic chemical evolution, *New Journal of Physics* **8** (2006) 195.

[Sakharov 1967b] A. D. Sakharov, Pis'ma Zh. Eksp. Teor. Fiz. 5 (1967) 32 [JETP Lett. 5 (1967) 24].

[Schifmann 2012] M. Shifmann, Reflections and impressionistic portrait at the conference, Frontiers beyond the standard model, arXiv:1211.0004v1 31th October 2012.

[Schilpp 1979] P. A. Schilpp, A. Einstein als Philosoph und Wissenschaftler, Vieweg, Braunschweig, 1979.

[Schneider 2006] P. Schneider, Extragalactic astronomy and cosmology, Springer, 2006.

[Selleri 1990] F. Selleri, Die Debatte um die Quantentheorie, Vieweg, 1990.

[Selleri 1996] F. Selleri, The inertial transformations and the relativity principle. Found. Phys. 26 (1996) 641–664.

[Selleri 1998] F. Selleri (ed.), Open questions in relativistic physics, Apeiron Montreal, 1998.

[Selleri 1996] F. Selleri, The liberation of time, Paper in O. Nawrot (Editor) Hevelius, Science-Technology-Philosophy, Scientific Society of Gdansk, Publishing House, SCIENTIA, Gdansk, Poland, (2004).

[Selleri 2004b] Recovering the Lorentz ether, Apeiron, 11 (2004).

[Shaeer 2001] J. R. Abo-Shaeer, C. Raman, J. Vogels, and W. Ketterle, Observation of vortex lattices in Bose–Einstein condensates, Science, 292 (2001) 5516.

[Shapiro 1968] Shapiro et al., Phys. Rev. Lett. 20 (1968) 1265–1269 und Phys. Rev. Lett. 26 (1971) 1132–1135.

[Shapiro 1976] Verification of the principle of equivalence for massive bodies in. Phys. Rev. Lett. 36 (1976) 555–558.

[Sheth 2003] Sheth K., Regan M. W., Scoville Z., and Strubbe L. E., Barred galaxies at z > 0.7: NICMOS Hubble Deep Field-North observations, Astrophys. J. 592 (2003) L13–L16.

[Shih 2003] Y. Shih, IEEE Quant. Elect. 9 (2003) 1455.

[Shih 2004] Y. Shih, Beyond the Heisenberg uncertainty, Special Issue J. Mod. Opt. 2004.

[Shklovsky 1964] I. S. Shklovsky, Soviet Astron. 7 (1964) 748.

[Silk2001] J. Silk, Supermassive black holes and galaxy formation, Space Sci. Rev. 100 (2001) 41.

[Simonyi 2001] K. Simonyi, Kulturgeschichte der Physik, Verlag Harry Deutsch, Frankfurt, 2001.

[Smoot 2006] G. Smoot, Interview with Die Welt, 23 December, 2006.

[Sommerfeld 1916] A. Sommerfeld: Zur Quantentheorie der Spectrallinien (I + II). In: Annalen der Physik. 51 (1916) 1–94.

[Spark & Gallagher, 2000, S. 72], L. S. Spark and J. S. Gallagher, Galaxies in the universe, Cambridge University Press, 2000.

[Sparnaay 1958] M. J. Sparnaay, Physica (Utrecht) 24 (1958) 751.

[Spergel 2007] D. N. Spergel et al., Wilkinson microwave anisotropy probe (WMAP) three year results: Implications for cosmology, APJS 170 (2007) 337–408.

[Spurio 2015] M. Spurio, Particles and astrophysics: A multi-messenger approach, Springer, 2015.

[Starobinskii 1973] A. A. Starobinskii, Amplification of waves during reflection from a rotating black hole, JETP 37 (1973) 28.

[Steimle 1998] B. Steimle, Metamorphosen der Äthertheorien, Dissertation Universität Gießen, DHS 2577, 1998.

[Steinhardt & Turek 2006] P. J. Steinhardt & N. Turek 2006, Why the Cosmological Constant Is Small and Positive, Science 312 (2006) 1180.

[Stenger 2000] V. J. Stenger, Natural explanation for the anthropic coincidences, Philo. 3 (2000) 50–67.

[Stodolna 2013] A. S. Stodolna, A. Rouzé F. Lépine, S. Cohen, F. Robicheaux, A. Gijsbert-sen, J. H. Jungmann, C. Bordas, and M. J. J. Vrakking, Phys. Rev. Lett. 110 (2013) 213001.

[Strekalov 1995] D. V. Strekalov et al., Phys. Rev. Lett. 74 (1995) 3600.

[Su 2001a] C. C. Su, Reinterpretation of the Michelson-Morley experiment based on the GPS Sagnac correction, Europhys. Lett. 56 (2001) 170.

[Suhl 1965] H. Suhl, Inertial Mass of a Moving Fluxoid, Phys. Rev. Lett. 14 (1965) 226.

[Super Nova Projekt 2010] Supernova cosmology project, Updated data of 2010, File: http://super nova.lbl.gov/Union/figures/SCPUnion2.1_mu_vs_z.txt, Download 13 August, 2016.

[Swenson 1950] C. Swenson, The liquid-solid transformation in helium near absolute zero, Phys. Rev. 79 (1950) 626.

[Tajmar 2007] M. Tajmar et al., Measurement of gravitomagnetic and acceleration fields around rotating superconductors, Space Technology and Applications International Forum, AIP, 978-0-7354-0386-4 (2007).

[Taylor 1934] G. I. Taylor, The mechanism of plastic deformation of crystals, Proc Roy. Soc A. 145 (1934) 362.

[Tegmark 1998] M. Tegmark, Is "the theory of everything" merely the ultimate ensemble theory? Ann. Phys. 270 (1998) 1–51.

[Thiessen 2009] D. Thiessen, Kritik der speziellen und allgemeinen Relativitätstheorie, Books on Demand, Norderstedt, 2009.

[Thirring 1991] Introduction in "The Stability of Matter: From Atoms to Stars", Selecta of Elliott H. Lieb, edited by W. Thirring, Springer-Verlag, Berlin Heidelberg, 1991.

[Thouless 1998] David J. Thouless, Topological quantum numbers in nonrelativistic physics, World Scientific Publishing, 1998.

[Tipler 1994] F. J. Tipler, Die Physik der Unsterblichkeit, Piper, Munich, 1994.

[Trujillo et al. 2005] I. Trujillo, C. Carretero, and S. G. Patiri, Detection of the effect of cosmological large-scale structure on the orientation of galaxies, Astrophys. J. Lett. 640 (2006) L111.

[Turyshev & Anderson 2004] S. Turyshev and J. Anderson, Pioneer anomaly put to the test, Phys. World, September / 2004.

[Unruh 1976] W. G. Unruh, Phys. Rev. D 14 (1976) 870.

[Urri & Padovani 1995] C. M. Urry and P. Pado-vani, PASP 107 (1995) 803.

[Vaas 2006] R. Vaas, Ist das Universum uns auf den Leib geschneidert? Bild der Wissenschaft 8(2006) 34.

[Vaucouleurs 1974] G. de Vaucouleurs, In: The formation and dynamics of galaxies, ed. J. R. Shakeshaft, Dordrecht, 1974.

[Ventura & Pines 2012] J. E. Ventura and D. Pines (ed.), Neutron stars: Theory and observation, Springer, 2012.

[Verhulst 1994] J. Verhulst, Der Glanz von Kopenhagen, 1994, 17.

[Vigier 1990] J. P. Vigier, Evidence for non-zero mass photons associated with a vacuum-induced dissipative redshift mechanism, IEEE Trans. On Plasma Science, 18 (1990) 64.

[Vigier 1997] J. P. Vigier, Relativistic interpretation of the small ether drift velocity detected by Michelson, Morley and Miller, Apeiron 4 (1997) 71.

[Vilenkin 2006] A. Vilenkin, The Vacuum Energy Crisis, Science 312 (2006) 1148.

[Völcker 2010] D. Völcker, Mehr Erfolg in Physik: Optik, Magnetismus, Elektrizitätslehre, Atomphysik, Mentor Verlag, 2010.

[Vollhardt & Wölfle 1990] D. Vollhardt and P. Wölfle, The superfluid phases of helium 3, Taylor & Francis, London, 1990.

[Volovik 2000] G. E. Volovik, Vortices observed and to be observed, J. Low Temp. Phys. 121 (2000) 357–366.

[Volovik 2003] G. E. Volovik, The universe in a helium droplet, Oxford Science Publications, 2003.

[Wallace 2009] D. Wallace, Gravity, entropy, and cosmology: In search of clarity, Br. J. Philos. Sci. 61 (2010) 513.

[Walter 2004] F. Walter et al., Resolved molecular gas in a quasar host galaxy at redshift z = 6.42, Astrophys. J. 615 (2004) L17–L20.

[Weinberg 1992] S. Weinberg, Der Traum von der Einheit des Universums, Bertelsmann, 1992.

[Weinberg 2012] S. Weinberg, The crisis of big science, The New York Review, 10 May, 2012, http://www.nybooks.com/articles/archives/2012/may/10/crisis-big-science

[Weizsäcker 1958] C. F. von Weizsäcker, Zum Weltbild der Physik, Hirzel Verlag, Stuttgart, 1958.

[Weizsäcker 1985] C. F. von Weizsäcker, Aufbau der Physik, Hanser Verlag, Munich, Vienna 1985.

[Wendel 2013] H. J. Wendel, Das Abgrenzungsproblem, in Logik der Forschung, Ed. By H. Keuth, Akademie Verlag, Berlin, 2013.

[West & Blakeslee 2000] M. J. West & J. P. Blakeslee, The principal axis of the Virgo Cluster, Astrophys. J. 543 (2000) L27.

[Weyl 1918] H. Weyl, Sitzsungber. Preuss. Akad. Wiss. (1918) 46.

[Wheeler 1975] J. Wheeler, The nature of scientific discovery, Owen Gingerich (ed.) Washington Smithsonian Press, 1975.

[Whitford & Rich, 1983] A. E. Whitford and R. M. Rich, Ap. J. 274 (1983) 723.

[Wilson 1979] K. G. Wilson, Die Renormierungsgruppe, Spektrum der Wissenschaft 10 (1979).

[Wilson 2000] A. S. Wilson et al., Astronom. J. 120 (2000) 1325.

[Wiltshire 2007] D. L. Wiltshire, "Cosmic clocks, cosmic variance and cosmicaverages," New Journal of Physics, 9 (2007) 377.

[Wolchover 2012] N. Wolchover, Supersymmetry fails test, forcing physics to seek new ideas, Scientific American, 11/2012.

[Wu 1957] Wu, C. S., Ambler, E., Hayward, R. W., Hoppes, D. D., Hudson, R. P., Experimental test of parity conservation in beta decay, Phys. Rev. 105 (1957) 1413–1415.

[Yin 2013] Juan Yin et al., Phys. Rev. Lett. 110 (2013) 260407.

[Yoon 2013] Y. Yoon, Problems with Mannheim's conformal gravity program, Phys. Rev. D 88 (2013) 027504.

[Zeh 1989] H. D. Zeh, The physical basis of the direction of time, Springer, Berlin, Heidelberg, 1989.

[Zeilinger 2003] A. Zeilinger, Einsteins Schleier, Die neue Welt der Quantenphysik, C. H. Beck, 2003.

[Zel'dovich 1971] Ya. B. Zeldovich, Generation of waves by a rotating body, JETP Lett. 14 (1971) 180.

[Zhabotinsky 1964] Zhabotinsky, A. M., Periodical oxidation of malonic acid in solution: A study of the Belousov reaction kinetics, Biofizika, 9 (1964) 306–11.

[Zwicky 1929] F. Zwicky, Proc. Nat. Ac. Sci. 15 (1929) 773.

Illustration sources

Figure 1   Composite graphic: From top left to bottom right: 1: W. Schaap Sloan Great Wall, Wikipedia CC BY-SA 3.0; 2: NASA, nasaima-ges.org, SDSS, Atlas of the Universe and R. Harris; 3: Adapted from NASA: MilkyWay-full-annotated.jpg, public domain; 4: Harman Smith and Laura Generosa, http://www.nasa.gov, public domain; 5: Image Science and Analysis Laboratory, NASA-Johnson Space Center, public domain; 7: Zephyris, Wikipedia, CC BY-SA 3.0; 8: Dhatfield, Wikimedia public domain.

Figure 2   left: Scott Bauer, Agricultural Research Service, United States Department of Agriculture, public domain; center: Serge Melki, Wikimedia, CC 2.0 generic; right: image derived: FleurdegivreL.jpg: Annick Monnier Wikimedia CC BY-SA 3.0.

Figure 4   According to [Chaplin 2008].

Figure 7   a) & b) M. Lenius & C. Volkert, Institute for Material Physics, Göttingen University, data processing: M. Tiegel; c) & d) Ch. Borchers, Institute for Material Physics, Göttingen University; e) and f) A. Hubert & R. Schäfer, *Magnetic Domains: The Analysis of Magnetic Microstructures*, Springer (1998), reprinted with permission of Springer.

Figure 10   a) With permission of H. Pniok, Wikimedia CC by-nc-nd 3.0, c) Ch. Borchers, Institute of Material Physics, University of Göttingen; d) T. Johansen, University of Oslo, with permission.

Figure 11   Photos Alexey Kljatov, license for printing purchased at www.shutterstock.com.

Figure 12   Top left: Astronomy Education at the University of Nebraska-Lincoln Web Site, http://astro.unl.edu/video/demonstrationvideos/movies/convection.mp4, with permission; right: NASA, http://www.nasa.gov/images/content/365724main_convection-STS-516.jpg.

Figure 13   left: Yevgeny Pashnin, Aurora_ Borelis_22Jan2004, Wikimedia CC by-sa.

Figure 14   Images: Bottom and top bottom left, H. H. Rotermund, FHI Berlin; Top right [Meissen 2001].

Figure 16   left: according to [Foulkes 2001].

Figure 18   NIST/JILA/CU-Boulder – NIST Image, public domain.

Figure 20   according to [Swenson 1950]; **Figure 21**: top left: according to W. Ketterle, *MIT Physics Annual* 2001, with permission of W. Ketterle; right: from J. R. Abo-Shaeer et al. *Science*, 292 (2001) Is. 5516 Reprinted with permission of AAAS (American Association for the Advancement of Science).

Figure 22   centre: "This Month in Physics History – January 1938: Discovery of Superfluidity", *APS News* January 2006, vol. 15, no. 1.

Figure 23   Reprinted with permission of Macmillan Publishers Ltd: I. Bloch et al. *Nature* 403 (2000) 166.

Figure 29   Right: T. H. Johansen, University of Oslo, with permission.

Figure 30   according to [Vollhardt & Wölfle 1990].

Figure 33   below: from S. K. Lamoreaux, *Phys. Rev. Lett.* 78 (1997) 5, reprinted with permission of APS.

Figure 34   G. Breitenbach, S. Schiller, and J. Mlynek, with permission of G. Breitenbach. http://www.gerdbreitenbach.de.

https://doi.org/10.1515/9783110644203-011

Figure 36   a) according to [Rauch 2000]; b) reprinted with permission of A. Zeilinger, R. Gähler, G. C. Shull, W., Treimer, and W. Mampe, *Rev. Mod. Phys.* 60 (1988) 1067, Copyright 1988 by American Physical Society.

Figure 37   a) and b) Reprinted with permission of A. Zeilinger, R. Gähler, G. C. Shull, W., Treimer, and W. Mampe, *Rev. Mod. Phys.* 60 (1988) 1067, Copyright 1988 by American Physical Society.

Figure 38   top left and right: reprinted with permission of Macmillan Publishers Ltd: M. Arndt, O. Nairz, J. Vos-Andreae, C. Keller, G. van der Zouw et al., *Nature* 401 (1999) 680, Copyright (1999); bottom left: Reprinted with permission of Macmillan Publishers Ltd: L. Hackermuller, K. Hornberger, B. Brezger, A. Zeilinger and M. Arndt, *Nature* 427 (2004) 711; C60 Macromolecule Inset: Sponk Wikimedia, CC Attribution-Share alike 3.0.

Figure 39   C. Philippidis, C. Dewdney and B. Hiley, Nuovo Cimento B 52, 15 (1979); reprinted with permission of Springer.

Figure 40   according to [Aspect et al. 1981, 1982].

Figure 41   Reprinted with permission of D.V. Strekalov et al, *Phy. Rev. Lett.* 74 (1995) 3600; Copyright (1995) by American Physical Society.

Figure 42   according to [Haken 1987].

Figure 45   Reprinted with permission of A. S. Stodolna, A. Rouzée F. Lépine, S. Cohen, F. Robicheaux, A. Gijsbertsen, J. H. Jungmann, C. Bordas, and M. J. J. Vrakking, *Phys. Rev. Lett.* 110, 213001 (2013). Copyright (2013) by American Physical Society.

Figure 48   according to [Feynman 1985, p.115–117].

Figure 49   according to [Feynman 1989 p. 97–99].

Figure 52   a) [Guye 1921] public domain.

Figure 53   (a) NOAA, Wikimedia GPS-constellation-3D-NOAA, public domain; (b) From C. W. Chou, D. B. Hume, T. Rosenband, D. J. Wineland, Science 329 (2010) 1630, Reprinted with permission of the American Association for the Advancement of Science (AAAS).

Figure 54   [Michelson and Morley 1887] public domain.

Figure 57   according to [Genz 1994].

Figure 58   a) ESO, Planetary Orbits, http://www.eso.org, CC Attr. 4.0 Int. Lic; b) Wikipedia, Precessing_Kepler_orbit_280frames_e0.6_smaller Wikimedia CC Attribution 3.0.

Figure 59   b) [Eddington 1919]; (c) Nasa, public domain; (d) Space Telescope Science Institute, STScI, [Nasa 1999].

Figure 60   a) Stefania de Luca, public domain;b) B. Famaey, S. S. McGaugh, Living Reviews in Relativity 15 (2012) 10, Creative Commons Licence CC BY-NC-ND 3.0 DE; c) Reprinted from P. D. Mannheim, Progress in Particle and Nuclear Physics, 56 (2006) 340 with permission from Elsevier.

Figure 61   b) Bibliotheque nationale de France, public domain.

Figure 64   Atlas Detektor, photo by Frank Hommes, public domain.

Figure 67   Standard Model of Elementary Particles, Wikimedia CC BY 3.0.

Figure 71   a) [CERN 2004] (b) [CERN 2005] both with the permission of CERN.

Figure 72   ATLAS experiment at CERN, reprinted with permission of CERN.

Figure 73   a) [Bethke 2007] published under CC BY NC ND; c) according to H. Georgi, in "Teilchen Felder Symmetrien", *Spektrum der Wissenschaft*, 1986.

Figure 77   a) Hertzsprung-Russel diagram, public domain; b) Image; stars_lifecycle_full.jpg, NASA and the Night Sky Network.

Figure 78	a) Binding energy curve, Wikimedia Commons public domain; b) Solar System Abundances, Wikimedia CC BY-SA 3.0.
Figure 79	a) and b): Anglo-Australian Observatory, with permission from David Malin; c): J. Pun (NASA/GSFC), R. Kirshner (CfA) and NASA, p.d.
Figure 80	[Rolfs & Rodney 1988] Permission to reprint by University of Chicago Press.
Figure 81	a) according to Pulsar, NASA, http://imagine.gsfc.nasa.gov.
Figure 82	a) Chandra X-ray Observatory, http://chandra.harvard.edu , NASA/CXC/ SAO; b) J. Hester and A. Loll, Arizona State University, NASA, ESA.
Figure 83	The Hubble Tuning Fork, http://hubblesite.org, NASA, public domain.
Figure 84	a) NASA/ESA; b) J. Blakeslee, Washington State University; c) NASA/ESA; d) NASA and ESA.
Figure 85	From [Lutz 1991].
Figure 86	Image courtesy of NRAO/AUI, Alan Bridle & Robert Laing [Martel 1999].
Figure 87	a) Centaurus A, Radio Galaxies, Wikipedia free GNU licence; b) from J. J. Condon and S. M. Ransom, National Radio Astronomy Observatory (NRAO).
Figure 88	a) NRAO/AUI 1999; b) NRAO/AUI 1999; c) with the permission of W. C. Keel, University of Alabama; (d) VLA, NRAO.
Figure 89	a) X-rays: NASA/CXC/CfA/W. Forman et al.; Optical: DSS; b) The Hubble Heritage Team STScI/AURA & NASA/ESA; c) F. Duccio Macchetto/NASA/ESA; d) National Radio Astronomy Observatory / Associated Universities.
Figure 90	Hubble Space Telescope (HST) image from NGC 4261, HST/NASA/ESA.
Figure 91	a) NASA/STScI; b) R. C. Thomson, & C.D. Mackay, Cambridge, UK; A. E. Wright, ATNF, Parkes, Australia; c) Multi-Element Radio Linked Interferometer Network (MERLIN); d) NASA/CXC/SAO/H. Marshall et al.
Figure 92	according to [Volovik 2003].
Figure 95	a) *The Two Micron All Sky Survey 2MASS*, T. H. Jarrett, J. Carpenter, & R. Hurt; b) Sloan Digital Sky Survey, (SDSS).
Figure 96	Gabriel Pérez Díaz, Instituto de Astrofísica de Canarias (Servicio MultiMedia), Permission to reprint of IAC.
Figure 98	b) Reprint with permission of Macmillan Publishers Ltd: J. R. Bond et al. "How filaments are woven into the cosmic web", Nature 380 (1996) 603.
Figure 100	Composite graphics using images from: 1st line Tiziana Di Matteo, Carnegie Mellon University, USA; 2nd line: NASA, ESA, F. Summers (STScI); ESO/ WFI (Optical) MPIfR/ESO/APEX/A.Weiss et al. (Submillimetre) NASA/CXC/CfA/R. Kraft et al. (X-ray); NASA/CXC/ASU/J. Hester et al.; 4th line, right: Copyright 1998–2016 by Sea and Sky, www.seasky.org, Used with permission.

Jooss: 1.6; 1.9–1.11; 3; 5–6; 8–9; 10 b); 12 bottom left; 13 right, according to [Alfvén & Arrhenius 1976]; 15; 16 right; 17; 19; 21 bottom left; 22 left and right, 24–28; 29 left; 31–32; 33 top; 35; 43–44; 46–47; 50–51; 52 b) and c), data from [Bertozzi 1964]; 55–56; 58 c) and d); 59 a); 61 a); 62–63; 65–66; 68; 69–70 according to [Greiner 1985]; 73 b) based on data from [Achard 2005, Burkhardt 2011]; 74–76; 81 b) based on results from [Ventura & Pines 2012]; 93–94; 97; 98 a); 99; 100 3. Line.

Index

https://doi.org/10.1515/9783110644203-012